NATURAL GROUNDWATER FLOW

Wouter Zijl
Marek Nawalany

Non intratur in veritate nisi per caritatem

LEWIS PUBLISHERS
Boca Raton Ann Arbor London Tokyo

This book has been written with financial support of the Netherlands Organization for Applied Scientific Research TNO.

Library of Congress Cataloging-in-Publication Data

Zijl, Wouter.
Natural groundwater flow / authors, Wouter Zijl and Marek Nawalany.
p. cm.
Includes bibliographical references and index.
ISBN 0-87371-868-2
1. Groundwater flow. I. Nawalany, Marek. II. Title.
TC176.Z55 1993
551.49--dc20 92-35372
CIP

Direct all inquiries to CRC Press, Inc., 2000 Corporate Blvd., N.W., Boca Raton, Florida, 33431.

PRINTED IN THE UNITED STATES OF AMERICA
1 2 3 4 5 6 7 8 9 0

Printed on acid-free paper

About the Authors

Wouter Zijl currently is a senior research fellow at the Netherlands Organization for Applied Scientific Research (TNO), working on environmental physics, especially in relation to flow and transport phenomena. He earned a B.Sc. in electrical engineering and an M.Sc. in theoretical and applied physics. His Ph.D. was obtained from Eindhoven University of Technology, the Netherlands, where he worked on heat and mass transfer in multi-phase flow. After 4 years of research management work at a Dutch-German consortium for energy technology, he joined the TNO Institute of Applied Geoscience in 1982. He is employed part-time at Amsterdam Free University in the faculties of earth sciences and mathematics for basic and applied research and educational tasks; he is part-time professor at the Brussels Free University; and he collaborates with a number of other universities. Recently, he has also been working in the relatively new field of environmental ethics, applying the general theory of systems as an 'interface' between environmental science and ethics.

Marek Nawalany currently is professor at, and director of, the Institute of Environmental Engineering, Warsaw University of Technology, Poland. He earned an M.Sc. in theoretical physics, an M.Sc. in applied mathematics, and an M.Sc. in hydrology and water resources management. His Ph.D. was obtained from the Academy of Mining and Metallurgy at Cracow, Poland, where he worked on the applications of nuclear tracers in geohydrology. Since 1975 he has worked at the Institute of Environmental Engineering on a wide range of topics including basic and applied research, software development, education, and management, all related to surface and groundwater hydrology and environmental engineering. He is vice president of the International Commission on Ground Water of the International Association of Hydrological Sciences. For the field of waste disposal management he is advisor of the Polish steel industry, and he is an advisor to the TNO Institute of Applied Geoscience, Delft, the Netherlands. He has great interest in systems theory applied to the analysis of water pollution with the aid of mathematical and numerical models for flow and transport phenomena.

CONTENTS

NATURAL GROUNDWATER FLOW

Chapter 1

Introduction

1.1. The scope of the book

The present book is devoted to the theory of groundwater flow, i.e., to the kinematics and dynamics of water in the pores of the subsurface. Groundwater flow finds its application in geohydrology, which is a branch of hydrology. Hydrology is devoted to the study of water in the cosmos, and especially on earth. Hydrology is a branch of geoscience, or earth science, in general. The scope of this book is on the mathematico-physical basis of the theory of groundwater flow.

Earth scientists, including geohydrologists, do not generally rely on the mathematical symbolism of physics to any great extent. The earth's 'reality' is so capricious and unpredictable that too much faith in mathematico-physical formulations would obviously be fallacious. Indeed, if the only choice was between 'true' and 'false,' then every such schematization of the geo reality would be 'false.' However, this is not the only choice, and it is important in all geo-scientific discussions to bear in mind that between 'true' and 'false' there is a large area of possibility, varying from the highly-probable to the highly-improbable. Consequently, we are entitled to consider all those aspects where the probability can be high.

Another important consideration is the general lack of geo-scientific data. Progress is being made, but there is a chronic and widespread shortage of measured data. Consequently, the most effective recourse is to develop a 'most-likely' model of the earth's structure and processes based on the data available. At the same time, this model can be complemented by using symbolic mathematical relationships based on 'natural laws' which are well-established in other scientific disciplines such as physics, chemistry, biology, etc. With the help of such mathematical symbolism based on natural laws, the measured data from a few space-time points $(\boldsymbol{x}, t)$ can be used more effectively in a kind of physico-mathematical way of interpolation, as it were.

Moreover, the analytical mathematical expressions can provide greater insight into subterranean geophysical processes and contribute to a more realistic conceptual model of these processes. Such conceptual models are important not only for estimations of global orders of magnitude but also for constructing more-detailed numerical mathematical models which, with appropriate caution, can be used to predict probable future trends.

In this book the importance of analytical mathematical models, and the order of magnitude estimates based on them, will be demonstrated mainly on the basis of groundwater flow, both on the small, local scale and on the large, regional scale. The transport of dissolved particles will also be discussed, so the term 'water' will not refer solely to pure water, but will include aqueous mixtures such as water-salt solutions. In general, the term 'fluid' can relate to both liquids and gases, or any mixture of the two, but in this book the terms 'fluid' and 'water' are used synonymously. The main concern is to emphasize the close relationship between mathematical symbolism and physical reality, and the derivation of numerical orders of magnitude for real-life situations.

The use of model codes, based on numerical mathematical approximation methods, to obtain detailed numerical results for field situations will be dealt with fully starting from Chapter 7. However, by way of introduction, it may be appropriate at this stage to include some extracts from the farewell speech of Prof.Dr.Ir. G. Vossers at the Eindhoven University of Technology, Eindhoven, 8 Sept. 1989, as follows:

> *Sometimes one method temporarily seems to render others superfluous. The current capabilities of present-day computation equipment, and the expectations for the future, have led the practitioners of numerical flow science into this temptation. But even if all the expected calculable mathematical possibilities were to be achieved, this dream will not itself be realized until well into the 21st century. The chaotic turbulence of reality will ensure that.*

Again,

> *Contacts with other disciplines are challenging because they can lead to new applications; they are also sobering and frustrating because one often sees fluid mechanics very inexpertly applied. The current offer of mathematical programs is an easy temptation for uses beyond their original*

intention, and often results in rubbish. I have occasionally advanced the theory that 50% of the calculus in the field of fluid mechanics is probably wrong because insufficient thought went into the problem formulation of the laws of conservation. Presumably the remaining 50% also contains a number of numerical and programming errors. In this respect, the applied information technology is no blessing.

The fallacies observed by Dr. Vossers will be avoided by concentrating on the 'laws of conservation,' the consequences of those laws, and some of their applications to extend our understanding. In other words, the first concerns will be with the pre-modeling requirements, the conceptual and model structuring which precedes numerical modeling.

On the other hand, the computer has made it possible to apply numerical mathematics in the construction of model codes aimed to simulate the behavior of complex systems that take numerous factors into consideration. Furthermore, the study of complex patterns leads to the discovery that there are unanticipated connections or overlappings of portions of many different disciplines. According to the physicist Heinz Pagels, 'problems in neuro-science, anthropology, population biology, learning theory, cognitive science, nonlinear dynamics, physics, and cosmology (to name but a few fields) have overlapping components' (see Pagels, 1988, p. 36). Pagels argues that the divisions among the sciences in the past have been influenced by the available tools for investigation and that the computer will lead to a more complex, less segmented approach. Physicists are taking the lead in the use of these new modeling tools. Since, historically speaking, physics played so large a role in modeling the present disciplinary narrowness, there is reason to hope that its new vision will lead to changes in other disciplines as well. Hopefully, this will open the way to a willingness to approach the geo-scientific problems in a real interdisciplinary way, without being oblivious to the enormous amount of knowledge stored in the various disciplines.

The positive aspects observed by Pagels strongly motivate us to include chapters on numerical mathematical modeling to enhance our understanding of it.

1.2. Natural environment and the role of water

Since the theory of groundwater flow is generally applied in the much broader context of earth sciences like hydrology, ecohydrology, ecology, geochemistry and geophysics applied to responsible management and use of the environment, it is appropriate to briefly introduce a very general description of nature. When thinking about 'nature,' i.e., the cosmos generally and the earth in particular, it is usually accepted that the various parts of the natural system are linked to each other by a complex network of many feedback loops. Furthermore, certain events are taking place in a way not uniquely determined by causes prior to that event. In the 'language' of ecology, this is a form of 'self-determination.' The intricacies of the feedback loops combined with the self-determination give rise to new properties not contained in the component parts of the system. To express that the whole is more than the sum of its parts, the term 'holism' has become fashionable, and such an intricate self-regulatory system is then called an 'organism' or an 'ecological unit.'

The recognition of causes preceding an event and uniquely determining it, i.e., excluding self-determination, is called determinism. A scientific description is essentially a deterministic description, however, sometimes complemented by a probabilistic description to account for events not uniquely pre-determined by prior causes. Quantum physics, i.e., the description of the laws governing the motion of elementary particles like electrons, protons, etc. can be interpreted as a demonstration of the nondeterministic character of ultimate reality, in which the random events are conditioned by a probability distribution. (Such an interpretation of quantum theory is advocated by Sir Karl Raimund Popper (1982, pp. 173-177).) Accordingly, any prediction is necessarily accompanied by a certain amount of uncertainty. The future has aspects which are really open, and in that sense there is real freedom, especially but not exclusively for humans, to determine their own future. So, in principle, mankind can decide to counter the present deterioration of the natural environment leading to the extinction of many species of life.

One of the best known organic/holistic theories about the earth is the Gaia Hypothesis of James Lovelock (1982, 1988) which stresses the close interrelationship between life on earth and the chemical composition of the earth's atmosphere. Fed by the sun, life and the atmosphere form a feedback system which, within limits, has a great measure of stability and order.

The Gaia Hypothesis fits in well with the earlier 'philosophy of organism' of the mathematician and philosopher Alfred North Whitehead and his like-minded companions. This philosophy, well-known in the U.S. and also to a growing number of ecologists in Europe, strongly emphasizes the existence of mutual bonds in the cosmos, and especially in the earth's nature and human culture. Originated in 1925 and receiving more attention with the publication of the book, *Science and the Modern World* (Whitehead, 1967), this philosophy is now known as 'process-relational philosophy.' In contrast to the Gaia Hypothesis, the process-relational philosophy recognizes the validity of a description of nature with some orientation to 'goals' because the higher species of animal, and man especially, can direct their activities toward a goal, leading to purposeful actions. Applications of the process-relational philosophy in biology, ecology and the environment are given by Birch and Cobb (1981). Further developments, with emphasis on redirecting the economy toward the environment, the community, and a sustainable future, are presented by Daly and Cobb (1990). In this context the work by Von Bertalanffy and Laszlo on the general theory of systems should also be mentioned (see Laszlo, 1972, 1983).

Similarities with the work of Whitehead will also be found in the works of Ilya Prigogine (Nobel prizewinner for Chemistry, 1974) on the dynamics of dissipative systems and the irreversibility of time (see Prigogine and Stengers, 1979). The intricate feedback loops mean that 'nature' has to be considered a nonlinear system. In dissipative (i.e., energy consuming) nonlinear systems unstable states exist. At these unstable states, the bifurcations, the system 'chooses' one of two possible avenues of further development. Near a bifurcation, random fluctuations play a decisive role in making the 'choice'; between bifurcations, the system operates more along deterministic lines. In this way nature produces 'order out of chaos.'

A well-known earth-scientific example of a nonlinear chaotic system is the weather. The unstable states, beyond which weather forecasts cannot be reliably extended, impose a fundamental limit to forecasting the long-range time-evolution of the weather. Of course, this limit can hardly be overcome by applying more detailed models and larger computers. A consequence for the hydrological sciences is that rainfall, which is strongly associated with weather, will exhibit a chaotic character (see Rodríquez-Iturbe *et al.*, 1989). In general, it can be said that the recognition and proper modeling of the most relevant feedback loops in nature will become a very important challenge in earth science as a whole.

The ongoing chaotic process in nature can be considered a permanent source of creativity which provides order as well as novelty. It creates new from old, and the eternal continuity of renewal and decay, birth and death, with its consequences for nature in general and life in particular, is one of the bases for religious reflection. No matter how one views religion, it is well to appreciate that in a sense there is a great similarity between earth science and religious thought; both struggle with the question of the relationship between empirical, rational and speculative aspects of a theory, albeit with more awareness of it in philosophical religious reflection (see Griffin, 1990).

By an empirical theory we mean a theory being adequately applicable to all the facts of experience. Therefore, an empirical theory is based on the impartial observation of all of what is observable. Returning to the earth sciences, empirical means based on field observations without denying or ignoring some facts of observation. This results in studies of particular areas, such as the Veluwe study in the Netherlands or the Kempisch Plateau study in Flanders, Belgium.

By a rational theory we mean a self-consistent theory, having no aspects that contradict other aspects or logical deductions therefrom. A rational theory consists of a set of general pronouncements based on a limited number of hard-core empirical truths and supported by logic and mathematics. For the subject matter of this book, the mechanics of groundwater flow, this means pronouncements based largely on the basic conservation laws (conservation of mass and momentum), which are considered as the hard-core empirical truths, and supported by analytical mathematics based on the continuum hypotheses. The results obtained by analytical mathematics are area-independent, and are of general validity.

By a speculative theory we mean a set of hypotheses about how 'things' are related to other 'things.' It is the power of hypotheses to illuminate previously unrecognized facts and/or to show how seemingly contradictory facts can be made compatible. It is the 'fertility' of the theory to understand how this complex web of interrelationships in itself. In the earth sciences, this often means that speculative theories take the form of action programs to achieve particular objectives, and which can be adapted as necessary in the course of completion (see also Popper, 1982, pp. 210-211).

In formulating earth-scientific theories it is important to use the correct mixture of empirical, rational and speculative aspects, and this is achieved with the 'hydrological systems' theory (see Engelen and Jones, 1986) which includes the 'flow systems analysis' described in Chapter 5. Hydrological systems theory is a practice-oriented organic/holistic theory which is centered on the role of water moving in nature, including groundwater. To ensure an even balance between empirical, rational and speculative thought, the rational aspects of the flow systems analysis are strongly emphasized in this book.

Theories on part-whole relationships have been developed since man started philosophical reflection. At present, opinions differ greatly; see Birch and Cobb (1981). However, it can be safely stated that, just as a doctor must know the functions of the various organs (the parts) to understand the human body (the whole), the hydrologist must know the various elements and processes of the geostructure (the parts) to understand nature (the whole). Water in nature, carrying various dissolved particles and materials, is one such component part, and the flow of water is an important process in the integral picture (see Figure 1). While bearing in mind its relationship to the whole of nature, we must be aware that water acts as a major transport medium and, in the form of fresh water, is a store of valuable raw materials. Consequently, fresh water occupies a central position in the attention currently being given to threats to nature and the environment. Indeed, the flow of water in nature is comparable in importance to the circulation of blood in animate life, and its importance for life cannot easily be overestimated.

Having said that, it must be added that this book does not consider all aspects of water movement in nature, but is restricted to the flow of *groundwater*, that is, water in the liquid phase which resides in the pores of the subsurface (see Figure 2). The pores are considered to be saturated with water, and the water may contain solutes. Especially fresh groundwater is a very important store of 'drinking water' for many types of life, including human life. Therefore, all our knowledge about the physical extension and movement of fresh groundwater will be required to arrive at a sustainable development.

A further limitation is that this study relates mainly to natural groundwater flow, i.e., flow due to gravity, and is not much concerned with flow caused by extraction or injection wells. This unconventional emphasis has been adopted since most of the available literature is devoted to the flow caused by wells, while the natural flow of groundwater remains relatively neglected despite its obvious and considerable importance in determining

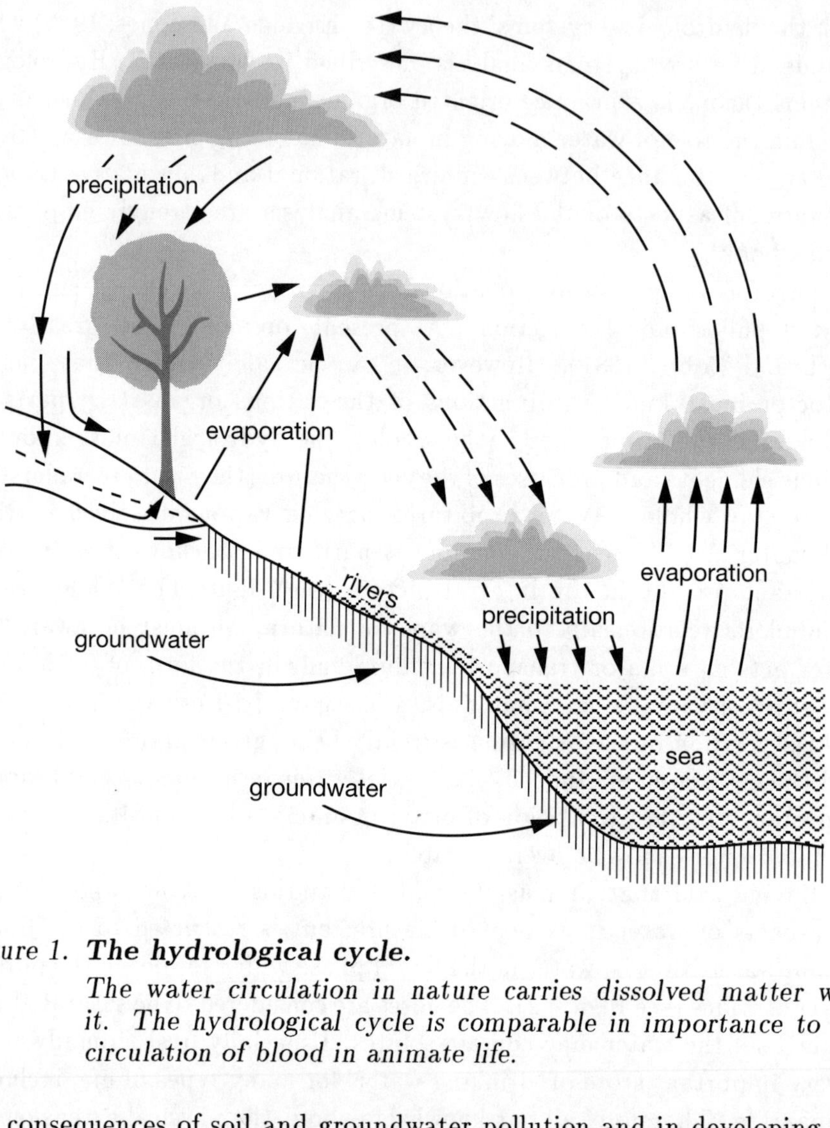

Figure 1. ***The hydrological cycle.***
The water circulation in nature carries dissolved matter with it. The hydrological cycle is comparable in importance to the circulation of blood in animate life.

the consequences of soil and groundwater pollution and in developing appropriate management strategies. Natural groundwater flow is more or less negligible on the local scale in the neighborhood of an extraction well, but it is the natural flow which predominates on the larger regional scale. Since natural flow determines the distribution of dissolved pollutants, the magnitude and direction of the groundwater flow velocity are particularly important. This is in contrast to the flow around wells, where the lowering of the groundwater table and its piezometric head are more important. The study of groundwater velocity on a regional scale leads to quite a different

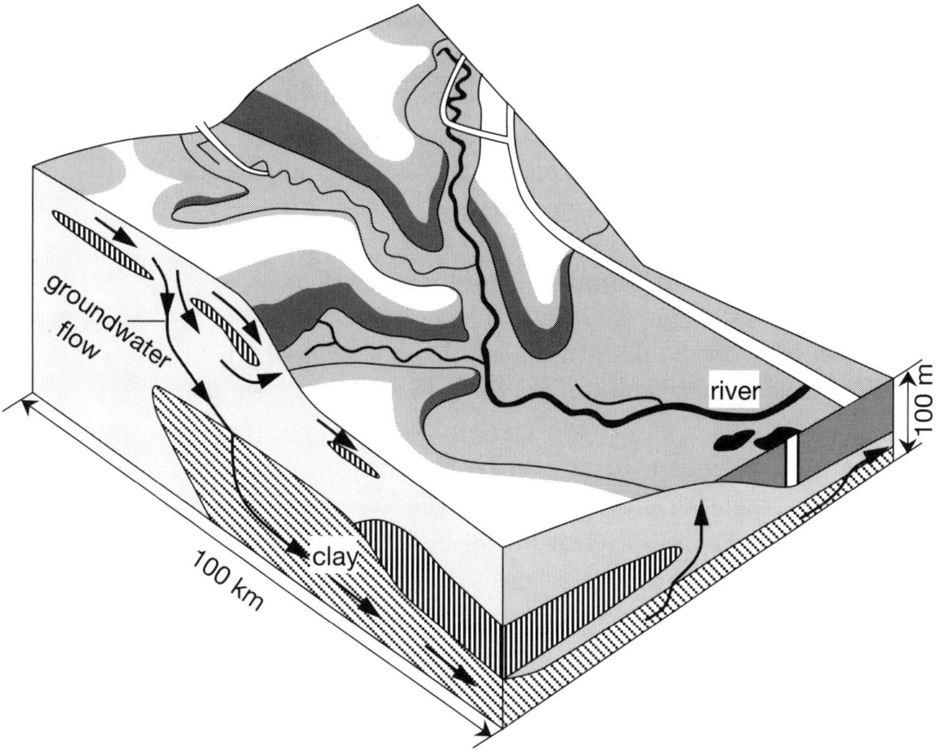

Figure 2. **Natural groundwater flow.**
Groundwater is the water flowing in the water-saturated pores of the subsurface. The great difference between the horizontal and vertical scales (a factor of 100 to 1000) is quite normal for natural groundwater flow systems. Therefore, in three-dimensional figures and in two-dimensional cross-sections the vertical scale is generally blown-up with respect to the horizontal scale.

approach to the theory, which is the justification for this book alongside other literature on the subject of groundwater flow.

A brief discussion of soil moisture, i.e., the liquid water which, together with air and water vapor, fills the pores, is included to define the depth below the ground level at which the zone saturated with groundwater begins. However, although water in its solid phase (ice) can sometimes play an important role, its effects are not discussed, and all further reference to water or groundwater or fluid relates to water in its liquid phase.

Three depth levels, shallow, middle and deep, are important for different reasons. Shallow groundwater is a transport medium which may carry pollutants to other regions or to the surface water. The groundwater at middle depths is largely fresh water which is suitable for drinking and irrigation. The deep groundwater contains carbon dioxide which influences the metamorphosis of subterranean rocks, and its flow is important in the exploration and production of ores and of natural oil and gas.

1.3. The role of mathematical analysis

This book is not a book on mathematics, but a book on the *application* of basic classical mathematics to groundwater flow. To gain insight into the scale-behavior of flow processes and the scale-dependent interpretation of computations, it is necessary to be familiar with a number of concepts which are mathematically derived from the hard-core empirical fundamentals of fluid mechanics in porous media. There has always been a chronic shortage of data in the earth sciences, but a conceptual model of the processes is needed nonetheless. It is needed to enable interpolations and even extrapolations of the here-and-now to become perceptible. To ensure that these interpolations and extrapolations are as reliable as possible, they should be based on universally-accepted hard-core physical laws like conservation of mass, momentum and energy, and physical theorems derived from them. These theorems are usually expressed in mathematical form with symbols that closely relate to perceived reality, e.g., velocity is denoted by the symbol $\boldsymbol{v}$, pressure by the symbol p, etc. In this context, the words of mathematician and philosopher Alfred North Whitehead (1982) are appropriate:

All science as it grows to perfection becomes mathematical in its ideas.

and

By the aid of symbolism, we can make transitions in reasoning almost mechanically by the eye, which otherwise would call into play the higher faculties of the brain.

The importance of the mathematico-physical system of symbols is also recognized by experts in the field of cognitive processes. For example, Raj Reddy, President of the American Association of Artificial Intelligence (AAAI), in 1988 summarized the fundamental principles of Artifical Intelligence (AI) as follows (see Reddy, 1988):

1. Limited rationality implies opportunistic searching.
2. *Physical symbol systems are necessary and adequate for intelligent activity.*
3. An expert knows 70,000 things plus or minus a binary order of magnitude.
4. Searching compensates for lack of knowledge.
5. Knowledge eliminates the need to search.

However, while systems of physical symbols may be adequate for intelligent activity in an artificial sense, it is not claimed here that they are entirely adequate in a real sense. Indeed, it should be clearly understood that some aspects of reality will inevitably be lost in constructing conceptual models by means of mathematical symbols. The acceptance of such a model as reality itself is what Whitehead (1967) calls the 'fallacy of misplaced concreteness.' Nash *et al.* (1990) write:

> *Often without adequate backgrounds in physics and chemistry, but with strong mathematical bias, they (i.e., the engineering hydrologists) may have tended to neglect the possibility of applying established scientific principles, and the necessity for careful observations. They may have tended to rely excessively on the mathematical analysis of 'data,' i.e., observations, usually made by other persons, often for other purposes, which were 'given' to the engineering hydrologist in the hope and expectation that he might detect some germ of information or expression of a relationship relevant to an immediate practical problem. Thus hydrology came to be practised as if it were a basic science, whose relationships were to be found exclusively by the formulation and testing of hypotheses against observations, rather than equally by the application of established principles of extant natural sciences. This approach, though providing many interesting challenges, which often left room for an almost Renaissance exercise of ingenuity in problem solving, was not, however, a sound basis on which to build an applied science.*
>
> *More significantly, however, engineering hydrologists undoubtedly fell into the error of preferring the means to the end and of learning, modifying and adapting the techniques of analysis often to a degree of sophistication far beyond what would be justified by nature of the observations to which these tools were applied. Much of this intellectual effort was entirely irrelevant to hydrological science and divorced from the practice of the hydrological profession.*

The above citation clearly points to the 'fallacy of misplaced concreteness.' Nevertheless, provided they are used with due caution, physically based mathematical models can be most useful.

It will be shown that the systems of physical symbols are essential for intelligent action, especially in this era of computer models and knowledge systems. Furthermore, it is hoped that the use of the symbolic language of the mathematician can permeate to the conceptual models we construct in our own consciousness from the reality we perceive. Indeed, there are various concepts (like flow system, net-infiltration, net-exfiltration) which cannot be suitably defined in any other way, and can be understood only through a mathematical symbolic system of thought within which the definitions are meaningful.

1.4. SI units and conventional units

Physical quantities such as length, time, pressure, potential, etc. are equivalent to the product of a numerical value, i.e., a pure number, and a unit:

$$\text{physical quantity} = \text{numerical value} \times \text{unit} \tag{1.4.1}$$

For a physical quantity denoted by symbol q this relation is usually represented in the form $q = \{q\}\,[q]$, where $\{q\}$ stands for the numerical value and $[q]$ stands for the unit. For instance, pressure (physical quantity) = 10 000 (numerical value) pascal (unit); or $p = 10\ 000\ Pa$ in symbols, where $\{p\} = 10\ 000$ and $[p] = Pa$.

In this book the International System of Units with the international abbreviation SI is used. One of the reasons is that SI units form a coherent system. A coherent system of units consists of various base units (among which are the meter, the kilogram, and the second in the SI system) which are defined independently and not derived from each other, but relate to each other to provide a complete quantitative description of nature. For example, the length unit meter relates to the time unit second to give the derived unit for velocity, meter-per-second.

To express this coherence, the formulae which describe physical relations do not contain any needless conversion numbers, so velocity = length/time and not something like velocity = 3.6 length/time, say.

Moreover, the SI system of units is comprehensive and usable in all subject disciplines, so the SI system is used in all subsequent chapters to strengthen the association with other physical subject disciplines. In this context there is an apt quotation from Appelo (1992):

Here, basic physics and chemistry offer the tools for investigation. Their application may protect you from violating principles that have proven their validity in much broader disciplines than ours.

However, it will also be appropriate to indicate the relationship with the traditional units used in geohydrological practice, and to do so the presentation is slightly different from that usually adopted in geohydrological literature. Remarks on units are included in the text between the symbols (# ... #); for example:

SI unit of pressure: pascal $[Pa]$; conventional unit of pressure: meter water $[mH_2O]$. This is noted briefly as: $p\ [Pa;\ mH_2O]$ is the pressure. Conversion: $10^{-2}\ mH_2O = 1\ cmH_2O \approx 100\ Pa = 1\ hPa = 1\ mbar$ (see Section 2.7 for a more detailed discussion on conversion). In the past the millibar $[mbar]$ was used instead of the hectopascal $[hPa]$, where $1\ mbar =$ $= 1\ hPa$. In concession to geohydrological practice where the meter water (in short, the meter) is used dogmatically as the unit of pressure, it is convenient to use the derived SI unit of pressure, the decibar $[dbar]$, where $1\ dbar = 10^4\ Pa \approx 1\ mH_2O$.

SI unit of time: second $[s]$; conventional unit of time: day $[d]$. This is noted briefly as: $t\ [s;\ d]$ is the time. Conversion: $1\ d = 86{,}400\ s$.

SI unit of length: meter $[m]$. This is also the conventional unit.

For more details on quantities, units and dimensional analysis, please refer to Silberberg and McKetta (1953). For an explanation of the SI system of units in relation to earth sciences, see JPT (1982).

1.5. Classification of flow types in the porous subsurface

The porous subsurface contains two phases, air (with water vapor) and water (with dissolved particles), and *flow behavior* can be classified in three main zones: (1°) the atmospheric zone, (2°) the two-phase zone, and (3°) the saturated zone.

(1°) In the atmospheric zone the gaseous phase (air and water vapor) flows while the liquid phase remains stagnant.
(2°) In the two-phase zone, the gaseous and liquid phases both flow.
(3°) In the saturated zone, the liquid phase flows while the gaseous phase, if present, remains stagnant.

It will be noted that water appears in all three zones: in both gaseous and liquid phases in the atmospheric and two-phase zones, and in the liquid phase in the saturated zone. Water can also appear in the solid phase at temperatures below freezing point, but this aspect will be ignored for present purposes.

It will also be noted that the atmospheric zone and the two-phase zone are usually classified together as the unsaturated zone in earlier terminologies, but greater distinction is needed for the more valid assessment of flow behavior introduced in the next section.

1.6. Flow in the unsaturated zone

In the atmospheric zone it is only the air and water vapor that can flow; the liquid phase remains stagnant. Consequently, the fluid mechanics of the gaseous phase can be described very simply. The flow of air and vapor is such that the air pressure $p_1(\boldsymbol{x}, t)$ is equal everywhere to the atmospheric pressure $p_{atm}(t)$ $[Pa;\ dbar \approx mH_2O]$, where t $[s;\ d]$ is the time and $\boldsymbol{x}$ $[m]$ is the location vector in three-dimensional Euclidean space.

Having introduced three-dimensional vectors, it is appropriate to explain here something more about their notation. In this book both two-dimensional and three-dimensional vectors are printed in bold type as is conventional in the literature. It follows from the context whether a two-dimensional or a three-dimensional vector is dealt with, but in most cases we deal with three-dimensional vectors. The three-dimensional position vector $\boldsymbol{x}$ has been given the three coordinates $(x_1, x_2, x_3) = (x, y, z)$ in a right-handed (i.e., clockwise-oriented) orthogonal Cartesian coordinate system. Since rows and columns of orthogonal Cartesian coordinates have similar mathematical properties as vectors, they will be equivalenced for the sake of shorthand notation, i.e., $\boldsymbol{x} = (x, y, z) = x_i$ $(i = 1, 2, 3)$. However, we should be very careful, since equivalencing vectors and their rows (or columns) of Cartesian coordinates is not allowed when changes in coordinate systems are involved. As an example, when in a coordinate system the vector $\boldsymbol{x}$ has coordinates (x, y, z), equivalenced as $\boldsymbol{x} = (x, y, z)$, and in another coordinate system the same vector $\boldsymbol{x}$ has coordinates (x', y', z'), equivalenced as $\boldsymbol{x} = (x', y', z')$, it is not allowed to equivalence $\boldsymbol{x} = (x, y, z) = (x', y', z')$.

The assumption is that $p_1(\boldsymbol{x}, t) = p_{atm}(t)$ is justified by the consideration that the viscosity of air is very much smaller than that of water (a ratio of about 5×10^{-5}) and that the density of air is also much smaller than that of water (a ratio of about 10^{-3}). The atmospheric pressure varies with time and has a mean value of about 1013 hPa = 10.13 $dbar$ = 10.33 mH_2O = = 1033 cmH_2O; variations in location can generally be neglected.

Below the atmospheric zone lies the two-phase zone where the air (gaseous phase) and the water (liquid phase) can flow separately. Hence the name two-phase zone. The air pressure can be assumed to be the same as the atmospheric pressure $p_{atm}(t)$. However, because of the surface tension on the small water-air interfaces in the pores, the water pressure $p(\boldsymbol{x}, t)$ in the two-phase zone will be smaller than the atmospheric pressure $p_{atm}(t)$, i.e., $p(\boldsymbol{x}, t) < p_{atm}(t)$.

Taken together, the atmospheric zone and the two-phase zone constitute what is usually known as the unsaturated zone. Directly below that lies the saturated zone.

1.7. Flow in the saturated zone

When the zones are characterized on the basis of flow behavior, the saturated zone is defined as that part of the subsurface where air no longer flows as a separate gaseous phase, and only the liquid phase can flow. Dissolved air can be carried in the water, of course, and any remaining pockets of air which may be present will have a pressure higher than atmospheric.

The interface between the saturated and unsaturated zones is known as the groundwater table or the phreatic plane. From the theory of flow in the unsaturated zone it appears that the surface tension in this plane is zero, therefore the water pressure here is atmospheric. But below the groundwater table, the water pressure increases to values much greater than atmospheric pressure.

The significance of these observations lies in the conclusion that the groundwater table $z = -h_f(x, y, t)$ defines the plane where the water pressure is atmospheric. This is known as the dynamic boundary condition:

$$p(x,\ y,\ -h_f(x,y,t),\ t) = p_{atm}(t) \tag{1.7.1}$$

where $p(\boldsymbol{x},t) = p(x,y,z,t)$ is the water pressure.

1.8. The two fundamental problem formulations

Two fundamental questions arise in any consideration of the saturated zone:

(1°) How do spatial variations in the phreatic plane take place with the passage of time?

(2°) What effects do the spatial variations in the groundwater table have on groundwater flow and the associated transport of solutes?

In general, a sensible procedure is to separate complex relationships into distinct but obviously linked conceptual modules so that one can concentrate on the solution of one aspect of the overall problem without being distracted by details of other aspects. However, the ways in which the various modules relate to each other, and the boundaries of the modules themselves, will need to be chosen wisely.

The distinctions made in considering the subsurface flow field as consisting of saturated and unsaturated zones is one example of modular thinking, and another is the distinction between the two sub-problems (1°) and (2°) above. The 'interface' between the conceptual modules in the latter example is the space-varying and time-evolving groundwater table, and the choice of this particular 'interface' is meaningful for several reasons. One reason is that the kinematic boundary condition on the water table (the boundary condition that describes phreatic storage) can be taken correctly with the description. Other reasons are given in Section 4.3.

Problem formulation (1°) is part of classic geohydrology, but problem formulation (2°) has only been realized within the framework of the 'flow systems analysis' mentioned in Section 1.2 (see Tóth, 1963) (see Figure 3). Indeed, the concept of 'flow systems analysis' is sometimes also limited to the study of type (2°) problems to indicate that type (1°) problems existed long before work was started on the more recent type (2°) formulations.

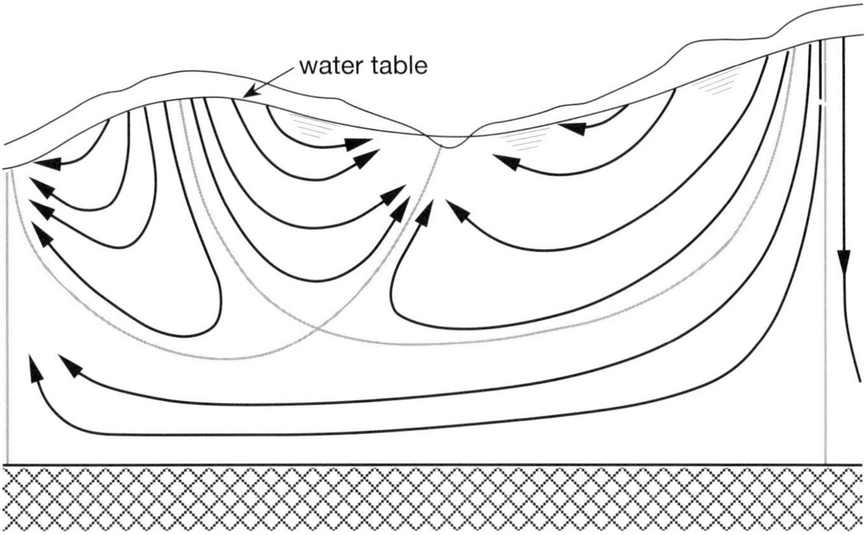

Figure 3. **Problem definition** *(2°).*
Cross-section of a homogeneous (but anisotropic) subsurface with a two-dimensional example of groundwater flow. On the water table the water pressure is atmospheric. Therefore, the spatial variations in the water table are causing the groundwater motion. In reality the groundwater flow is generally three-dimensional, giving rise to more complex flow patterns. Note that the deeper groundwater flow extends over a larger horizontal scale than the shallow groundwater flow.

1.9. The time-evolution of the water table

The first problem formulation, the evolution in time of the groundwater table, belongs to the classic domain of geohydrology and is concerned not only with the flow of water through the saturated zone but also with precipitation and evaporation, and with surface water and the unsaturated zone. The measurement of water levels in observation wells and the execution of physical and stochastic model studies are all ways of gaining insight into the question: how is the spatially varying movement of the groundwater table brought about? Answering this difficult question is a major activity of most groundwater studies, and involves the need for a groundwater measuring network and access to a variety of modeling methods and techniques. Studies in relation to drought, flooding, and damage to farming caused by moisture shortage or excess, etc. are all based on this problem formulation, but will not be discussed further in this book.

1.10. Flow and transport in the subsurface

Subsequent discussion will be concerned solely with the second problem formulation, the question of flow and transport caused by a given, assumed-known evolution-in-time of the spatial variations in the groundwater table. With a specified evolution of the groundwater table (the phreatic plane) $z = -h_f(x, y, t)$ it is possible to compute what the subsurface potential field $\phi(\boldsymbol{x}, t) = \phi(x, y, z, t)$ looks like at every point in space-time. The potential field can be determined by piezometric head measurements in observation wells (see Section 2.6), but it is only the velocity field $\boldsymbol{v}(\boldsymbol{x}, t) = (v_x(x, y, z, t),\ v_y(x, y, z, t),\ v_z(x, y, z, t))$ which is important for subsurface transport. Therefore, the problem formulation is directed primarily to answering the question of what the velocity field $\boldsymbol{v}(\boldsymbol{x}, t)$ below the groundwater table will look like as a result of a specified evolution of the groundwater table. This velocity field can be determined with the premise that the conductivity and porosity is known at every point in the subsurface. With this space-time determination of the velocity field, the stream paths and residence times can be determined and the transport equation can be solved. These last aspects are discussed briefly in Chapter 6.

If the spatial three-dimensional velocity field is computed at every time-point, then obviously the vertical component of the groundwater flow at the groundwater table is also defined at every time-point. In this way, the evolution in time of the infiltration (recharge) and exfiltration (discharge) patterns can also be defined (though, strictly speaking, these patterns are defined by the normal [perpendicular] flow component on the groundwater table rather than the vertical flow component on the groundwater table). Neglecting nonphreatic storage, there will be as much water infiltration as exfiltration if there are no well-extractions. In principle, it is possible that a stream-line (a line where the direction in every point at time t is the same as the direction of the velocity vector field) could be closed instead of open between infiltration and exfiltration points. This internal circulation can occur with free convection, or density-driven flow, but if it can be neglected, then nearly every stream-line will start at an infiltration point and end at an exfiltration point (though a few will begin or end at a subsurface stagnation point). For a steady (i.e., time-independent) flow, almost every water particle that infiltrates will eventually exfiltrate, apart from the few that go to stagnation points. However, assuming that the flow field generally is unsteady (i.e., time-dependent), it is possible in principle that some water particles never return to the groundwater table, and even that the point where a particle infiltrates coincides with the point where it exfiltrates. However, lateral transport from infiltration to exfiltration generally takes place, and is of particular importance if the infiltrating water is polluted say, by poor farming practices, since neighboring nature areas can be polluted in this way. Consequently, with the increasing concern for the protection of the environment, natural groundwater flow is receiving attention along with the artificially induced flow to wells that has been frequently studied in the past.

1.11. Scale-dependent interpretations

A simple physical system might be defined as one that obeys simple basic laws like, for instance, the continuity equation and Darcy's law for the description of groundwater flow. But simplicity in the laws does not necessarily imply simplicity in the outcomes. We discussed already the 'fallacy of misplaced concreteness,' meaning that too much confidence can be given in the ultimate truth of the basic laws. Apart from this fallacy, it might well be true that the action of the simple basic laws working over large spatial dimensions and long periods of time will give rise to complex structures and events not directly 'visible' from these simple laws. It is, therefore, necessary to conceive of different laws at different spatio-temporal scales of description. An example is Darcy's law holding on the laboratory scale (the small scale). However, on the regional scale (the large scale) the Dupuit equations hold, which can be generalized to a 'large-scale Darcy's law.' However, the parameters in the (local-scale) Darcy's law (the [local-scale] conductivities), have only a rather loose relationship with the parameters in the Dupuit equations (the transmissivities) and in the large-scale Darcy's law (the large-scale conductivities). A practical consequence of this is that field measurements of local-scale conductivities have to be performed in ways completely different from field measurements of large-scale conductivities. The theoretical derivation (and practical lack of derivation) of large-scale laws from local-scale laws will mainly be addressed in Chapter 4.

It is also possible that different scale phenomena interact. This plays, for instance, a role when groundwater is transporting dissolved matter from one place to another. Indeed, if this is the subject of interest, awareness of the different scales involved is of primordial importance. In particular, the meaning of stream paths calculated on the basis of the scale-dependent laws should be questioned. The significance of stream paths along which water is flowing from infiltration point to exfiltration point is not easy to interpret for three reasons. Firstly, the conductivity for subsurface water flow is spatially heterogeneous; secondly, the groundwater table varies with time; and thirdly, the small-scale spatial variations in the groundwater table are never known. The small-scale spatial variations and the fluctuations over time at short time scales are usually averaged to give a large-scale impression of the groundwater table, both in time and space. This obviously will have considerable effect on the infiltration and exfiltration zones which are computed from it. And there will be even greater ultimate effect on the stream paths which are computed in turn from these zones. Plainly, with the data available at present, there can be no great correspondence between the computations and reality, and considerable experience and insight will be needed to arrive at a valid interpretation of the results. Nevertheless, especially with the application of mathematics, it is a worthwhile undertaking which leads to increasingly more certain knowledge of the future groundwater flow patterns and their ecological consequences. These aspects will mainly be discussed in Chapter 5.

Chapter 2

Basic Equations

2.1. The first principles: conservation laws

This section provides a summary of the two most important principles describing groundwater flow: the continuity equation and Darcy's law. In this book these principles will be considered as the basic level upon which the subsequent examinations are based. However, these basic principles are derivable from principles valid on a deeper level; the Navier-Stokes equations supplemented with some additional conditions. The Navier-Stokes equations describe the flow of fluids (gases and liquids) such as air and surface water which are not contained in porous media. The flow in the pores of a porous medium also satisfies the Navier-Stokes equations, of course, and since this flow is very slow (more precisely, has a very low Reynolds number), it also satisfies the steady Stokes equations as a particular case. These deeper-level principles are in turn derivable from still more deeper-level principles such as the kinetic theory of gases and liquids, etc., and this hierarchy of levels extends to a metaphysics, a deepest level that describes the ultimate basic events of the entire cosmos (see Popper, 1982, pp. 210-211). In the ecological model described in Section 1.2, an event at the higher level can be explained for the most part by events at the lower level, but the event at the lower level cannot be explained fully without reference to the event at the higher level. An example of this will be given in Section 4.8, where a medium-scale Darcy's law is derived from the small-scale Darcy's law by making assumptions about the flow pattern on the large scale. Accordingly, such derivations from deeper levels cannot be considered as a strict proof of the continuity equation and Darcy's law. These laws can be well-proved by experiment, but the possibility of such derivations demonstrates the relatedness of nature at all levels. More practically, these derivations provide important information on the range in which the principles are valid and the range where corrections are needed.

However, we need not delve further into this material here. More details of these derivations can be found in Whitaker (1986) and an introduction to the basic lines of thought is presented in Appendix A.

An extensive presentation of the basic equations, together with numerical solution methods and computer model codes based on them, is given in the book by Bear and Verruijt (1987). Again, a discussion of the basic equations based on stochastic methods, and taking into account the heterogeneity of the subsurface, is presented by Dagan (1989). Except for a brief discussion in Section 3.5, no stochastic methods are used in this book, but much attention is given to the heterogeneity of the subsurface. This heterogeneity means that the equations, and particularly their parameters, are different for different spatial scales and sometimes also for different time scales.

The dependency of Darcy's law, and, in particular, of the conductivity, on the spatial scale is discussed in Section 4.8. The dependency of the porosity and the length of dispersion on the spatial scale as well as on the time scale are discussed in Section 6.4.

2.2. Conservation of mass

The first fundamental principle is the conservation of mass, expressed mathematically in the continuity equation:

$$\partial/\partial t(\rho\theta) + \mathrm{div}(\rho \boldsymbol{q}) = 0. \tag{2.2.1}$$

Where ρ is the density of the fluid (water with dissolved mass), θ is the effective porosity of the water in the porous subsurface, and $\boldsymbol{q} = (q_x, q_y, q_z)$ is the volumetric flow rate of the flowing fluid (a vector), also known as the Darcy velocity.

SI unit density $[kg{\cdot}m^{-3}]$; unit porosity [–] (dimensionless); SI unit volumetric flow rate $[m^3{\cdot}m^{-2}{\cdot}s^{-1} = m{\cdot}s^{-1}]$; conventional unit volumetric flow rate $[m^3{\cdot}m^{-2}{\cdot}d^{-1} = m{\cdot}d^{-1}]$.

In general, the porosity of the porous medium is the volume fraction in pores but, in the continuity equation above, the porosity effective for flow is used. This *effective* porosity is the volume fraction of the porous medium that contains *flowing* fluid, excluding stagnant fluid. The effective porosity equals the porosity of the porous medium at locations in the saturated zone where air does not appear as a separate nonflowing phase, and where all the water participates in the flow, with none left trapped in crevices and hollows. See also Section 2.10, where nonflowing water in the unsaturated zone reduces the above-presented effective porosity to even smaller values. The question how much hardly flowing, almost stagnant water contributes to the large-scale effective porosity is briefly addressed in Section 6.4.

In the context of the continuity equation it is appropriate to explain briefly the meaning of the vector differential operators div(•) and **grad**(•) and of the scalar product of two vectors.

The divergence $\operatorname{div}\boldsymbol{a}$ is a scalar function. For a continuously differentiable vector function $\boldsymbol{a}(\boldsymbol{x},t)$ the divergence can be obtained by partial differentiation of that vector function. In an orthogonal Cartesian system of coordinates, the three components of the vector $\boldsymbol{a}$ are given by $(a_1, a_2, a_3) = (a_x, a_y, a_z)$. It has already been noted that rows and columns of orthogonal Cartesian components have properties similar to vectors; therefore, the vector $\boldsymbol{a}$ will also be denoted as $\boldsymbol{a} = (a_x, a_y, a_z) = a_i$ $(i = 1, 2, 3)$. The divergence is then given as:

$$\operatorname{div}\boldsymbol{a} = \partial a_x/\partial x + \partial a_y/\partial y + \partial a_z/\partial z = \sum_{i=1}^{3} \partial a_i/\partial x_i. \tag{2.2.2}$$

Other conventional notations for the divergence are:

$$\operatorname{div}\boldsymbol{a} = \nabla\boldsymbol{\cdot}\boldsymbol{a} = \partial_i a_i \tag{2.2.3}$$
(summation over the twice-occurring index $i = 1, 2, 3$),

where $\partial_i = \partial/\partial x_i$.

The notation $\nabla\boldsymbol{\cdot}\boldsymbol{a}$ has some resemblance with the scalar product of two vectors. The scalar product $\boldsymbol{a}\boldsymbol{\cdot}\boldsymbol{b}$ of two vectors $\boldsymbol{a} = (a_x, a_y, a_z) = a_i$ $(i = 1, 2, 3)$ and $\boldsymbol{b} = (b_x, b_y, b_z) = b_i$ $(i = 1, 2, 3)$ is defined as the scalar:

$$\boldsymbol{a}\boldsymbol{\cdot}\boldsymbol{b} = (a_x b_x + a_y b_y + a_z b_z) = a_i b_i \tag{2.2.4}$$
(summation over the twice-occurring index i = 1,2,3).

The gradient $\mathbf{grad}\,\alpha$ is a vector function. For a continuously differentiable scalar function $\alpha(\boldsymbol{x},t)$, the gradient can be obtained by partial differentiation of that scalar function, and is so defined that its three orthogonal Cartesian components are given by:

$$\mathbf{grad}\,\alpha = (\partial\alpha/\partial x,\ \partial\alpha/\partial y,\ \partial\alpha/\partial z) = \partial\alpha/\partial x_i \quad (i = 1,2,3). \tag{2.2.5}$$

Other common notations of the gradient are:

$$\mathbf{grad}\,\alpha = \nabla\alpha = \partial_i\alpha \quad (i = 1,2,3). \tag{2.2.6}$$

However, for geohydrological applications, where the parameters of the subsurface may change discontinuously at some planes, it is important to note that weak forms of the divergence and gradient can be defined in which $\boldsymbol{a}$ and α need not be continuously differentiable, but where $\boldsymbol{n}\boldsymbol{\cdot}\boldsymbol{a}$ ($\boldsymbol{n}$ is the unit vector normal to the plane of discontinuity) and α need to be continuous only; see, for example, Morse and Feshbach (1953, part I, p. 35), and Butkov (1973, p. 26).

The continuity equation (2.2.1) can also be written as:

$$(\theta/\rho)(\partial\rho/\partial t + \boldsymbol{v}\boldsymbol{\cdot}\,\mathbf{grad}\,\rho) + \partial\theta/\partial t + \operatorname{div}\boldsymbol{q} = 0 \tag{2.2.7}$$

where $\boldsymbol{v} = \boldsymbol{q}/\theta$ is the mean fluid velocity in the pore space; this velocity is also known as the transport velocity.

Let us now consider the situation where the fluid density ρ depends only on the fluid pressure p $[Pa;\ dbar \approx mH_2O]$ and the mass fraction $\omega[-]$ of dissolved material such as salt; i.e., $\rho = \rho(p,\omega)$. The porosity is assumed to be exclusively dependent on the intergranular pressure p_i and the position vector $\boldsymbol{x}$; in other words, $\theta = \theta(\boldsymbol{x}, p_i)$, where the intergranular pressure is defined as minus the intergranular stress. The intergranular stress $-p_i$ plus the fluid pressure p are equal to the total stress σ caused by the weight of the 'overburden' plus the atmospheric pressure; i.e., $-p_i + p = \sigma$. The assumption $\theta = \theta(\boldsymbol{x}, p_i)$ is valid if the solid grains are not deformable so that changes in porosity are solely caused by rearrangements of the solids. For more details on elastic effects in the subsurface see, for instance, Green and Wang (1990). Now, the continuity equation (2.2.7) can be written as:

$$\begin{aligned}&[(\theta/\rho)\ \partial\rho/\partial p + \partial\theta/\partial p_i]\ \partial p/\partial t + (\theta/\rho)\ \partial\rho/\partial p\ \boldsymbol{v}\boldsymbol{\cdot}\,\mathbf{grad}\,p +\\ &\quad + (\theta/\rho)\ \partial\rho/\partial\omega\ (\partial\omega/\partial t + \boldsymbol{v}\boldsymbol{\cdot}\,\mathbf{grad}\,\omega) + \operatorname{div}\boldsymbol{q} =\\ &\quad = (\partial\theta/\partial p_i)\ \partial\sigma/\partial t.\end{aligned} \tag{2.2.8}$$

In general, the velocity $\boldsymbol{v}$ and the associated pressure gradient $\mathbf{grad}\, p$ associated by Darcy's law (see Section 2.7) are small enough for the term $(\theta/\rho)\, \partial\rho/\partial p\ \boldsymbol{v}\boldsymbol{\cdot}\,\mathbf{grad}\, p$ to be negligibly small in relation to the other terms. For the transport of dissolved material, the convection-dispersion equation holds:

$$\partial\omega/\partial t + \boldsymbol{v}\boldsymbol{\cdot}\,\mathbf{grad}\,\omega = \mathrm{div}(\underline{\boldsymbol{D}}\boldsymbol{\cdot}\,\mathbf{grad}\,\omega) + b \tag{2.2.9}$$

where $\underline{\boldsymbol{D}}$ is the three-dimensional dispersion tensor $[m^2 \cdot s^{-1};\ m^2 \cdot d^{-1}]$ and $b\ [s^{-1};\ d^{-1}]$ represents possible sorption and decay (see Sections 6.2 and 6.3). For a conservative material like salt, $b = 0$. If the dispersion is negligible also, then the right-hand term of equation (2.2.9) will equal zero. Based on these considerations, it will generally be reasonable to assume that the term $(\theta/\rho)\, \partial\rho/\partial\omega\ (\partial\omega/\partial t + \boldsymbol{v}\boldsymbol{\cdot}\,\mathbf{grad}\,\omega)$ in the continuity equation (2.2.8) will also be negligibly small. Thus the continuity equation (2.2.1) will be simplified to:

$$s\ \partial p/\partial t + \mathrm{div}\,\boldsymbol{q} = (\partial\theta/\partial p_i)\ \partial\sigma/\partial t \tag{2.2.10}$$

in which

$$s = (\theta/\rho)\ \partial\rho/\partial p + \partial\theta/\partial p_i \tag{2.2.11}$$

is the specific storage coefficient $[Pa^{-1};\ dbar^{-1} \approx mH_2O^{-1}]$.

2.3. The continuity equation for incompressible flow

The compressibility $(1/\rho)\ \partial\rho/\partial p$ of water at a temperature of 283 K (10^oC) is $0.4789 \times 10^{-9}\, Pa^{-1} \approx 0.5 \times 10^{-9}\, Pa^{-1}$ (see Weast, F-5, 1975). From this, with a porosity of $\theta = 0.3$, it follows that $(\theta/\rho)\ \partial\rho/\partial p \approx 0.15 \times 10^{-9}\, Pa^{-1}$. Here we are dealing with a completely elastic and reversible process of compression and decompression of water. If the water pressure increases by an amount dp then, with constant total stress (the stress caused by the weight of the 'overburden' plus the atmospheric pressure), the intergranular stress will decrease by the same amount dp, and will result in an increase of the porosity θ. It is often accepted that this process also appears to be elastic, and it is sometimes assumed that the term $\partial\theta/\partial p_i$ has the same order of magnitude as the term $(\theta/\rho)\,\partial\rho/\partial p$; so $\partial\theta/\partial p_i \approx 0.15\times 10^{-9}\,Pa^{-1}$ (see, for instance, Todd, 1980). In that case the specific elastic storage coefficient s will have the following order of magnitude:

$$\begin{aligned} s &= (\theta/\rho)\,\partial\rho/\partial p + \partial\theta/\partial p_i \approx 0.3\times 10^{-9}\,Pa^{-1} = \\ &= 3\times 10^{-6}\,dbar^{-1} \approx 3\times 10^{-6}\;\; mH_2O^{-1}. \end{aligned} \tag{2.3.1}$$

This last numeric is also mentioned by Todd (1980). However, much larger values are usually mentioned for the term $\partial\theta/\partial p_i$ especially for poorly permeable (clayey) soils, and larger values also occur when the pores contain gas as a separate phase, so that values can be found to be 100 times greater than mentioned above. However, this aspect will not be explored further in this book.

The elastic compressibility of a 'mixed' fluid-solid porous medium is always known as elastic storage in geohydrology. Elastic storage should not be confused with the phreatic storage, to be discussed in Section 2.10, which relates to fluctuations of the groundwater table in time. If the compressibility of the subsurface is not elastic, but plastic or irreversible, it is referred to as nonelastic (specific) storage.

Looking ahead briefly to the discussions of Darcy's law in Sections 2.4 and 2.7, it can be postulated, from a combination of Darcy's law and the continuity equation with elastic storage, that there follows a relationship between the characteristic length l_c and the characteristic time t_c. For instance, neglecting changes in total stress, and ignoring the fact that Darcy's law describes flow with respect to the elastically *moving* solids, we find the following diffusion equation for a homogeneous and isotropic porous medium:

$$(s/k)\,\partial\phi/\partial t = \nabla^2\phi \tag{2.3.2}$$

where $\nabla^2\phi = \nabla\cdot\nabla\phi = \mathrm{div}\,\mathbf{grad}\,\phi = \partial^2\phi/\partial x^2 + \partial^2\phi/\partial y^2 + \partial^2\phi/\partial z^2$ represents the Laplacean of the potential ϕ, and k is the conductivity (see Section 2.7). Exact solutions of equation (2.3.2) are known for a variety of situations (see Carlslaw and Jaeger, 1959). The relationship between l_c and t_c is important in these solutions, and is given by:

$$t_c = (s/k)\,l_c^2. \tag{2.3.3}$$

Consider, for example, a well-permeable layer with a thickness of $l_c = 100\ m$ and a conductivity of $k = 3\ m^2{\cdot}dbar^{-1}{\cdot}d^{-1}$, then $s/k = 10^{-6}\,d{\cdot}m^{-2}$ and $t_c = 10^{-2}\,d \approx 15\ min$. Thus the time needed for a pressure wave to be transmitted through this thickness of permeable layer is about a few times 15 minutes. If we next consider the example of a thinner but less-permeable layer with a thickness of 1 m and a conductivity of $3 \times 10^{-6}\,m^2{\cdot}dbar^{-1}{\cdot}d^{-1}$, then $s/k = 1\ d{\cdot}m^{-2}$ and $t_c = 1\ d$. This second example shows that it would take a few days for a pressure wave to be transmitted through a poorly permeable layer.

In general, the elastic storage term $s\ \partial p/\partial t$ in the continuity equation can be neglected after a time lapse exceeding the characteristic time t_c by a factor of say, 3 or 4, and the term *incompressible flow* is then used. However, this term does not necessarily imply steady flow; it means only that observed changes take place over a much longer period than the time scale of the term t_c for elastic storage. The elastic storage time scales are usually negligible by comparison with the time scales relevant for the transport of materials, ranging from a few months to several years, so elastic storage will be ignored from now on. Furthermore, phenomena like compaction and related land subsidence, or variations in the total stress and other forms of nonphreatic storage, including plastic deformation, will also be ignored.

In mathematical terms, this produces the well-known simplified continuity equation:

$$\operatorname{div} \boldsymbol{q} = \partial q_x/\partial x + \partial q_y/\partial y + \partial q_z/\partial z = 0. \tag{2.3.4}$$

This equation states that, at every point in time, the volume of water per unit of time flowing into a fixed volume of the subsurface will be equal to the volume of water per unit of time flowing out of that volume of subsurface. Obviously, this will not hold if there are wells in this subsurface volume, but well extraction is a separate problem and one that has been dealt with at length in the geohydrological literature. Consequently, the flow near wells will be considered only in Sections 3.2, 8.4 and 10.2, and will hardly be considered further. Therefore, subsequent problem formulation will assume there are no wells in the subsurface, and the evolution in time of spatial variations in the phreatic plane will be as given. The subsurface flow of water can then be studied in relation to local differences in the groundwater table, without forced flow due to extraction. In other words, the study will be devoted to natural groundwater flow, and nothing else.

2.4. Conservation of momentum

The second fundamental principle is the conservation of momentum as expressed mathematically by Darcy's law. According to Darcy's law, there is at each point $\boldsymbol{x}$ in the subsurface a *linear* relationship between the volumetric flow rate $\boldsymbol{q} = (q_x, q_y, q_z) = q_i$ $(i = 1,2,3)$ (a vector) $[m{\cdot}s^{-1};\ m{\cdot}d^{-1}]$ and the driving force per unit volume $\mathbf{grad}\, p - \rho \boldsymbol{g} = (\partial p/\partial x - \rho g_x,$ $\partial p/\partial y - \rho g_y, \partial p/\partial z - \rho g_z) = \partial_i p - \rho g_i$ $(i = 1,2,3)$ (a vector) $[N{\cdot}m^{-3}]$.

The 'point' concept should not be taken too literally; here it relates to a small volume-element, say 1 cm^3, which contains a great many grains (say more than 100) of the porous subsurface. Such a volume-element is called a representative elementary volume or REV.

The relationship is linear because the coefficients which link the components of the volumetric flow rate to the components of the driving force per unit volume are independent of the volumetric flow rate and the driving force per unit volume. The coupling-coefficients (the conductivities referred to in Section 2.7) are generally dependent on other quantities such as location, time, temperature, and mass fraction of solutes. However, before expressing Darcy's law as a formula, let us consider the concepts of potential and piezometric head.

2.5. The potential for an irrotational driving force

If the driving force per unit volume is irrotational, it can be written as a potential gradient $\mathbf{grad}\,\phi$, which means that:

$$\begin{aligned}\mathbf{grad}\, p - \rho \boldsymbol{g} &= \mathbf{grad}\,\phi = \\ &= (\partial\phi/\partial x,\ \partial\phi/\partial y,\ \partial\phi/\partial z) = \partial_i \phi \quad (i = 1,2,3)\end{aligned} \tag{2.5.1}$$

where $\phi(\boldsymbol{x}, t)$ is the potential.

SI unit potential $[Pa]$; conventional unit potential $[mH_2O \approx dbar]$; the potential can also be given as energy per unit volume, 1 $Pa = 1\ J{\cdot}m^{-3}$. SI unit potential gradient $[Pa{\cdot}m^{-1}]$; conventional unit potential gradient $[mH_2O{\cdot}m^{-1}]$; conversion: 1 $mH_2O{\cdot}m^{-1} \approx 1\ dbar{\cdot}m^{-1}$.

The meaning of irrotationality can be explained as follows: The curl, $\mathbf{curl}\,\boldsymbol{a}$, is a vector function. For a continuously differentiable vector function $\boldsymbol{a} = (a_x, a_y, a_z) = a_i$ $(i = 1,2,3)$, the curl can be obtained by partial differentiation of that vector function. In a right-handed orthogonal Cartesian coordinate system the curl is defined in such a way that:

$$\begin{aligned}\mathbf{curl}\,\boldsymbol{a} &= (\partial a_z/\partial y - \partial a_y/\partial z,\ \partial a_x/\partial z - \partial a_z/\partial x,\\ &\qquad \partial a_y/\partial x - \partial a_x/\partial y) =\\ &= \sum_{j=1}^{3}\sum_{k=1}^{3} \epsilon_{ijk}\ \partial a_k/\partial x_j \quad (i = 1,2,3).\end{aligned} \tag{2.5.2}$$

The permutation-tensor ϵ_{ijk} is defined here as:

$$\begin{aligned}\epsilon_{ijk} &= +1 \text{ if } ijk \text{ are cyclic,} &&\text{i.e., } \epsilon_{123} = \epsilon_{231} = \epsilon_{312} = +1\\ \epsilon_{ijk} &= -1 \text{ if } ijk \text{ are not cyclic,} &&\text{i.e., } \epsilon_{132} = \epsilon_{321} = \epsilon_{213} = -1\\ \epsilon_{ijk} &= 0, \text{ if there are two equal indices present.}\end{aligned} \tag{2.5.3}$$

(In fact, $\mathbf{curl}\,\boldsymbol{a}$ is not a 'true' vector. It is an axial vector having the property that its definition changes sign if we change from a right-handed to a left-handed coordinate system; see Morse and Feshbach, 1953, part I, p. 11 and p. 29.)

Other usual notations of the curl are:

$$\mathbf{curl}\,\boldsymbol{a} = \mathbf{rot}\,\boldsymbol{a} = \nabla \times \boldsymbol{a} = \epsilon_{ijk}\,\partial_j\,a_k \tag{2.5.4}$$

(summation over the twice-occurring indices j and k).

For the weak-form definition of the curl for noncontinuously differentiable $\boldsymbol{a}$ at a plane where $\boldsymbol{n} \times \boldsymbol{a}$ is continuous ($\boldsymbol{n}$ is the unit normal vector on the plane of discontinuity) see, for example, Morse and Feshbach (1953, part I, p. 43), and Butkov (1973, p. 26).

To explain the notation $\nabla \times \boldsymbol{a}$, we show some similarity between the curl of a vector and the vector product of two vectors. In a right-handed orthogonal Cartesian system of coordinates, the vector product $\boldsymbol{a} \times \boldsymbol{b}$ of two vectors $\boldsymbol{a} = (a_x, a_y, a_z) = a_i$ $(i = 1,2,3)$ and $\boldsymbol{b} = (b_x, b_y, b_z) = b_i$

$(i = 1,2,3)$ is defined as the vector:

$$\boldsymbol{a} \times \boldsymbol{b} = (a_y b_z - a_z b_y,\ a_z b_x - a_x b_z,\ a_x b_y - a_y b_x) = \\ = \epsilon_{ijk}\, a_j\, b_k \qquad (2.5.5)$$

(summation over the twice-occurring indices j and k).

This vector product results in a vector that stands perpendicular on the vectors $\boldsymbol{a}$ and $\boldsymbol{b}$, i.e., $\boldsymbol{a}\boldsymbol{\cdot}(\boldsymbol{a} \times \boldsymbol{b}) = \boldsymbol{b}\boldsymbol{\cdot}(\boldsymbol{a} \times \boldsymbol{b}) = 0$. However, the curl of a vector $\boldsymbol{a}$ does *not* generally stand perpendicularly on the vector $\boldsymbol{a}$.

The curl gets its practical *usefulness* from the following two identities:

(1) $\mathrm{div}\,\mathbf{curl}\,\boldsymbol{a} = 0$ for every (sufficiently smooth) vector function $\boldsymbol{a}$.
(2) $\mathbf{curl}\,\mathbf{grad}\,\alpha = \mathbf{0}$ for every (sufficiently smooth) scalar function α.

The reverse is also true:

(1) Every solenoidal (i.e., divergence-free) vector function can be written as the curl of another vector function.
(2) Every irrotational vector function can be written as the gradient of a scalar function.

The latter proposition (2) is used to introduce the potential ϕ:

If $\mathbf{curl}(\mathbf{grad}\,p - \rho\boldsymbol{g}) = \boldsymbol{g} \times \mathbf{grad}\,\rho = \mathbf{0}$, then $\mathbf{grad}\,p - \rho\boldsymbol{g} = \mathbf{grad}\,\phi$.

For water with a spatially constant density $\rho_0(t)$ $[kg{\cdot}m^{-3}]$ the potential is defined by the well-known expression:

$$\phi(x_h, y_h, z_v, t) = p(x_h, y_h, z_v, t) - \rho_0(t)\, g\, z_v \qquad (2.5.6)$$

where $p(x_h, y_h, z_v, t)$ is the water pressure $[Pa;\ dbar \approx mH_2O]$, $g = |\boldsymbol{g}| = \sqrt{(\boldsymbol{g}\cdot\boldsymbol{g})}$ is the absolute value of the gravitational acceleration $[m{\cdot}s^{-2}]$, $\rho_0(t)\ g$ is the specific gravity $[Pa{\cdot}m^{-1};\ dbar{\cdot}m^{-1} \approx mH_2O{\cdot}m^{-1}]$, x_h and y_h are the exactly horizontal coordinates; and z_v is the exactly vertical coordinate directed positively downward $[m]$. The term $\rho_0\, g\, z_v$ $[Pa = J{\cdot}m^{-3};\ dbar \approx mH_2O]$ is sometimes called the potential energy (per unit volume).

$\rho_0 \approx 1000\ kg{\cdot}m^{-3}$, $g \approx 10\ m{\cdot}s^{-2}$ on earth, from which it follows that $\rho_0\, g \approx 1\ dbar{\cdot}m^{-1}$; $\rho_0\, g = 1\ mH_2O{\cdot}m^{-1}$ (exact at 10^oC at sea-level at latitude 52^oN.)

If the density of the water is solely dependent on the vertical coordinate (and the time), i.e., if $\rho = \rho(z_v, t)$, then expression (2.5.6) for the potential can be generalized as:

$$\phi(x_h, y_h, z_v, t) = p(x_h, y_h, z_v, t) - \int_0^{z_v} \rho(z_v', t)\; g\; \mathrm{d}z_v'. \tag{2.5.7}$$

However, if the density of the water depends also on the horizontal coordinates x_h and y_h, i.e., when $\rho = \rho(x_h, y_h, z_v, t)$, then the driving force per unit volume is not irrotational. In that case, the introduction of a potential is still possible, but then the potential gradient is not the only driving force. Therefore, in that case it is often simpler to deal directly with the water pressure rather than a potential. Dependency of the density on the horizontal coordinates x_h and y_h gives rise to what is called density-driven flow or free convection, which is important when combinations of fresh and saline water are studied. Free convection will be considered further in Appendix C.

2.6. The piezometric head

In an observation well at a horizontal location (x_h, y_h) with a filter depth z_v, water with density $\rho(z_v, t)$ will rise to a height $h(x_h, y_h, z_v, t)$ measured with respect to the horizontal plane $z_v = 0$ (i.e., $z_v = -h(x_h, y_h, z_v, t)$ is the vertical coordinate of the top of the water column in the observation well). When the water pressure is in equilibrium with the weight of the water column plus the atmospheric pressure, then:

$$p(x_h, y_h, z_v, t) = \rho(z_v, t)\; g\; [h(x_h, y_h, z_v, t) + z_v] + p_{atm}(t). \tag{2.6.1}$$

Substitution of equation (2.6.1) into equation (2.5.7) shows that the potential at the horizontal location (x_h, y_h) and filter depth z_v is given by:

$$\begin{aligned} \phi(x_h, y_h, z_v, t) &= p_{atm}(t) + \rho(z_v, t)\; g\; h(x_h, y_h, z_v, t) + \\ &+ \rho(z_v, t)\; g\; z_v - \int_0^{z_v} \rho(z_v', t)\; g\; \mathrm{d}z_v'. \end{aligned} \tag{2.6.2}$$

Expression (2.6.2) is often used to calculate the potentials needed in Darcy's law (for a horizontally stratified water density) to solve problems with both fresh and saline water, where the piezometric heads $h(x_h, y_h, z_v, t)$ in the fresh and the saline water are measured in observation wells.

If the density of the water is constant all over, i.e., when $\rho(z_v, t) = \rho_0(t)$, expression (2.6.2) for the potential is simplified to the well-known formula:

$$\phi(x, y, z, t) = p_{atm}(t) + \rho_0(t)\, g\, h(x, y, z, t). \tag{2.6.3}$$

In this last formula, the horizontal and vertical coordinates x, y and z are not necessarily the exactly horizontal-vertical coordinates of the location vector $\boldsymbol{x}$, but may belong to a system of coordinates which is rotated in relation to the exact horizontal-vertical system.

2.7. Darcy's law and the conductivity tensor

The most general linear relationship between the volumetric flow rate $\boldsymbol{q}$ and the driving force per unit volume $\mathbf{grad}\, p - \rho \boldsymbol{g}$ is given by Darcy's law:

$$\boldsymbol{q} = -\underline{\boldsymbol{k}} \cdot (\mathbf{grad}\, p - \rho \boldsymbol{g}) \tag{2.7.1}$$

where $\underline{\boldsymbol{k}}$ is a three-dimensional tensor. (Note: tensors are indicated by underlined symbols in bold typeface.)

In the irrotational approximation Darcy's law (2.7.1) becomes:

$$\boldsymbol{q} = -\underline{\boldsymbol{k}} \cdot \mathbf{grad}\, \phi. \tag{2.7.2}$$

Written in orthogonal Cartesian components, we find from equation (2.7.2) that:

$$q_x = -k_{xx}\, \partial\phi/\partial x - k_{xy}\, \partial\phi/\partial y - k_{xz}\, \partial\phi/\partial z \tag{2.7.3a}$$

$$q_y = -k_{yx}\, \partial\phi/\partial x - k_{yy}\, \partial\phi/\partial y - k_{yz}\, \partial\phi/\partial z \tag{2.7.3b}$$

$$q_z = -k_{zx}\, \partial\phi/\partial x - k_{zy}\, \partial\phi/\partial y - k_{zz}\, \partial\phi/\partial z \tag{2.7.3c}$$

or briefly:

$$q_i = -\sum_{j=1}^{3} k_{ij}\ \partial\phi/\partial x_j = k_{ij}\ \partial\phi/\partial x_j \quad (i = 1,2,3) \tag{2.7.4}$$

(summation over the twice-occurring index $j = 1,2,3$).

In the same way as with vectors, the tensor $\underline{\boldsymbol{k}}$ with orthogonal Cartesian components k_{ij} $(i,j = 1,2,3)$ can be denoted by a matrix as:

$$\underline{\boldsymbol{k}} = \begin{bmatrix} k_{xx} & k_{xy} & k_{xz} \\ k_{yx} & k_{yy} & k_{yz} \\ k_{zx} & k_{zy} & k_{zz} \end{bmatrix} = k_{ij} \quad (i,j = 1,2,3). \tag{2.7.5}$$

The above form of Darcy's law (2.7.2), together with the continuity equation (2.3.4) ($\operatorname{div} \boldsymbol{q} = 0$), can be derived from the steady Stokes equations which describe flow in the pore space (see Appendix A). From the same derivation it also follows that the tensor $\underline{\boldsymbol{k}} = k_{ij}$ $(i,j = 1,2,3)$ is symmetrical, i.e., $k_{ij} = k_{ji}$, and positive definite, i.e., $a_i\, k_{ij}\, a_j > 0$ (summation over the twice-occurring indices i and j) for a_i and a_j not equal to 0. It should be noted that the symmetry property of the conductivity tensor $\underline{\boldsymbol{k}}(\boldsymbol{x},t)$ holds only on the scale of spatial 'points' $\boldsymbol{x}$, or, more accurately, in the limit of infinitely small representative elementary volume elements $dxdydz$. When averaging $\underline{\boldsymbol{k}}(\boldsymbol{x},t)$ over larger volumes with a heterogeneous distribution of $\underline{\boldsymbol{k}}(\boldsymbol{x},t)$-values, the symmetry is generally lost (see Appendix D).

The six Cartesian components $k_{xx}(x,y,z,t)$, $k_{yy}(x,y,z,t)$, $k_{zz}(x,y,z,t)$, $k_{xy}(x,y,z,t) = k_{yx}(x,y,z,t)$, $k_{xz}(x,y,z,t) = k_{zx}(x,y,z,t)$, $k_{yz}(x,y,z,t)$ $= k_{zy}(x,y,z,t)$ in Darcy's law will be called here the conductivities; in petroleum reservoir engineering, however, they are called the mobilities.

SI unit conductivity $[m^2 \cdot Pa^{-1} \cdot s^{-1}]$; conventional unit conductivity $[m^2 \cdot mH_2O^{-1} \cdot d^{-1}]$; conversion $1\ m^2 \cdot mH_2O^{-1} \cdot d^{-1} \approx 1\ m^2 \cdot dbar^{-1} \cdot$ $\cdot d^{-1} = 1/86400\ m^2 \cdot dbar^{-1} \cdot s^{-1}$.

If there is no stagnant air as a separate phase in the pore space, the conductivity is equal to the intrinsic permeability κ $[m^2;\ \mu m^2 = darcy =$ $= D]$ divided by the dynamic viscosity μ $[Pa \cdot s;\ dbar \cdot d \approx mH_2O \cdot d]$. The intrinsic permeability is a property of only the porous medium, whereas the dynamic viscosity is a property of only the fluid. Consequently, the conductivity is a property of the fluid-solid porous 'mixture.' However, if stagnant air does appear as a separate phase in the pore space, then the above expression for the conductivity must be multiplied by the relative permeability κ_r [–], a number less than one.

The intrinsic permeability can change in time due to compaction/decompaction and metamorphosis, though these changes are very slow, and the time-dependence of k_{ij} may well be ignored for most practical purposes. The spatial variability, however, plays a large part in all practical situations.

The dynamic viscosity of water is strongly temperature-dependent; the conductivity increases by 50% in the temperature range of 10^oC to 26^oC (see Weast, 1975, F-49).

Conductivity, as defined here, is independent of the gravitational acceleration $\boldsymbol{g}$. A definition independent of the gravitational acceleration is preferable from a physical point of view. It is especially advantageous in centrifuges where $\boldsymbol{g}$ is increased, and in drop-towers and spacecraft where $\boldsymbol{g}$ is decreased.

In virtually all textbooks on geohydrology, a gravity-dependent definition of the conductivity is given. There the customary unit of potential mH_2O is taken equal to the head of fresh water (with uniform density ρ_0) in meters. This is a gravity-dependent definition since this head depends on the gravitational acceleration g. In that case 1 mH_2O = 1 m and the conventional unit of conductivity is $m^2 \cdot mH_2O^{-1} \cdot d^{-1} = m^2 \cdot m^{-1} \cdot d^{-1} = = m \cdot d^{-1}$. The conductivity defined in this latter way is generally called the hydraulic conductivity. In a similar way the conventional unit of potential gradient is $mH_2O \cdot m^{-1} = m \cdot m^{-1} = 1$ (dimensionless).

The more conventional, but equivalent way of introducing the fresh water head ϕ_{head} is to relate it to the earlier-defined potential ϕ by $\phi_{head} = = \phi/(\rho_0\, g)$. Then the hydraulic conductivity $\underline{\boldsymbol{k}}_{hyd}$ is related to the earlier-defined conductivity $\underline{\boldsymbol{k}}$ by $\underline{\boldsymbol{k}}_{hyd} = \rho_0\, g\, \underline{\boldsymbol{k}}$.

In this context it is appropriate to recall the words of the renowned physicist, Lord Rayleigh:

On the other hand, engineers, who might make much more use of it (the rules of dimensional analysis) than they have done, employ a notation which tends to obscure it. I refer to the manner in which gravity is treated. When the question under consideration depends essentially upon gravity, the symbol of gravity (g) makes no appearance, but when gravity *does not enter into the question at all, g obtrudes itself conspicuously.*

(see Silberberg and McKetta, 1953, part II).

The above-mentioned gravity-dependence in the units of potential and conductivity can be avoided when defining 1 mH_2O as the head of fresh water measured precisely at 10^oC, i.e., $\rho_0 = 999.728\ kg{\bullet}m^{-3}$ (see Weast, 1975, F-5) at sea-level at latitude 52^oN, i.e., $g = 9.81247\ m{\bullet}s^{-2}$ (see Weast, 1975, F-195). Then 1 $mH_2O = 0.9809801\ dbar$, and equivalencing 1 mH_2O to 1 m is no longer allowed. #

2.8. Anisotropy

It is practically impossible to know all six values of k_{ij} from one point to another in the subsurface. The normal practice is to change over to data reduction with the help of plausible assumptions.

In Section 4.6, the principal axes of a symmetrical tensor are introduced for a two-dimensional tensor $\underline{\boldsymbol{T}} = T_{ij}$ $(i,j = 1,2)$. However, the theory for two-dimensional tensors can be extended to three-dimensional tensors $\underline{\boldsymbol{T}} = T_{ij}$ $(i,j = 1,2,3)$. From this extension it follows that, since k_{ij} is symmetrical at every point in the subsurface, there are three mutually perpendicular principal axes (directions 1, 2 and 3 with curvilinear orthogonal coordinates ξ_1, ξ_2 and ξ_3) in which Darcy's law can be written as:

$$q_1 = -k_1\ h_1^{-1}\ \partial\phi/\partial\xi_1 \tag{2.8.1a}$$

$$q_2 = -k_2\ h_2^{-1}\ \partial\phi/\partial\xi_2 \tag{2.8.1b}$$

$$q_3 = -k_3\ h_3^{-1}\ \partial\phi/\partial\xi_3 \tag{2.8.1c}$$

where h_1, h_2 and h_3 are the scale factors belonging to the curvilinear orthogonal coordinate system. For an explanation of the concept of curvilinear orthogonal coordinates see Morse and Feshbach (1953, part I, pp. 21-31).

To obtain data reduction, a much-used assumption now is that the principal axes of the conductivity tensor coincide with those of a Cartesian system of coordinates (with coordinates $\xi_1 = x$, $\xi_2 = y$, $\xi_3 = z$ and scale factors $h_1 = h_2 = h_3 = 1$; not necessarily exactly horizontal and vertical coordinates). These axes are supposed to coincide with those along the stratigraphic pattern (directions x and y) and perpendicular to the stratigraphic pattern (direction z). For the sake of further data reduction, it is customary to assume that the two conductivities in the direction of the stratigraphy are equal, i.e., $k_1 = k_2 = k_h$. The two remaining conductivities $k_h(x,y,z,t)$ and $k_z(x,y,z,t)$ are in fact dependent on the position (x,y,z) in the subsurface, and possibly on the time t. Thus, the following simplified Darcy equations will always be used from now on:

$$q_x(x,y,z,t) = -k_h(x,y,z,t)\ \partial\phi(x,y,z,t)/\partial x \tag{2.8.2a}$$

$$q_y(x,y,z,t) = -k_h(x,y,z,t)\ \partial\phi(x,y,z,t)/\partial y \tag{2.8.2b}$$

$$q_z(x,y,z,t) = -k_z(x,y,z,t)\ \partial\phi(x,y,z,t)/\partial z \tag{2.8.2c}$$

It is noted here that almost all results derived on the basis of equations (2.8.2) can simply be extended to situations $k_1 = k_h$, $k_2 = c\,k_h$, where c is a constant. This can be done by defining a coordinate $y' = y/\sqrt{c}$ and a volumetric flow rate $q_{y'}(x,y',z,t) = q_y(x,y,z,t)/\sqrt{c}$. Then Darcy's law (2.8.2) and the continuity equation (2.3.4) ($\operatorname{div} \boldsymbol{q} = 0$) hold for $q_x(x,y',z,t)$, $q_{y'}(x,y',z,t)$ and $q_z(x,y',z,t)$ in the (x,y',z) coordinate system.

As a clarifying example, the even-more-simplified problem of the *perfectly layered* subsurface will be considered. A perfectly layered subsurface is understood to be a subsurface where the conductivities change not in the lateral but in the vertical direction, where 'vertical' means perpendicular to the lateral, but not exactly horizontal direction. The conductivities are then given by: $k_h = k_h(z,t)$, $k_z = k_z(z,t)$. In reality, the subsurface is never perfectly layered; there is always dependence on the lateral position (x,y). Nevertheless, the subsurface often does have a layered character, especially in sedimentary basins, and it is relatively easy to describe the flow mathematically and obtain much qualitative insight into the character of the flow in a perfectly layered subsurface, as will be demonstrated later.

2.9. The dynamic boundary condition on the water table

The top boundary of the groundwater flow domain is given by the groundwater table $z = -h_f(x, y, t)$. Based on the dynamic boundary condition $p(x,\ y,\ -h_f(x, y, t),\ t) = p_{atm}(t)$ discussed earlier, it follows from equation (2.5.7) that the boundary condition for the potential is:

$$\phi(x,\ y,\ -h_f(x, y, t),\ t) = p_{atm}(t) - \int_0^{-h_f} \rho(z_v', t)\ g\ \mathrm{d}z_v' \qquad (2.9.1)$$

where the integration is over the exactly vertical coordinate z_v. Conversion from coordinate system (x, y, z) to coordinate system (x_h, y_h, z_v) is always possible, but will not be elaborated further here.

Note, however, that the magnitude of the gravitational acceleration g plays a part in the boundary condition given above. In conventional units, where the unit of $\rho g = mH_2O \cdot m^{-1}$ and, moreover, where $mH_2O = m$ is assumed, the dependence on gravitational acceleration is rendered invisible.

In many situations, it can be assumed that, in the interval between $z_v = 0$ and $z_v = -h_f(x, y, t)$, the density of the water is spatially constant, i.e., $\rho(z_v, t) = \rho_0(t)$. In that case, the dynamic boundary condition (2.9.1) reduces to:

$$\phi(x,\ y,\ -h_f(x, y, t),\ t) = p_{atm}(t) + \rho_0(t)\ g\ h_f(x, y, t). \qquad (2.9.2)$$

2.10. The kinematic boundary condition on the water table

The coordinates $(\boldsymbol{x}, t) = (x, y, z, t)$ of the spatially variable and fluctuating-in-time groundwater table are given by the general relationship $F(\boldsymbol{x}, t) =$ $= F(x, y, z, t) = 0$. Such a relationship describes a plane in three-dimensional space at every point in time. The unit direction vector $\boldsymbol{n}$ normal to this plane is given by:

$$\boldsymbol{n} = \textbf{grad}\ F / |\textbf{grad}\ F| \qquad (2.10.1a)$$

or, in Cartesian coordinates:

$$(n_x, n_y, n_z) = (\partial F/\partial x, \partial F/\partial y, \partial F/\partial z)/ \\ /[(\partial F/\partial x)^2 + (\partial F/\partial y)^2 + (\partial F/\partial z)^2]^{1/2}. \tag{2.10.1b}$$

The equation $F(\boldsymbol{x}, t) = 0$ generally gives a multi-valued relation between the lateral location (x, y) and the 'vertical' (but not necessarily exactly vertical) depth z. On physical grounds, at least one such relationship is known to exist. This relationship is written as:

$$z = -h_f(x, y, t). \tag{2.10.2}$$

The index f denotes a choice of one particular solution if $F(\boldsymbol{x}, t)$ gives more values for z at the same location (x, y); thus other solutions $z = = -h_g(x, y, t)$, $z = -h_h(x, y, t)$, etc. are also possible at the same location (x, y). These last two solutions, with the indices g and h, could be two *apparent* water tables of a saturated rain-lens above the phreatic plane, for instance, while the first value, with index f, describes the phreatic plane or (non-apparent) water table. The expression $z = -h_f(x, y, t)$ can obviously be written as:

$$F_f(x, y, z, t) = F(x,\ y,\ -h_f(x, y, t),\ t) = \\ = z + h_f(x, y, t) = 0. \tag{2.10.3}$$

According to equation (2.10.1b), the above formulation (2.10.3) provides the basis for the following expression for the normal unit vector $\boldsymbol{n}_f$ on the water table:

$$\boldsymbol{n}_f = (\partial h_f/\partial x,\ \partial h_f/\partial y,\ 1)/ \\ /[(\partial h_f/\partial x)^2 + (\partial h_f/\partial y)^2 + 1]^{1/2}. \tag{2.10.4}$$

The velocity of motion of the location vector $\boldsymbol{x}_f(t) = (x,\ y,\ -h_f(x, y, t))$ on the water table at a fixed, motionless lateral location (x, y) is given by:

$$\partial \boldsymbol{x}_f/\partial t = (0,\ 0,\ -\partial h_f/\partial t). \tag{2.10.5}$$

Thus, by combining equations (2.10.4) and (2.10.5), the normal component of the velocity of motion of the water table $\boldsymbol{n}_f \cdot \partial \boldsymbol{x}_f / \partial t$ is given by:

$$\begin{aligned} \boldsymbol{n}_f \cdot \partial \boldsymbol{x}_f / \partial t &= -(\partial h_f / \partial t) / \\ & /[(\partial h_f / \partial x)^2 + (\partial h_f / \partial y)^2 + 1]^{1/2}. \end{aligned} \tag{2.10.6}$$

The above velocity of displacement perpendicular to the groundwater table is equal to the normal component of the transport velocity of the groundwater at the level of the groundwater table $\boldsymbol{n}_f \cdot \boldsymbol{v}$, reduced by the normal component of the groundwater replenishment velocity from the unsaturated zone.

With the help of expression (2.10.4) we find:

$$\begin{aligned} \boldsymbol{n}_f \cdot \boldsymbol{v} &= (v_x \partial h_f / \partial x + v_y \partial h_f / \partial y + v_z) / \\ & /[(\partial h_f / \partial x)^2 + (\partial h_f / \partial y)^2 + 1]^{1/2}. \end{aligned} \tag{2.10.7}$$

The kinematic boundary condition follows then from equations (2.10.6) and (2.10.7) as:

$$\theta \; \partial h_f / \partial t + q_x \; \partial h_f / \partial x + q_y \; \partial h_f / \partial y + q_z = P_e. \tag{2.10.8}$$

Here the vector $\boldsymbol{q}/\theta = (q_x/\theta, q_y/\theta, q_z/\theta) = \boldsymbol{v} = (v_x, v_y, v_z)$ $[m{\cdot}s^{-1}; m{\cdot}d^{-1}]$ is the transport velocity at the location of the water table; θ $[-]$ is the porosity at the location of the groundwater table; and P_e $[m{\cdot}s^{-1}; m{\cdot}d^{-1}]$ is the component of the groundwater replenishment from the unsaturated zone perpendicular to the groundwater table, multiplied by the term $[(\partial h_f / \partial x)^2 + (\partial h_f / \partial y)^2 + 1]^{1/2}$.

Since we may reasonably assume that flow in the unsaturated zone is in the vertical direction only, it follows from the above-given definition of P_e and from expression (2.10.4) that, in that case, it is a good approximation to say that P_e is the groundwater replenishment. This replenishment is usually called the 'natural groundwater recharge,' or the 'effective precipitation,' but these expressions could be somewhat misleading for present purposes.

When the groundwater table moves up or down, most of the unsaturated zone will move up or down with it. However, there will always be a percentage of water (often ca. 5%) that does not flow, particularly in the upper parts of the unsaturated zone (i.e., in the atmospheric zone), which therefore provides a contribution P_e' to P_e that is not the result of effective precipitation. This contribution is proportional to $\partial h_f/\partial t$ and equal to $\theta_0\ \partial h_f/\partial t$, where θ_0 is the volume fraction of nonflowing water. When included in the kinematic boundary condition (2.10.8) given above, it results in an effective porosity $\theta - \theta_0$. To avoid confusion with the effective porosity introduced in Section 2.2, this effective porosity $\theta - \theta_0$ is also known as the phreatic storage coefficient. Consequently, in calculations based on the phreatic storage coefficient $\theta - \theta_0$, P_e will be related only to the natural groundwater recharge, or effective precipitation, whereas calculations based on the 'ordinary' effective porosity θ (see Section 2.2) will also include in P_e the contribution of water that is stagnant in the unsaturated zone.

In reality, the situation is even more complicated. Most of the water in the unsaturated zone will flow, but only very slowly. Consequently, this flow will not respond to rapid changes of the groundwater table (which means small values for $\theta - \theta_0$), though it will respond to slow changes (which means $\theta_0 \approx 5\%$). This makes the phreatic storage coefficient a frequency-dependent or time-dependent flow property. Since a flow-dependent coefficient may be considered as a weak point in a theory, we observe here a draw-back of the separation of the description of flow in the saturated zone from the description of flow in the unsaturated zone.

As we have seen, the kinematic boundary condition (2.10.8) describes the evolution-in-time of the groundwater table; in other words, it describes the phreatic storage. However, it is a far from trivial operation to solve the kinematic boundary condition with a model for flow in the subsurface in combination with a model for P_e. In the first place, the moving position of the top-boundary and the terms $q_x\ \partial h_f/\partial x$ and $q_y\ \partial h_f/\partial y$ make the flow problem nonlinear; in the second place, the setting-up of a good, practical yet manageable model for the 'natural groundwater recharge' is a difficult proposition.

The task of defining the evolution-in-time of spatial variations in the groundwater table $h_f(x, y, t)$ is a classic problem in geohydrology, and one on which much has been published. It should be noted that the problem is drastically simplified by combining the kinematic boundary condition (2.10.8) with the Dupuit approximation (see Section 4.2). This combination results in the two-dimensional Dupuit-Forchheimer approximation (see Section 4.3) which is nearly always brought into use for numerical models with a description of phreatic storage, but claims that the computations are 'completely three-dimensional' will need to be taken 'with a pinch of salt' so far as this aspect is concerned.

The description of the evolution of $h_f(x, y, t)$ will hardly be considered in this book, which means that terms such as $\partial h_f/\partial t$ and $\partial\phi/\partial t$ do not appear in the mathematical formulations. However, that does not mean that the flow problem is steady. Indeed, the dynamic boundary condition for $\phi(x,\ y,\ -h_f(x, y, t),\ t)$ on the top plane, and also the position of the top plane $z = -h_f(x, y, t)$, form a time-dependent boundary condition for the problem to be considered.

2.11. Projection of the top-boundary conditions

By way of the simplification that is needed to progress with analytical mathematical methods, the dynamic and kinematic boundary conditions are projected on the plane $z = 0$. The dynamic boundary condition (2.9.1) then becomes:

$$\phi(x, y, 0, t) = p_{atm}(t) - \int_0^{-h_f} \rho(z_v', t)\ g\ \mathrm{d}z_v' \tag{2.11.1}$$

and the kinematic boundary condition (2.10.8) becomes:

$$\theta\ \partial h_f/\partial t + q_x(0)\ \partial h_f/\partial x + q_y(0)\ \partial h_f/\partial y + q_z(0) = P_e. \tag{2.11.2}$$

The above two projected boundary conditions (2.11.1) and (2.11.2) hold on $z = 0$; thus $q_x(0) = q_x(x, y, 0, t)$, $q_y(0) = q_y(x, y, 0, t)$ and $q_z(0) =$ $= q_z(x, y, 0, t)$ in the above formulae. This is a permissible simplification if the groundwater basin is sufficiently thick.
It is also worth noting that, in the computation of $h_f(x, y, t)$ with the help of the projected kinematic boundary condition (2.11.2), it is sometimes used in its linearized form:

$$\begin{aligned} \theta(x,\ y,\ -h_f(x, y, t),\ t)\ \partial h_f(x, y, t)/\partial t + q_z(x, y, 0, t) &= \\ = P_e(x, y, t). & \end{aligned} \tag{2.11.3}$$

An essentially nonlinear description of the groundwater table can be linearized in this way. In the well-known Dupuit-Forchheimer approximation, the assumption of a transmissivity independent of the groundwater table is equivalent to projection and linearization of the kinematic boundary condition (see Section 4.3).

2.12. Summary of the natural flow formulation

The lower boundary of the domain is chosen to be on the plane $z = d$, where the boundary condition $q_z(x, y, d, t) = 0$. However, the choice of the depth d, the 'impermeable' base, is far from trivial, and a criterion for the determination of d will be derived in Sections 5.1 and 5.2.

The flow problem consists of the field equations (continuity and Darcy's law), the dynamic boundary condition at $z = -h_f(x, y, t)$ (or $z = 0$ as an approximation) and the boundary condition on the impermeable base which, being *linear*, leads to a relatively simple mathematical analysis. This analysis is relatively simple because the superposition principle is valid for linear problems.

N.B. The definition of the evolution-in-time of the spatial variations in the groundwater table is a *nonlinear* problem for which the superposition principle is *not* valid.

Following this detailed explanation of the fundamentals, we can now continue with further elaboration.

Chapter 3

Perfectly and Nonperfectly Layered Basins

3.1. Four Laplace-type equations for perfectly layered basins

Inserting Darcy's law in the continuity equation leads to the well-known Laplace-type equation for the potential ϕ:

$$\begin{aligned}\partial/\partial x(k_h\,\partial\phi/\partial x)\ &+ \partial/\partial y(k_h\,\partial\phi/\partial y)+\\ &+ \partial/\partial z(k_z\,\partial\phi/\partial z) = 0.\end{aligned} \tag{3.1.1}$$

Equation (3.1.1) states that the vector with Cartesian components $(k_h\,\partial\phi/\partial x, k_h\,\partial\phi/\partial y, k_z\,\partial\phi/\partial z)$ is solenoidal (solenoidal means divergence-free).

If $k_h = k_z = k$ with constant k (i.e., $\mathbf{grad}\,k = \mathbf{0}$), then equation (3.1.1) reduces to the Laplace equation. The Laplace equation may be considered as the simplest partial differential equation of mathematical physics. In combination with appropriate boundary conditions on a closed boundary, it can be proved that a unique solution exists, which depends continuously on the parameters in the boundary conditions. Furthermore, an abundance of mathematical properties of the solution is known (see, for example, Morse and Feshbach, 1953, parts I and II; and Butkov, 1973). The Laplace-type equation (3.1.1) with positive coefficients $k_h(x, y, z)$ and $k_z(x, y, z)$ varying in space may be considered as the next-to-simplest partial differential equation for which many mathematical properties are known as well.

It is somewhat less well-known that three more Laplace-type equations can be derived for $e_x = q_x/k_h = -\partial\phi/\partial x$, $e_y = q_y/k_h = -\partial\phi/\partial y$ and $q_z = -k_z\,\partial\phi/\partial z$ for a *perfectly layered* subsurface, as follows:

$$\begin{aligned}\partial/\partial x(k_h\,\partial e_x/\partial x)\ &+ \partial/\partial y(k_h\,\partial e_x/\partial y)+\\ &+ \partial/\partial z(k_z\,\partial e_x/\partial z) = 0\end{aligned} \tag{3.1.2a}$$

$$\partial/\partial x(k_h\,\partial e_y/\partial x)\;+\partial/\partial y(k_h\,\partial e_y/\partial y)+ \\ +\,\partial/\partial z(k_z\,\partial e_y/\partial z)=0 \tag{3.1.2b}$$

$$\partial/\partial x(k_z^{-1}\,\partial q_z/\partial x)\;+\partial/\partial y(k_z^{-1}\,\partial q_z/\partial y)+ \\ +\,\partial/\partial z(k_h^{-1}\,\partial q_z/\partial z)=0 \tag{3.1.2c}$$

(For a derivation see Appendix B.)

The three equations (3.1.2) state that in a perfectly layered subsurface the three vectors $(k_h\,\partial e_x/\partial x,\ k_h\,\partial e_x/\partial y,\ k_z\,\partial e_x/\partial z)$, $(k_h\,\partial e_y/\partial x,\ k_h\,\partial e_y/\partial y,\ k_z\,\partial e_y/\partial z)$ and $(k_z^{-1}\,\partial q_z/\partial x,\ k_z^{-1}\,\partial q_z/\partial y,\ k_h^{-1}\,\partial q_z/\partial z)$ are solenoidal (divergence-free).

The boundary conditions for a *perfectly layered* subsurface are:

$$\text{at } z=0: \begin{cases} \phi=f, \quad e_x=-\partial f/\partial x, \quad e_y=-\partial f/\partial y \\ \partial q_z/\partial z=k_h(0)(\partial^2 f/\partial x^2+\partial^2 f/\partial y^2) \end{cases} \tag{3.1.3a}$$

$$\text{at } z=d: \begin{cases} \partial\phi/\partial z=0, \quad \partial e_x/\partial z=0, \\ \partial e_y/\partial z=0, \quad q_z=0 \end{cases} \tag{3.1.3b}$$

where $f=f(x,y,t)$ is the prescribed potential on the top plane given by the dynamic boundary condition (2.11.1). For a perfectly layered basin, therefore, there are four uncoupled Laplace-type problems for the four quantities ϕ, e_x, e_y and q_z. There is a single unique solution to each problem, and that solution is continuous in the boundary conditions and the parameters. In other words, we are dealing with four well-posed problems.

Time does not appear explicitly in these four Laplace-type problems, but it is an implicit parameter in the phreatic boundary condition for the potential, and also possibly in the time-dependency of the conductivities. Consequently, the flow field can be solved in any space-point at time-point t, independent of the solutions at other points in time.

Perfectly layered subsurfaces often play a role in the analysis of flow problems with rotational symmetry around a well. In that case, the equations (3.1.1) and (3.1.2) become:

$$\partial/\partial r(k_h\,r\,\partial\phi/\partial r)+\partial/\partial z(k_z\,r\,\partial\phi/\partial z)=0 \tag{3.1.4a}$$

$$\partial/\partial r(k_h\,r^{-1}\,\partial a_r/\partial r)+\partial/\partial z(k_z\,r^{-1}\,\partial a_r/\partial z)=0 \tag{3.1.4b}$$

$$\partial/\partial r(k_z^{-1}\, r\, \partial q_z/\partial r) + \partial/\partial z(k_h^{-1}\, r\, \partial q_z/\partial z) = 0 \tag{3.1.4c}$$

where r and z are the lateral and vertical cylindrical coordinates, and $a_r = 2\pi\, r\, e_r$, where e_r is the component of (e_x, e_y) in the radial direction. By this definition $k_h\, a_r$ is the flow rate per unit length of z through a ring with radius r around the axis of symmetry.

The boundary conditions (3.1.3) are now written as:

$$\text{at } z = 0 : \begin{cases} \phi = f, \quad a_r = -2\pi\, r\, \partial f/\partial r \\ \partial q_z/\partial z = k_h(0)\, r^{-1}\, \partial/\partial r(r\, \partial f/\partial r) \end{cases} \tag{3.1.5a}$$

$$\text{at } z = d : \{\partial\phi/\partial z = 0, \quad \partial a_r/\partial z = 0, \quad q_z = 0. \tag{3.1.5b}$$

The above equation (3.1.4a) for the potential ϕ is used by Koefoed (1979) to compute the flow to (or from) a point-source in perfectly layered infinitely deep basins, with piecewise constant conductivities per layer. It is true that Koefoed's work is concerned with geo-electrical methods, not with groundwater flow, but his problem-posing is identical to that for groundwater flowing to one point-source. An extraction well can be thought of as constructed from a collection of extraction points for which all the exact solutions may be added. This procedure is practical and manageable, and leads to precise results (see Section 3.2).

The equations (3.1.4b) and (3.1.4c) for the flux components a_r and q_z offer no advantages for *exact* solutions, but they are useful for *approximations* (see Sections 4.2 and 5.3).

Formulations with a_r and q_z certainly seem to offer advantages in combination with free convection, or density-driven flow. (Free convection leads to extra terms which are dealt with in Appendix C.)

3.2. Introduction to flow near wells

The four Laplace-type equations (3.1.1), (3.1.2) and, hence, the three equations (3.1.4) hold under the condition that $k_h(z)$ and $k_z(z)$ are continuously differentiable functions of z, i.e., they hold if the derivatives $\partial k_h(z)/\partial z$ and $\partial k_z(z)/\partial z$ exist and are continuous functions of z. However, the meaning of these Laplace-type equations can be extended to weak forms in situations where the conductivities $k_h(z)$ and $k_z(z)$ are discontinuous at some discrete locations $z = \delta_i$, $i = 1, 2, \ldots, n$. In a perfectly layered subsurface with piecewise constant conductivities, the weak forms of equations (3.1.4) have the following meaning:

$$\begin{cases} \alpha\, r^{-1}\, \partial/\partial r(r\, \partial\phi/\partial r) + \partial^2\phi/\partial z^2 = 0 \quad \text{in each layer} \\ \phi \ \text{is continuous at each interface}\ z = \delta_i \\ k_z\, \partial\phi/\partial z \ \text{is continuous at each interface}\ z = \delta_i \end{cases} \tag{3.2.1}$$

$$\begin{cases} \alpha\, r\, \partial/\partial r(r^{-1}\, \partial a_r/\partial r) + \partial^2 a_r/\partial z^2 = 0 \quad \text{in each layer} \\ a_r \ \text{is continuous at each interface}\ z = \delta_i \\ k_z\, \partial a_r/\partial z \ \text{is continuous at each interface}\ z = \delta_i \end{cases} \tag{3.2.2}$$

$$\begin{cases} \alpha\, r^{-1}\, \partial/\partial r(r\, \partial q_z/\partial r) + \partial^2 q_z/\partial z^2 = 0 \quad \text{in each layer} \\ q_z \ \text{is continuous at each interface}\ z = \delta_i \\ k_h^{-1}\, \partial q_z/\partial z \ \text{is continuous at each interface}\ z = \delta_i \end{cases} \tag{3.2.3}$$

where $\alpha = k_h/k_z$ is the anisotropy factor which is a constant for each layer.

The solution for a point-extraction at location $r = 0$ and $z = z_p$ in an infinitely extended homogeneous medium (one infinitely extended layer) is given by the following solutions of the above equations (3.2.1), (3.2.2) and (3.2.3):

$$\phi(r,z) = -Q \Big/ \left[4\pi k \sqrt{\{r^2 + \alpha(z - z_p)^2\}}\right] \tag{3.2.4a}$$

$$a_r(r,z) = -Qr^2 \Big/ \left[2k \sqrt{\{r^2 + \alpha(z - z_p)^2\}^3}\right] \tag{3.2.4b}$$

$$q_z(r,z) = -\sqrt{\alpha}\, Q(z - z_p) \Big/ \left[4\pi \sqrt{\{r^2 + \alpha(z - z_p)^2\}^3}\right] \tag{3.2.4c}$$

where Q $[m^3{\cdot}s^{-1};\ m^3{\cdot}d^{-1}]$ is the production rate extracted by the point-source $k = \sqrt{(k_h k_z)}$, and $\sqrt{r^2 + (z - z_p)^2}$ is the distance of the point with cylindrical coordinates r and z to the point-source at location $r = 0$, $z = z_p$.

The above solutions (3.2.4) for the point-extraction can also be written as a linear combination of Bessel functions of r and exponentials of z. For instance, equation (3.2.4a) can be combined with the following expression (see Watson, 1966, Section 13.2; Rikitake *et al.*, 1987, p. 112):

$$\int_0^\infty J_0(wr)\exp(-w\left|z' - z'_p\right|)\,dw = 1 \Big/ \sqrt{r^2 + (z' - z'_p)^2} \qquad (3.2.5)$$

where $z' = z\sqrt{\alpha}$ and $z'_p = z_p\sqrt{\alpha}$.

Not only the above solution for the point-extraction in one layer, but all possible solutions to the partial differential systems (3.2.1), (3.2.2) and (3.2.3) for respectively ϕ, a_r and q_z in a multi-layer subsurface can be expressed as a linear combination in the form of integrals from $w = 0$ to $w \to \infty$ of the functions $\phi_w(r, z') = J_0(wr)\exp(\pm wz')$, $a_{rw}(r, z') = -J_1(wr)\exp(\pm wz')$ and $q_{zw}(r, z') = J_0(wr)\exp(\pm wz')$, for $0 \leq w < \infty$, where w $[m^{-1}]$ is the wave number. This principle is known as the Hankel transformation (see Rikitake *et al.*, 1987, p. 13; Dunford and Schwartz, 1963, p. 978).

In these Hankel transformations, J_0 is the zeroth-order Bessel function and J_1 is the first-order Bessel function. The zeroth-order Bessel function has a wavy character, rather like the cosine, but with decreasing amplitude for increasing argument. The first-order Bessel function also has a wavy character, rather like the sine, and again with decreasing amplitude for increasing argument. For more details, see Abramowitz and Stegun (1972, pp. 358-433).

The general solutions $\phi_w(r, z') = J_0(wr)\exp(\pm wz')$ in the layers, combined with the continuity requirements for ϕ and $k_z\,\partial\phi/\partial z$ at the interfaces, lead to recurrence relations connecting the solutions in the different layers to each other. Combined with the requirement of no flow at infinite depth $z \to \infty$ and with the top boundary condition of zero potential $\phi_w(r, 0) = 0$, the exact solution can be found for each separate mode with wave number w. The final solution for a point-extraction is then obtained by superposition of all these part-solutions, i.e., by integration over the wave number w. After some logarithmic transformations of variables, this integration leads to a convolution integral that can be solved numerically by the use of log-scale Fourier theory and Fast Fourier Transform (see Van Veldhuizen *et al.*, 1992b). Koefoed and Ghosh were the first to develop such an approach and they applied it to find solutions in the context of geophysical exploration with the geo-electrical method (see Koefoed, 1979; and Ghosh, 1970). However, Koefoed's and Ghosh's approach can as well be extended to groundwater flow, since the governing equations for electrical fields and groundwater flow are the same. For the application to flow near a well, as discussed here, the well is considered as a set of neighboring points along a line for which the various solutions are superimposed on each other to obtain the total solution. For more details about the many extensions to Koefoed's and Ghosh's theory necessary for application to flow near wells, see Nieuwenhuizen (1992).

Since this book is concerned with natural groundwater flow, we shall not elaborate further on flow to wells, but there is one aspect of this example which will be of considerable use later. Differentiation of the integral (3.2.5) with respect to z', then taking the limit $z' \to z'_p$, and making use of the fact that the inverse of a Hankel transformation is again the same Hankel transformation, we find that (see Rikitake *et al.*, 1987, p. 13):

$$\int_0^\infty 1 \; J_0(wr) \; w \; \mathrm{d}w = 2\pi\delta(r) \tag{3.2.6a}$$

$$\int_0^\infty 2\pi\delta(r) \; J_0(wr) \; r \; \mathrm{d}r = 1 \tag{3.2.6b}$$

where $\delta(r)$ represents the delta function for the Hankel transformation. This delta function is defined as $\delta(r) = 0$ for $r > \epsilon$, $\delta(r) = 1/(\pi\epsilon^2)$ for $r \leq \epsilon$ and $\epsilon \to 0$ while the volume of the cylinder with radius ϵ and height $\delta(r)$ is kept 1. Equation (3.2.6a) says that the Hankel Transform of 1 is equal to $2\pi\delta(r)$ and equation (3.2.6b) says that the Hankel Transform of $2\pi\delta(r)$ is equal to 1.

The importance of the above expressions (3.2.6) is in their suggestion that any function $f(r)$ of r can be expressed as the integral of Bessel functions. This is indeed the case, and using the nth-order Bessel function $(n = 0, 1, 2, \ldots, \infty)$ this yields:

$$\int_0^\infty f^*(w) \; J_n(wr) \; w \; \mathrm{d}w = f(r) \tag{3.2.7a}$$

where

$$\int_0^\infty f(r) \; J_n(wr) \; r \; \mathrm{d}r = f^*(w). \tag{3.2.7b}$$

The above principle is known as a Hankel transformation, or Fourier-Bessel transformation, where $f^*(w)$ is the Hankel Transform of $f(r)$, and $f(r)$ is the Hankel Transform of $f^*(w)$. If we are not dealing specifically with rotational symmetry problems, then we can choose sines and cosines or complex e-powers instead of Bessel functions. This leads to the Fourier transformation, which is simpler to use than the Fourier-Bessel transformation, as will be explained in Section 3.6. The Fourier transformation will be used for the description of natural groundwater flow in Chapter 5.

One aspect of such integral transformations is important to note. From equations (3.2.6) it follows that 1 is the Hankel Transform of $2\pi\delta(r)$ and that $2\pi\delta(r)$ is the Hankel Transform of 1. This means that a function of r which is concentrated around the location $r = 0$, and is vanishingly small everywhere else, as is the case for the delta function, is spread out almost infinitely in the w-space. Also the inverse holds; a function of r which is spread out almost infinitely in the r-space, is concentrated around a 'location' w and is vanishingly small everywhere else in the w-space. Unlike the Hankel transformation, the inverse of a Fourier transformation is not again the same Fourier transformation, and also its delta function is slightly different from the delta function for the Hankel transformation. Nevertheless similar properties with respect to spreading and concentrating in the space and wave number domains hold for the Fourier transformation. The practical consequence is that for the study of very local phenomena, not repeating themselves at other locations, a great many Fourier modes need to be taken into account. We return to this point in Section 3.6.

Finally, a somewhat different approach to flow near wells is possible by expressing the general solution of the above partial differential equations (3.2.1), (3.2.2) and (3.2.3) for ϕ, a_r and q_z, respectively, as a linear combination of $\phi_u(r, z') = I_0(ur)\exp(iuz')$, $a_{ru}(r, z') = I_1(ur)\exp(iuz')$ and $q_{zu}(r, z') = I_0(ur)\exp(iuz')$, for $-\infty < u < +\infty$, where u $[m^{-1}]$ is the decrement and $i = \sqrt{-1}$.

In this general solution, I_0 is the zeroth-order modified Bessel function and I_1 is the first-order modified Bessel function. The modified Bessel functions have an exponentially-decreasing character with increasing argument, rather like the e-power. The z'-dependency is periodic with $\exp(iuz') = \cos(uz') + i\sin(uz')$. Using this form of the general solution, Maas (1987a) obtained solutions for flow near wells in a perfectly layered subsurface with piecewise constant conductivities (see also Maas, 1986; Maas, 1987b). The well in this case is not an assembly of discrete points along a line, as in Koefoed's approach, but a continuous vertical line over the interval $z_1 < z < z_2$. This means that wells deviating from a vertical line cannot be described in this way. Furthermore, in Maas's approach z must not become infinitely deep, which in practice implies limitation to shallow basins with an impervious base. Consequently, the Maas solution cannot be used in combination with the flow systems analysis (see Chapter 5), where the facility to study infinitely deep basins is needed.

3.3. The character of natural flow in perfectly layered basins

Since the four systems of equations (3.1.1) and (3.1.2) with boundary conditions (3.1.3) are mutually uncoupled, it means that q_z can be calculated independently of ϕ, e_x and e_y, and the same is true for the other three quantities. This lack of coupling says much about the character of flow in a perfectly layered subsurface.

From this we can see from the top-boundary conditions (3.1.3a) and the dynamic boundary condition (2.11.1) that q_z is driven exclusively by the sum of the second derivatives in the groundwater table $\partial^2 h_f/\partial x^2 + +\partial^2 h_f/\partial y^2$, or $r^{-1}\,\partial/\partial r(r\partial h_f/\partial r)$ in cylindrical symmetry (see equations (3.1.5a)). If this sum is equal to zero at every point of the top plane $z = 0$, then $q_z(x, y, z, t)$ is equal to zero in the whole flow field. A groundwater table where $\partial^2 h_f/\partial x^2 + \partial^2 h_f/\partial y^2 = 0$ is called a harmonic groundwater table, and a few examples of harmonic groundwater tables are:

$$h_f(x,y) = h_0 + c_1\,x + c_2\,y \quad \text{(plane groundwater table)} \tag{3.3.1}$$

$$h_f(x,y) = c_3 \ln(r) = c_3 \ln\left[\sqrt{(x^2+y^2)}\right], \qquad r \neq 0, \quad (x,y) \neq 0 \tag{3.3.2}$$

$$h_f(x,y) = c_4 \exp(-x) \sin y \quad \text{(and variants thereof)} \tag{3.3.3}$$

where h_0, c_1, c_2, and c_3 are constants (possibly dependent on t).

All linear combinations of the above harmonic functions are also harmonic. From these examples we see that exfiltration/infiltration patterns correlate with deviations from the harmonic character of the groundwater table, and do not depend solely on the phreatic plane being curved. On this basis we also see that it would be misleading to calculate the exfiltration/infiltration patterns $q_z(x, y, 0, t)$ from the curve of h_f relating to a two-dimensional cross-section of the subsurface.

From the uncoupled nature of the equation for $q_z(x, y, z, t)$ it follows that, with spatially constant input $q_z(x, y, 0, t) = N(t)$, the vertical transport velocity $v_z(x, y, z, t) = q_z(x, y, z, t)/\theta(z)$ will not change in the lateral directions, whereas the lateral velocities obviously do change in the lateral directions. We find that:

$$q_z(x, y, z, t) = N(t) \int_z^d k_h(z')\,\mathrm{d}z' \bigg/ \left[\int_0^d k_h(z')\,\mathrm{d}z'\right]. \tag{3.3.4}$$

This means that water in perfectly layered basins with spatially constant infiltrating top flow resides in lateral planes of the same water age. As a consequence, uniformly distributed pollutants, such as tritium originating from nuclear explosions in the atmosphere during the 1960s, are distributed in lateral planes in the depths of such basins.

On the other hand, e_x and e_y are driven, respectively, by the first derivatives $\partial h_f/\partial x$ and $\partial h_f/\partial y$ in the groundwater table. From the examples of harmonic groundwater tables given earlier, it will be clear that the lateral velocities can be large (large $\partial h_f/\partial x$, $\partial h_f/\partial y$), while at the same time the vertical velocity is small (small $\partial^2 h_f/\partial x^2 + \partial^2 h_f/\partial y^2$). Consequently, the small vertical velocity is often neglected by contrast with the large lateral velocity, but the vertical velocity can sometimes be important, especially if we need to know how deep a pollutant has penetrated in the subsurface, for example, or how much a fresh-saline interface has risen during the centuries.

As has already been mentioned, the four equations (3.1.1) and (3.1.2) may be interpreted as weak forms in situations where the conductivities $k_h(z)$ and $k_z(z)$ are discontinuous at some locations $z = \delta_i$, $i = 1, 2, \ldots, n$. Here we further illustrate the practical meaning of the weak form. Firstly, at the interfaces $z = \delta$ where k_h and k_z jump from $k_h(\delta-)$ and $k_z(\delta-)$ to $k_h(\delta+)$ and $k_z(\delta+)$, the four quantities ϕ, e_x, e_y and q_z are continuous, i.e., $\phi(\delta-) = \phi(\delta+)$, $e_x(\delta-) = e_x(\delta+)$, $e_y(\delta-) = e_y(\delta+)$ and $q_z(\delta-) = q_z(\delta+)$. Secondly, the four vectors $(k_h\,\partial\phi/\partial x,\ k_h\,\partial\phi/\partial y,\ k_z\,\partial\phi/\partial z)$, $(k_h\,\partial e_x/\partial x,\ k_h\,\partial e_x/\partial y,\ k_z\,\partial e_x/\partial z)$, $(k_h\,\partial e_y/\partial x,\ k_h\,\partial e_y/\partial y,\ k_z\,\partial e_y/\partial z)$ and $(k_z^{-1}\,\partial q_z/\partial x,\ k_z^{-1}\,\partial q_z/\partial y,\ k_h^{-1}\,\partial q_z/\partial z)$ are solenoidal (divergence-free), where the divergence may be interpreted in its weak form. This means that $k_z\partial\phi/\partial z$,

$k_z\partial e_x/\partial z$, $k_z\partial e_y/\partial z$, and $k_h^{-1}\,\partial q_z/\partial z$ are continuous. Thus, $k_h(\delta-)^{-1}\cdot$ $\cdot(\partial q_z/\partial z)_{(\delta-)} = k_h(\delta+)^{-1}(\partial q_z/\partial z)_{(\delta+)}$, for example, from which it follows that $(\partial q_z/\partial z)_{(\delta+)} = [k_h(\delta+)/k_h(\delta-)]\,(\partial q_z/\partial z)_{(\delta-)}$. If $[k_h(\delta+)/$ $/k_h(\delta-)] \to 0$, then above $z = \delta$, at $z = \delta-$, there is a well-permeable layer (an aquifer) and below $z = \delta$, at $z = \delta+$, there is a poorly permeable layer (an aquitard). Since under this condition $(\partial q_z/\partial z)_{(\delta+)} \to 0$, then, for a sufficiently thin aquitard, this means that q_z remains constant over the thickness of the aquitard. The assumption that q_z is approximately constant over the thickness of an aquitard is very popular in geohydrology. This approximation leads to $q_z = (\phi_u - \phi_l)/c$, where ϕ_u and ϕ_l are, respectively, the potential above and below the poorly permeable layer, and c is the resistance of that layer (see Appendix C). Note that this approximation does *not* mean that the flow through a poorly permeable layer is predominantly vertical. The latter approximation is frequently applied in combination with the Dupuit approximation (see Section 4.2), applied to the well-permeable layers. The reverse is also true; q_z changes sharply (is strongly damped) over the thickness of an aquifer.

Similar arguments for ϕ, e_x and e_y give the well-known result that ϕ, e_x and e_y change very little over the thickness of an aquifer, and that ϕ, e_x and e_y change considerably (are strongly damped) over the thickness of an aquitard. The approximation that ϕ, e_x and e_y are constant over the thickness of an aquifer is very popular in geohydrology. This approximation is the Dupuit approximation that will be treated extensively in Section 4.2.

3.4. Extension to lateral heterogeneities

The flow behavior acquires a totally different character in places where lateral heterogeneities in the conductivities occur. There the equations for the quantities e_x, e_y and q_z are indeed coupled, and the reasoning for the exfiltration/infiltration patterns of harmonic groundwater tables, etc. given in Section 3.3, will no longer apply. The study of coupled flow behavior near lateral heterogeneities can provide insights which are both interesting and of practical importance.

When dealing with a laterally piecewise perfectly layered subsurface, the Laplace-type equations (3.1.2) for e_x, e_y and q_z will be valid for each perfectly layered piece of the subsurface. With the aid of continuity conditions at the interfaces between two different perfectly layered areas, the solutions in both areas can be linked together. However, this approach will not be discussed further.

For a more restricted, but still interesting class of heterogeneous subterranean conductivity distributions, a system of coupled equations can be written for e_x, e_y and q_z, which will be called the coupled Laplace-type equations because of the coupling. The conductivities must then satisfy:

$$k_h(x, y, z) = k_{h0}(z) \exp\ \lambda_h(x, y, z) \tag{3.4.1a}$$

$$k_z(x, y, z) = k_{z0}(z) \exp\ \lambda_z(x, y, z) \tag{3.4.1b}$$

where $k_{h0}(z)$ and $k_{z0}(z)$ may be discontinuous functions of z. However, the functions $\lambda_h(x, y, z)$ and $\lambda_z(x, y, z)$ must *not* be discontinuous but should be sufficiently 'smooth' (i.e., sufficiently often differentiable). Of course, the cases where $\lambda_h = \lambda_z = 0$ conform with the perfectly layered basins described in the previous sections.

The coupled Laplace-type equations are:

$$\begin{aligned} &\partial/\partial x(k_h\, \partial e_x/\partial x) + \partial/\partial y(k_h\, \partial e_x/\partial y) + \partial/\partial z(k_z\, \partial e_x/\partial z) + \\ &\quad + k_h\, e_x\, \partial^2 \lambda_h/\partial x^2 + k_h\, e_y\, \partial^2 \lambda_h/\partial y \partial x + q_z\, \partial^2 \lambda_z/\partial z \partial x + \\ &\quad - (\partial \xi/\partial x)\ \partial q_z/\partial z = 0 \end{aligned} \tag{3.4.2a}$$

$$\begin{aligned} &\partial/\partial x(k_h\, \partial e_y/\partial x) + \partial/\partial y(k_h\, \partial e_y/\partial y) + \partial/\partial z(k_z\, \partial e_y/\partial z) + \\ &\quad + k_h\, e_x\, \partial^2 \lambda_h/\partial x \partial y + k_h\, e_y\, \partial^2 \lambda_h/\partial y^2 + q_z\, \partial^2 \lambda_z/\partial z \partial y + \\ &\quad - (\partial \xi/\partial y)\ \partial q_z/\partial z = 0 \end{aligned} \tag{3.4.2b}$$

$$\begin{aligned} &\partial/\partial x(k_z^{-1}\, \partial q_z/\partial x) + \partial/\partial y(k_z^{-1}\, \partial q_z/\partial y) + \\ &\quad + \partial/\partial z(k_h^{-1}\, \partial q_z/\partial z) + \\ &\quad + e_x\, \partial^2 \lambda_h/\partial z \partial x + e_y\, \partial^2 \lambda_h/\partial z \partial y + \\ &\quad - k_h^{-1}\, q_z\, [\partial/\partial x(\alpha \partial \lambda_z/\partial x) + \partial/\partial y(\alpha \partial \lambda_z/\partial y)] + \\ &\quad + k_z^{-1}\ [(\partial \xi/\partial x)\ \partial q_z/\partial x + (\partial \xi/\partial y)\ \partial q_z/\partial y] = 0 \end{aligned} \tag{3.4.2c}$$

where:

$$\begin{aligned} \alpha(x, y, z) &= k_h(x, y, z)/k_z(x, y, z) = \\ &= [k_{h0}(z)/k_{z0}(z)] \exp[\lambda_h(x, y, z) - \lambda_z(x, y, z)] \end{aligned} \tag{3.4.3}$$

is the anisotropy factor, and where:

$$\xi(x,y,z) = \lambda_h(x,y,z) - \lambda_z(x,y,z). \tag{3.4.4}$$

In many practical problems the anisotropy factor α is taken only as a function of z, or even as a constant, for reasons of data reduction. In that case, the three equations (3.4.2a), (3.4.2b) and (3.4.2c) will be simplified because the last terms will be omitted and the next-to-last term in equation (3.4.2c) will be simplified.

The boundary conditions are:

$$\text{at } z=0: \begin{cases} \phi = f, \quad e_x = -\partial f/\partial x, \quad e_y = -\partial f/\partial y \\ \partial q_z/\partial z = \partial/\partial x(k_h\,\partial f/\partial x) + \partial/\partial y(k_h\,\partial f/\partial y) \end{cases} \tag{3.4.5a}$$

$$\text{at } z=d: \begin{cases} \partial\phi/\partial z = 0, \quad \partial e_x/\partial z = 0, \\ \partial e_y/\partial z = 0, \quad q_z = 0. \end{cases} \tag{3.4.5b}$$

For a derivation of the above equations and boundary conditions, see Appendix B. Now, it can be seen what 'sufficiently smooth' means: at least all the derivatives of the conductivities that are present in the equations must exist. For a perfectly layered subsurface the above-presented equations (3.4.2) with boundary conditions (3.4.5) simplify to the classic, uncoupled Laplace-type equations (3.1.2) with boundary conditions (3.1.3). From a mathematical point of view, the latter classic Laplace-type equations (3.1.2) are very simple, but application of the extended, coupled Laplace-type equations (3.4.2) may be rather complex for some applications. For situations where this is the case, an alternative will be presented in Section 4.7 (see equations (4.7.7)).

The method of series expansion (i.e., perturbation calculus, which will be introduced in Section 4.2) for perfectly layered porous media is not generally applicable to porous media with discontinuous lateral heterogeneities, but it is readily extendable to basins with sufficiently smooth lateral heterogeneities in the conductivities, as will be shown in Section 4.5. Therefore, in the following sections, the more in-depth treatment of perfectly layered basins should not be seen as a dead-end but as a means of enlightenment. In principle, the methodology can always be expanded or extended to basins with smooth lateral heterogeneities in the conductivities, and these characteristics give a good approximation of reality for a large class of practical geohydrological problems.

3.5. Directly calculated velocity components

To avoid confusion, and to maintain the independence and differences in character and scale of the lateral and vertical velocity components in groundwater basins, it is best to consider each velocity component separately in terms of its own equation and its own boundary conditions. As will be shown in the following chapters, this leads to a systematic construction of the analytical mathematical operations, and often it also has advantages for determining the flow field with numerical methods.

One advantage of applying Laplace-type equations (3.4.2) with boundary conditions (3.4.5) in numerical methods is that, in this way, the speed of convergence for grid-refinement is an order higher than when the velocity is obtained from the first calculated potential. For example, if the potential is computed by solving Laplace-type equation (3.1.1) with order of convergence $h^{3/2}$ (see Ababou, 1988, p. 337), then the order of convergence for the velocity calculated from this potential by applying Darcy's law (2.7.2) will be $h^{1/2}$, whereas in solving the Laplace-type equations (3.4.2) for e_x, e_y and q_z the order of convergence will be $h^{3/2}$ (where h is the magnitude of the spatial discretization). This is a great advantage if grid-refinement is necessary.

Another advantage is the precise determination of small vertical velocity components in perfectly layered subsurfaces. If these velocity components are determined from the differences between numerically calculated potentials at two near points, then large errors in the velocity components can arise from small numerical errors in the potentials (see Figure 4), e.g., through breaking-off the iterations too soon (see also Section 4.2). Such errors can be avoided when q_z is calculated directly either by equation (3.4.2c), or by the alternative equation (4.7.7d) which will be presented in Section 4.7 (see also the discussion on well-posedness in Section 4.3).

A further advantage of application of the Laplace-type equations (3.4.2), (3.4.5) in the numerical approximation method is that an impression of the accuracy associated with a particular discretization can be gained by using the numerical solutions for the residuals; see Section 4.7. It should be borne in mind that a numerical method gives approximate solutions, not exact ones, so it would be wise after a numerical approximation to ascertain the magnitude of the deviation of the approximate solutions with respect to

100	300	500	700	900	1100	1300	exact potential distribution
101	301	501	701	901	1101	1301	$q_x = -\partial\phi/\partial x = -200$
							$q_z = -\partial\phi/\partial z = -1$
102	302	502	702	902	1102	1302	
103	303	503	703	903	1103	1303	

100	299	505	695	901	1099	1300	inaccurately approximated potential distribution
98	300	500	702	900	1099	1297	$q_x = -\partial\phi/\partial x = -200 \pm 10$
							$q_z = -\partial\phi/\partial z = -1 \pm 10$
104	301	497	705	897	1102	1306	
100	299	503	705	907	1106	1302	

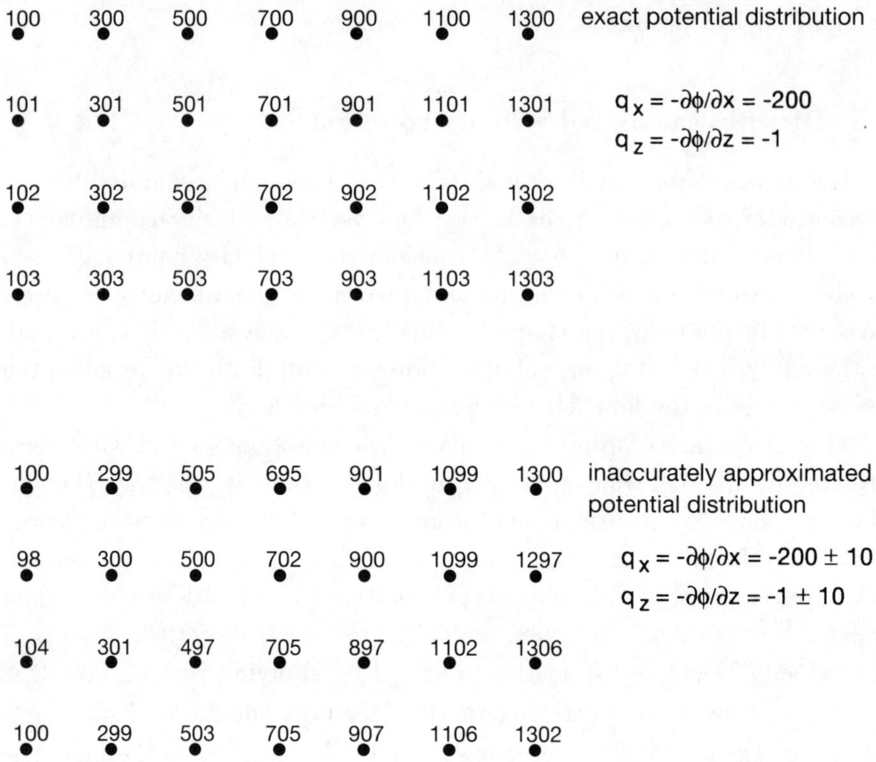

Figure 4. **Inaccurately calculated vertical flux.**
 Cross-section of a homogeneous subsurface with a two-dimensional example of groundwater flow. The components of the volumetric flow rate (the flux components) are determined from the potential distribution ϕ by $q_x = -\partial\phi/\partial x$ and $q_z = -\partial\phi/\partial z$. Figure 4a shows the calculation from the exact potential distribution. Figure 4b shows the calculation from an inaccurate approximation of the potential distribution; in that case the calculated vertical flux component is intolerably inaccurate. This example reflects reality: in perfectly layered (and inhomogeneous) subsurfaces, $\partial\phi/\partial z = 0$ is often a good approximation. However, determination of the vertical flux component by $q_z = -\partial\phi/\partial z$ results in $q_z = 0$, which is an intolerably inaccurate approximation.

the exact solutions. Based on this information, the need for (local) grid-refinement can then be decided. Such simple estimations are not possible in the calculation of velocities from numerically computed potentials. However, some numerical methods for solving the transport of dissolved matter require that $\boldsymbol{q}$ exactly satisfies $\operatorname{div} \boldsymbol{q} = 0$. This latter requirement can be met by applying the alternative equation (4.7.7d) presented in Section 4.7, rather than application of Laplace-type equation (3.4.2c).

Application to equation (3.1.1) for ϕ of the conventional finite-element method with conformal piecewise linear elements results in a potential field that is continuous everywhere; but in calculating the volumetric flow rate from this computed potential field by Darcy's law (2.7.2), the normal component of the calculated volumetric flow rate is discontinuous on the inter-element boundaries, which makes it difficult to obtain a reliable calculation of the flow paths. However, the same conventional finite-element method applied to the equations (3.4.2) for e_x, e_y and q_z results in a velocity field with continuous normal components of the volumetric flow rate on the inter-element boundaries.

Finally, there are also advantages in directly computing e_x, e_y and q_z in situations where density-driven flow occurs (see Appendix C).

To summarize, we can say that the direct calculation of e_x, e_y and q_z works perfectly for perfectly layered basins, but disadvantages arise with the presence of lateral heterogeneities. One disadvantage is that the lateral heterogeneities must be sufficiently smooth. Another disadvantage is that, in the presence of lateral heterogeneities, the resulting matrix is not positive-definite; this means that the usual iterative techniques (such as successive over-relaxation or preconditioned conjugate gradients) cannot be used. However, the difficulty can be avoided if the terms which make the matrix nonpositive-definite are calculated in advance from the potentials calculated by solving equation (3.1.1) for ϕ; in that case, the order of the speed of convergence will decrease a little in grid-refinement. Another, even simpler but less accurate alternative will be presented in Section 4.7. In conclusion, direct numerical computation of e_x, e_y and q_z is possible not only for perfectly layered, but also for 'approximately layered' subsurfaces. The model code FLOSA (Flow Systems Analysis) for three-dimensional flow provides the capability to calculate the potential and, from that, to compute the velocities for completely heterogeneous subsurfaces, as well as the ability to directly compute e_x, e_y and q_z for perfectly layered, and 'approximately layered' subsurfaces (see Nawalany, 1986a; 1986b) and see

Chapters 8, 9 and 10.

In this context, where applications of the Laplace-type equations for the three flux components are discussed, it is appropriate to mention Ababou's improvement of Gelhar's theory for porous media flow with randomly distributed values of the conductivity (see Gelhar, 1986); (see Ababou, 1988, pp. 211-220). Ababou improved Gelhar's theory thanks to direct determination of the volumetric flow rate $\boldsymbol{q}$ with the aid of the vector Laplace-type equation:

$$\boldsymbol{\nabla}^2\boldsymbol{q} - \mathbf{grad}\,\lambda\boldsymbol{\cdot}\underline{\mathbf{grad}\boldsymbol{q}} - \boldsymbol{q}\,\nabla^2\lambda + \boldsymbol{q}\boldsymbol{\cdot}\underline{\mathbf{grad}\,\mathbf{grad}}\,\lambda = \mathbf{0}. \tag{3.5.1}$$

Equation (3.5.1) is equivalent to equations (3.4.2). For convenience, it is given here for an isotropic subsurface, where the conductivity is given by $k(\boldsymbol{x}) = k_G \exp\lambda(\boldsymbol{x})$, but an extension to anisotropy is straightforward. In equation (3.5.1) $\underline{\mathbf{grad}\boldsymbol{q}} = \partial_i q_j$ $(i,j = 1,2,3)$ is a tensor; and the vector Laplacean $\nabla^2\boldsymbol{q} = \partial_j\partial_j q_i$ $(i = 1,2,3)$ is a vector (summation over the twice-occurring index j). It is observed that equation (3.5.1) is linear in the log conductivity $\lambda(\boldsymbol{x})$ (dimensionless).

Ababou's results are very instructive for the understanding of groundwater flow phenomena on large spatial scales. Since this is an important topic in this book, we briefly devote some attention to Ababou's improvement of Gelhar's theory, translated to the style of presentation adopted in this book.

First, the mean value, or expected value, $E[\psi(\boldsymbol{x})]$ of a function $\psi(\boldsymbol{x})$ is defined as the volume integral over all function values $\psi(\boldsymbol{x})$ in a domain, divided by the volume of that domain. In general this mean value is a function of the coordinates of the center of the integration domain and of the volume of that domain. But for a statistically homogeneous porous medium, introduced here as a simplification, the mean value of the log conductivity is constant (see Davis, 1986, p. 259). Under that condition the log conductivity is given by:

$$\lambda(\boldsymbol{x}) = \lambda_0 + \sigma\,\lambda_1(\boldsymbol{x}) \tag{3.5.2}$$

where σ is a nonnegative constant, and where λ_0 is the constant mean value of the log conductivity. This means that $E[\lambda(\boldsymbol{x})] = E(\lambda_0) = \lambda_0$ is constant and $E[\lambda_1(\boldsymbol{x})] = 0$. From now on, the mean log conductivity λ_0 is chosen equal to zero, i.e., $E[\lambda(\boldsymbol{x})] = \lambda_0 = 0$. Since the perturbation $\lambda_1(\boldsymbol{x})$

will be normed by requiring $E[\lambda_1(\boldsymbol{x})^2] = 1$, the dimensionless parameter σ accounts for the magnitude of the perturbation.

Making use of the statistical homogeneity of the log conductivity, substitution of expression (3.5.2) into equation (3.5.1) yields:

$$\boldsymbol{\nabla}^2 \boldsymbol{q} - \sigma\,(\mathbf{grad}\,\lambda_1 \cdot \underline{\mathbf{grad}}\boldsymbol{q} + \boldsymbol{q}\;\nabla^2\lambda_1 - \boldsymbol{q}\cdot\underline{\mathbf{grad}}\,\mathbf{grad}\,\lambda_1) = \mathbf{0}. \tag{3.5.3}$$

The occurrence of the 'magnitude parameter' σ in equation (3.5.3) suggests that solutions may be obtained by a perturbation method (see, for instance, Van Dyke, 1975; or Nayfeh, 1973). In the perturbation method the volumetric flow rate $\boldsymbol{q}(\boldsymbol{x})$ is written in the form of the following perturbation series in powers of σ:

$$\boldsymbol{q}(\boldsymbol{x}) = \boldsymbol{q}_0(\boldsymbol{x}) + \sigma\,\boldsymbol{q}_1(\boldsymbol{x}) + \sigma^2\,\boldsymbol{q}_2(\boldsymbol{x}) + \ldots \tag{3.5.4}$$

Substitution of the above series (3.5.4) into equation (3.5.3) leads to another power series in the parameter σ. Since the sum of the terms of this power series is equal to zero for any value of the parameter σ, each term in the series must be equal to zero. This leads to the following infinite hierarchy of equations:

$$\boldsymbol{\nabla}^2 \boldsymbol{q}_0 = \mathbf{0} \tag{3.5.5a}$$

$$\boldsymbol{\nabla}^2 \boldsymbol{q}_1 - \mathbf{grad}\,\lambda_1 \cdot \underline{\mathbf{grad}}\boldsymbol{q}_0 - \boldsymbol{q}_0\;\nabla^2\lambda_1 + \boldsymbol{q}_0\cdot\underline{\mathbf{grad}}\,\mathbf{grad}\,\lambda_1 = \mathbf{0} \tag{3.5.5b}$$

$$\boldsymbol{\nabla}^2 \boldsymbol{q}_2 - \mathbf{grad}\,\lambda_1 \cdot \underline{\mathbf{grad}}\boldsymbol{q}_1 - \boldsymbol{q}_1\;\nabla^2\lambda_1 + \boldsymbol{q}_1\cdot\underline{\mathbf{grad}}\,\mathbf{grad}\,\lambda_1 = \mathbf{0} \tag{3.5.5c}$$

$$\vdots$$

$$\begin{aligned}\boldsymbol{\nabla}^2 \boldsymbol{q}_n &- \mathbf{grad}\,\lambda_1 \cdot \underline{\mathbf{grad}}\boldsymbol{q}_{n-1} - \boldsymbol{q}_{n-1}\;\nabla^2\lambda_1 + \\ &+ \boldsymbol{q}_{n-1}\cdot\underline{\mathbf{grad}}\,\mathbf{grad}\,\lambda_1 = \mathbf{0}\end{aligned} \tag{3.5.5d}$$

for $n = 1, 2, \ldots, \infty$.

The zeroth-order equation (3.5.5a) is a vector Laplace equation and can be applied to obtain the zeroth-order approximation $\boldsymbol{q}_0(\boldsymbol{x})$. A further simplification is obtained by restricting to the zeroth-order solutions $\boldsymbol{q}_0 =$

= *constant*. Under this assumption the first-order equation (3.5.5b) simplifies to the following vector Poisson equation for the first-order correction $\boldsymbol{q}_1(\boldsymbol{x})$:

$$\boldsymbol{\nabla}^2 \boldsymbol{q}_1 - \boldsymbol{q}_0 \, \nabla^2 \lambda_1 + \boldsymbol{q}_0 \cdot \underline{\mathbf{grad}} \, \mathbf{grad} \, \lambda_1 = \mathbf{0}. \tag{3.5.6}$$

Taking the mean value of equation (3.5.6) shows that $E[\boldsymbol{q}_1(\boldsymbol{x})] = \mathbf{0}$ is a solution. On the other hand, from equations (3.5.5c) and (3.5.5d) it follows that $E[\boldsymbol{q}_2(\boldsymbol{x})] \neq \mathbf{0}, \ldots, E[\boldsymbol{q}_n(\boldsymbol{x})] \neq \mathbf{0}$. Consequently, in general $E[\boldsymbol{q}(\boldsymbol{x})] \neq \boldsymbol{q}_0$, but up to first-order accuracy $E[\boldsymbol{q}(\boldsymbol{x})] = \boldsymbol{q}_0$. In the context of random porous media theory, it turns out that also the first-order accurate approximation to $\boldsymbol{q}(\boldsymbol{x})$ is statistically homogeneous, but higher-order approximations are not.

Certain spatial distributions for $\lambda_1(\boldsymbol{x})$ with $E[\lambda_1(\boldsymbol{x})] = 0$ and $E[\lambda_1(\boldsymbol{x})^2] = 1$ can be assumed. Strictly speaking, the distribution of $\lambda(\boldsymbol{x})$ may be obtained in any possible way; for instance, by a deterministic simulation of the processes that, in the past, have led to the formation of sedimentary basins. Such a simulation finally yields a conductivity distribution that is not statistically homogeneous (see Stam *et al.*, 1989). However, another possibility is to assume random conductivity distributions. A well-known example of a random distribution is the normal, or Gaussian distribution with zero mean value and unit standard deviation with probability density $F(\lambda_1) = \exp(-\lambda_1^2/2)/\sqrt{(2\pi)}$; with this choice the standard deviation of the log conductivity λ is equal to the magnitude parameter σ. In any spatial point $\boldsymbol{x}$ the log conductivity perturbation $\lambda_1(\boldsymbol{x})$ may be considered as a realization of the above normal distribution. This realization may not be completely independent since the required differentiability of $\lambda_1(\boldsymbol{x})$ imposes some correlation between the values of λ_1 at points $\boldsymbol{x}$ and $\boldsymbol{x} + \mathrm{d}\boldsymbol{x}$. The distribution of $\lambda(\boldsymbol{x}) = \lambda_0 + \sigma\,\lambda_1(\boldsymbol{x})$ obtained in this way leads to a lognormal distribution of $k(\boldsymbol{x})$. The 'theory of breakage' leading to a lognormal distribution of grain size (see Davis, 1986, pp. 87-92) offers a plausible justification for the application of lognormal conductivity distributions within one specific formation type. However, apart from the correlations required by the differentiability requirement, the realizations at the different spatial points may also be correlated to account for mechanisms that have led to the formation of the sediment. This correlation can be obtained by the 'turning band method' (Ababou, 1988, pp. 64-73), and in this way statistically anisotropic random distributions can be generated.

We have seen that the choice of a statistically homogeneous log conductivity leads to statistical homogeneity of the first-order correct volumetric flow rate in such a way that its mean value is equal to the constant zeroth-order solution. This homogeneity makes it possible to apply relatively simple spectral analysis for infinitely extended domains. With the aid of spectral analysis, the first-order correct result for the random distribution of the volumetric flow rate $\boldsymbol{q}'(\boldsymbol{x}) = \boldsymbol{q}_0 + \sigma\, \boldsymbol{q}_1(\boldsymbol{x})$ has been determined by Ababou.

In a similar way, the first-order accurate result for the driving force per unit volume $\boldsymbol{e}'(\boldsymbol{x}) = \boldsymbol{e}_0 + \sigma\, \boldsymbol{e}_1(\boldsymbol{x})$ can be derived. For this purpose Ababou applies the Laplace-type equation for the potential $\phi(\boldsymbol{x})$:

$$\nabla^2\phi + \mathbf{grad}\,\lambda \boldsymbol{\cdot} \mathbf{grad}\,\phi = 0 \tag{3.5.7}$$

but the Laplace-type equation for $\boldsymbol{e}(\boldsymbol{x})$:

$$\boldsymbol{\nabla}^2\boldsymbol{e} + \mathbf{grad}\,\lambda \boldsymbol{\cdot} \underline{\mathbf{grad}\boldsymbol{e}} + \boldsymbol{e} \boldsymbol{\cdot} \underline{\mathbf{grad}}\,\mathbf{grad}\,\lambda = \mathbf{0} \tag{3.5.8}$$

could have been used equally well. Similar to equation (3.5.1), also equations (3.5.7) and (3.5.8) are linear in $\lambda(\boldsymbol{x})$. This is at variance with Darcy's law $\boldsymbol{q}(\boldsymbol{x}) = -k_G \exp[\lambda(\boldsymbol{x})]\, \mathbf{grad}\,\phi(\boldsymbol{x})$, which is strongly *nonlinear* in $\lambda(\boldsymbol{x})$. This nonlinearity introduces additional inaccuracy in the perturbation method truncated after the first-order terms, and this observation motivated Ababou not to apply Darcy's law in combination with equation (3.5.7), but to use the linear equation (3.5.1) in combination with equation (3.5.7). Of course, the first-order accurate approximations $\boldsymbol{q}'(\boldsymbol{x}) = \boldsymbol{q}_0 + {}+\sigma\, \boldsymbol{q}_1(\boldsymbol{x})$ and $\boldsymbol{e}'(\boldsymbol{x}) = \boldsymbol{e}_0 + \sigma\, \boldsymbol{e}_1(\boldsymbol{x})$ do not exactly satisfy Darcy's law $\boldsymbol{q}(\boldsymbol{x}) = k_G \exp[\sigma\, \lambda_1(\boldsymbol{x})]\, \boldsymbol{e}(\boldsymbol{x})$. But, with the aid of the series expansion $\exp[\sigma\, \lambda_1(\boldsymbol{x})] \approx 1 + \sigma\, \lambda_1(\boldsymbol{x}) + (\sigma^2/2)\, \lambda_1(\boldsymbol{x})^2$, the constant solution $\boldsymbol{e}_0$ of the vector Laplace equation for $\boldsymbol{e}_0(\boldsymbol{x})$ can be related to $\boldsymbol{q}_0$ in such a way that Darcy's law holds as accurately as possible for the first-order correct solutions $\boldsymbol{q}'(x)$ and $\boldsymbol{e}'(\boldsymbol{x})$ in the limit for small σ.

After having derived the first-order correct statistically homogeneous solutions $\boldsymbol{q}'(\boldsymbol{x})$ and $\boldsymbol{e}'(\boldsymbol{x})$, it is also possible to determine the constant mean values $E[\boldsymbol{q}'(\boldsymbol{x})] = \boldsymbol{q}_0$ and $E[\boldsymbol{e}'(\boldsymbol{x})] = \boldsymbol{e}_0$. Under the assumption of an 'ellipsoidally anisotropic' log conductivity distribution (see Ababou, 1988, p. 104), it follows that there are three mutually perpendicular directions (say the x-, y- and z-directions), the principal axes, along which $\boldsymbol{q}_0$ and

$\boldsymbol{e}_0$ have the same direction. In that case we may define the ratios $\langle k_x \rangle$, $\langle k_y \rangle$ and $\langle k_z \rangle$ by:

$$\langle k_x \rangle = q_{0x}/e_{0x} \tag{3.5.9a}$$

$$\langle k_y \rangle = q_{0y}/e_{0y} \tag{3.5.9b}$$

$$\langle k_z \rangle = q_{0z}/e_{0z} \tag{3.5.9c}$$

or:

$$\boldsymbol{q}_0 = \langle \underline{k} \rangle \cdot \boldsymbol{e}_0 \tag{3.5.9d}$$

where $\langle \underline{k} \rangle$ is a symmetric tensor. Defining $\langle \boldsymbol{q} \rangle = \boldsymbol{q}_0$ and the zeroth-order potential $\langle \phi \rangle(\boldsymbol{x})$ as linear in $\boldsymbol{x}$ by $\textbf{grad} \langle \phi \rangle(\boldsymbol{x}) = -\boldsymbol{e}_0$, equation (3.5.9d) is written as:

$$\langle \boldsymbol{q} \rangle = -\langle \underline{k} \rangle \cdot \textbf{grad} \langle \phi \rangle(\boldsymbol{x}). \tag{3.5.10}$$

Equation (3.5.10) for the 'domain scale' has the same form as Darcy's law (2.7.2) on the 'point scale.' Therefore it seems appropriate to call equation (3.5.10) a large-scale Darcy's law. Along the same line of thought, $\langle \underline{k} \rangle$ will be called the large-scale conductivity, or, as is customary in the theory of random porous media, the effective conductivity. In this book bold type brackets $\langle \cdot \rangle$ denote large-scale properties of a three-dimensional domain, which justifies the notations $\langle \boldsymbol{q} \rangle = \boldsymbol{q}_0$, $\langle \boldsymbol{e} \rangle = \boldsymbol{e}_0$ and $\langle \phi \rangle(\boldsymbol{x})$.

It is important to note here that the large-scale conductivity may deviate considerably from the conductivity k_G, i.e., the conductivity without perturbation. Furthermore, in this presentation we started with an *isotropic* porous medium, with scalar conductivity $k(\boldsymbol{x})$, on the local point scale. On the large scale, here the infinitely large scale, we end up with an *anisotropic* conductivity described by the large-scale conductivity tensor $\langle \underline{k} \rangle$. Consequently, anisotropy turns out to be largely a scale effect.

For the above-presented first-order correct approximation, Ababou (1988) derives expressions for the large-scale conductivity for several patterns of the lognormal conductivity distribution. For more details, especially with respect to the flux standard deviation and its consequences for dispersive transport of solutes, where Ababou's theory improves Gelhar's original theory, see Ababou (1988, pp. 211-220).

The validity of the first-order results is expected to give satisfactory results only when the log conductivity perturbation is sufficiently small to permit higher-order terms to be neglected. Here sufficiently small means that not only the amplitude σ, but also the *first and second derivatives* of the log conductivity must be sufficiently small, since these derivatives appear in the hierarchy of equations (3.5.5). It has already been mentioned that in the first-order correct theory the large-scale anisotropy is described by a *symmetric* tensor $<\underline{k}>$. However, large-scale conductivities are *generally nonsymmetric* (see Chapter 4). To avoid both the assumption of statistical homogeneity and the assumption of small derivatives in the log conductivity perturbation, another approach to the determination of large-scale conductivities will be presented in Chapter 4. Furthermore, random conductivity distributions will not be discussed further in this book.

3.6. Fourier analysis of spatial variations in the water table

The Hankel transformation, or Fourier-Bessel transformation, and its application to flow near wells, was introduced in Section 3.2. The somewhat simpler Fourier transformation and its applications will now be introduced here.

For insight into natural groundwater flow, it is important to express the spatial variations of the two-dimensional water table $h_f(x, y, t)$ at every time-point t as the inverse Fourier Transform of the Fourier Transform $h_f^*(w_x, w_y, t)$ of the water table $h_f(x, y, t)$ (see Rikitake *et al.*, 1987, pp. 12):

$$h_f(x,y,t) = \int_{-\infty}^{\infty}\int_{-\infty}^{\infty}\left[h_f^*(w_x, w_y, t)/(4\pi^2)\right] \cdot \\ \cdot \exp\left[i(w_x x + w_y y)\right] dw_x dw_y \tag{3.6.1}$$

where $i = \sqrt{-1}$; $h_f^*(w_x, w_y, t)$ is the (generally complex) Fourier Transform of $h_f(x, y, t)$; $w_x = 1/l_x$ $[m^{-1}]$ and $w_y = 1/l_y$ $[m^{-1}]$ are the wave numbers ($-\infty < w_x$, $w_y < +\infty$); the lengths $2\pi\,|l_x| = 2\pi/\,|w_x|$ $[m]$ and $2\pi\,|l_y| =$ $= 2\pi/\,|w_y|$ $[m]$ are the wavelengths in the x- and y-directions, respectively.

Fourier transforms $h_f^*(w_x, w_y, t)$ exist for the spatial variations in the groundwater table that are found in geohydrological practice, and their determination will be discussed later in this section. With a change-over to polar-coordinates, the Hankel transformation described in Section 3.2 can be derived from the two-dimensional Fourier transformation (3.6.1) (see Rikitake *et al.*, 1987, p. 12-13).

Using the familiar formula $\exp[i(w_x x + w_y y)] = \cos(w_x x + w_y y) + + i \sin(w_x x + w_y y)$ and with $h_f^*(w_x, w_y, t)/(4\pi^2) = a_f(w_x, w_y, t) + - i\, b_f(w_x, w_y, t)$, a_f and b_f real, the inverse Fourier Transform (3.6.1) will give the following expression:

$$\begin{aligned} h_f(x, y, t) &= \int_{-\infty}^{\infty}\int_{-\infty}^{\infty} [a_f(w_x, w_y, t)\cos(w_x x + w_y y) + \\ &\quad + b_f(w_x, w_y, t)\sin(w_x x + w_y y)]\,\mathrm{d}w_x \mathrm{d}w_y. \end{aligned} \tag{3.6.2}$$

Since the flow problem is linear, the four solutions for the scalar potential field and the vectorial velocity field can be determined separately for each Fourier component, or Fourier mode, h_{fw}:

$$\begin{aligned} h_{fw}(x, y, t) &= a_f(w_x, w_y, t)\cos(w_x x + w_y y) + \\ &\quad + b_f(w_x, w_y, t)\sin(w_x x + w_y y). \end{aligned} \tag{3.6.3}$$

Thereafter, all these solutions must be integrated over (w_x, w_y) to obtain the complete solution. This is an application of the superposition principle. Therefore, in the continuation of this discussion, we shall always look at one particular Fourier component.

The above-presented inverse Fourier transformation holds for infinitely extended regions where $-\infty < x < \infty$ and $-\infty < y < \infty$. We have already discussed in Section 3.2 that local phenomena, i.e., phenomena not repeating themselves periodically, give rise to an infinitely broad continuous spectrum in the w-space. Of course, real groundwater tables are 'infinitely' extended (i.e., generally have very large superregional dimensions) and are never periodic over finite distances, which means that any phenomenon is local. However, for practical reasons, we want to restrict ourselves to a finite, discrete spectrum valid in the neighborhood of the phenomena we want to study. This means that we want to consider water table maps with finite dimensions, in such a way that an 'infinite' number of mutually adjacent maps cover the whole 'infinitely extended' region. Then for each finite map the water table can be decomposed in much fewer Fourier modes than required for the whole infinite region. In this way, we have obtained a Fourier decomposition of the whole infinite region with wave numbers depending on which map is decomposed. Or, in other words, we want a location-dependent Fourier decomposition. The question is whether the concept of location-dependent wave numbers makes sense mathematically. Since the work of Meyer (a French mathematician) in 1986, mathematical research on wavelet theory has begun. Very popularly speaking, wavelets are a kind of 'Fourier modes' with location-dependent wave numbers. This is an improvement over the Fourier transformation where the basis functions (the complex exponentials $\exp[-i(w_x x + w_y y)]$) result in a perfectly concentrated Fourier Transform in the wave number domain (w_x, w_y), but spread all over the space domain (x, y); and vice versa for the inverse Fourier Transform. For a brief introduction to wavelet theory, see Strang (1989).

We will not attempt to mathematically justify the above-described 'finite map approach,' but will discuss it from an intuitive point of view. Let us consider such a finite rectangular water table map where x and y are bounded, i.e., $0 \leq x \leq L_x$ and $0 \leq y \leq L_y$. Instead of a Fourier integral, we now obtain a Fourier series with the longest wavelengths being the lengths L_x and L_y of the region, hence, with smallest wave numbers $2\pi/L_x$ and $2\pi/L_y$. In this Fourier series, all other wave numbers are integer multiples of these smallest wave numbers, i.e., $w_x = 2\pi n/L_x$ and $w_y = 2\pi m/L_y$ with $n, m = 1, 2, \ldots, \infty$. It is generally an infinite series. However, maps never exhibit all details of the water table; in general a map gives the water table only at N_x equidistant points x_i and N_y equidistant points y_j, i.e., at points $x_i = i\, L_x/N_x$, $i = 1, 2, \ldots, N_x$ and $y_j = j\, L_y/N_y$, $j = 1, 2, \ldots, N_y$. In that case the Fourier series may be truncated after $n = N_x/2$ and $m = N_y/2$.

A further simplification is possible if the map under consideration is a square with linear dimensions $L_x = L_y = L$ and with a discretization such that $N_x = N_y = N = 2^l$ with $l = 1, 2, 3, \ldots$. In that case the Fourier components can efficiently be calculated from the equidistantly-given water table values by the Fast Fourier Transform (FFT) (see Rikitake *et al.*, 1987, pp. 57-63). We divide the square water table map under consideration in N^2 smaller squares with linear dimensions L/N; the discretization points (x_i, y_j) are located in the centers of these squares. These squares are called the pixels of the water table map. Details smaller than the pixel dimensions cannot be presented on such a map. The longest wavelength occurring in the Fourier series is equal to the linear dimensions L of the region, and the smallest wavelength occurring in the Fourier series is equal to $2L/N$ which is two times the linear pixel dimension. This relationship between pixel dimension and smallest wavelength shows that there is a relationship between the linear dimensions of the smallest details presented on a water table map and the smallest wavelength occurring in the Fourier series belonging to that map. The more details in the map, the smaller the pixel size and, consequently, the smaller the smallest wavelength. The above method is relatively simple to handle, but there are some limitations inherent in its use; the wave numbers are determined by the map characteristics (map dimensions and pixel dimensions) and have no necessary relation to any wave form which might actually be present in the data. This arbitrary selection of wave numbers also excludes the possibility of extrapolation beyond the map boundaries; all water tables will begin to duplicate themselves in the x- and y-directions beyond the map boundaries. A way to overcome these limitations is described by James (1966), who also treates irregularly spaced data. However, if the map characteristics are chosen meaningfully with respect to the water table data, the above limitations can be overcome. Much of this has been worked out in a geological context by Harbaugh and Merriam (1968, pp. 125-155).

Both from a numerical and from a conceptual point of view, a finite Fourier series for a discretized finite region is obviously much easier to manage than an integral to infinity. Thus it is easier to describe the natural groundwater flow than the flow to a well which needs a Hankel integral running to infinity.

3.7. Decay of spatial Fourier modes with increasing depth

From the four Laplace-type equations (3.1.1) and (3.1.2) for ϕ, e_x, e_y and q_z for a perfectly layered basin, we have already learned in Section 3.3 a few things about the behavior of a solution, and especially about the damping of the potential and the flow velocities with increasing depth due to the transitional behavior between well- and poorly permeable layers (aquifers and aquitards). On closer examination, which will be presented in Chapter 5, we learn that the flow due to the Fourier components with short wavelengths is decaying much more strongly with increasing depth than flow due to the Fourier components with long wavelengths. We are here referring to the behavior of the Fourier components in space, not in time, as discussed in the previous section (Section 3.6). From this latter discussion we may then conclude that flow due to small-scale variations in the groundwater table is decaying much more strongly with increasing depth than flow due to large-scale variations in the water table. This fact relates the scale of the water table map, by virtue of its associated pixel size, to the depth below which we want to represent the groundwater flow. This scale-dependency will be discussed in more detail in Chapter 5.

We have already noted in Section 3.3 that ϕ, e_x and e_y are particularly damped with increasing depth in aquitards, and that q_z is damped most with increasing depth in aquifers. This has a bearing on the depth where the effects of the scale of spatial variations in the groundwater table – variations of scale associated with differing wavelengths – will penetrate the subsurface. Since the penetrating depth depends on the spatial scale in question, it means that any justifiable choice of the minimum depth for the 'impermeable base' d must also be scale-dependent. Theoretically the best choice would be $d \rightarrow \infty$, but this is obviously unrealistic for practical numerical model studies. This aspect will be discussed in more detail in Chapter 5.

Chapter 4

Large-Scale Flow Parameters: Transmissivity and Conductivity

4.1. Characteristic and dimensionless quantities

Characteristic quantities are brought in to transform the reasoning given in Section 3.7 into a qualitative criterion. Characteristic lengths and times and their interrelationship through the process of elastic storage were already introduced in Section 2.3. We now introduce more characteristic quantities:

The characteristic amplitude of a Fourier component is defined by $\phi_c = \rho_0\, g\sqrt{(a_f^2 + b_f^2)}$, where a_f and b_f are defined in Section 3.6; the lateral characteristic length of a Fourier component is defined by $l_c = 1/\sqrt{(w_x^2 + w_y^2)} = 1/\sqrt{(1/l_x^2 + 1/l_y^2)}$, where the wave numbers w_x and w_y are defined in Section 3.6; the vertical characteristic length is defined, independent of the Fourier component under consideration, as the depth of the basin d; the characteristic lateral conductivity is denoted by k_{hc} and the characteristic vertical conductivity is denoted by k_{zc}. It is not easy at this stage to give a meaningful definition of characteristic conductivities for heterogeneous porous media, so we shall return to it later with more extensive discussion in Section 4.4.

All quantities can be made dimensionless with the help of the characteristic quantities: $X = x/l_c$ and $Y = y/l_c$ are the dimensionless lateral coordinates; $Z = z/d$ is the dimensionless vertical coordinate; $K_h = k_h/k_{hc}$ and $K_z = k_z/k_{zc}$, respectively, are the dimensionless lateral and vertical conductivities; $F = f/\phi_c$ and $\Phi = \phi/\phi_c$ are dimensionless potentials; and $E_x = e_x\, l_c/\phi_c$, $E_y = e_y\, l_c/\phi_c$ and $Q_z = q_z\, l_c^2/(\phi_c\, d\, k_{hc})$ are the dimensionless 'fluxes.' In this way, the equations can be written in dimensionless form, and the dimensionless number $\epsilon = (d/l_c)^2\, (k_{hc}/k_{zc})$ turns out to be a very important parameter in these dimensionless equations.

The dimensionless Laplace-type or Poisson-type equations describing groundwater flow in the basin are:

$$\begin{aligned} &\epsilon\, \partial/\partial X(K_h\, \partial\Phi/\partial X) + \epsilon\, \partial/\partial Y(K_h\, \partial\Phi/\partial Y) + \\ &\quad + \partial/\partial Z(K_z\, \partial\Phi/\partial Z) = 0 \end{aligned} \tag{4.1.1a}$$

$$\begin{aligned} &\epsilon\, \partial/\partial X(K_h\, \partial E_x/\partial X) + \epsilon\, \partial/\partial Y(K_h\, \partial E_x/\partial Y) + \\ &\quad + \partial/\partial Z(K_z\, \partial E_x/\partial Z) + \epsilon\, K_h\, E_x\, \partial^2\lambda_h/\partial X^2 + \\ &\quad + \epsilon\, K_h\, E_y\, \partial^2\lambda_h/\partial Y\partial X + \epsilon\, Q_z\, \partial^2\lambda_z/\partial Z\partial X + \\ &\quad - \epsilon\, (\partial\xi/\partial X)\, \partial Q_z/\partial Z = 0 \end{aligned} \tag{4.1.1b}$$

$$\begin{aligned} &\epsilon\, \partial/\partial X(K_h\, \partial E_y/\partial X) + \epsilon\, \partial/\partial Y(K_h\, \partial E_y/\partial Y) + \\ &\quad + \partial/\partial Z(K_z\, \partial E_y/\partial Z) + \epsilon\, K_h\, E_x\, \partial^2\lambda_h/\partial X\partial Y + \\ &\quad + \epsilon\, K_h\, E_y\, \partial^2\lambda_h/\partial Y^2 + \epsilon\, Q_z\, \partial^2\lambda_z/\partial Z\partial Y + \\ &\quad - \epsilon\, (\partial\xi/\partial Y)\, \partial Q_z/\partial Z = 0 \end{aligned} \tag{4.1.1c}$$

$$\begin{aligned} &\epsilon\, \partial/\partial X(K_z^{-1}\, \partial Q_z/\partial X) + \epsilon\, \partial/\partial Y(K_z^{-1}\, \partial Q_z/\partial Y) + \\ &\quad + \partial/\partial Z(K_h^{-1}\, \partial Q_z/\partial Z) + E_x\, \partial^2\lambda_h/\partial Z\partial X + \\ &\quad + E_y\, \partial^2\lambda_z/\partial Z\partial Y + \\ &\quad - \epsilon\, K_h^{-1}\, Q_z\, [\partial/\partial X(A\, \partial\lambda_z/\partial X + \partial/\partial Y(A\, \partial\lambda_z/\partial Y)] + \\ &\quad + \epsilon\, K_z^{-1}\, [(\partial\xi/\partial X)\, \partial Q_z/\partial X + (\partial\xi/\partial Y)\, \partial Q_z/\partial Y] = 0 \end{aligned} \tag{4.1.1d}$$

where:

$$\begin{aligned} A(X,Y,Z) &= K_h(X,Y,Z)/K_z(X,Y,Z) = \\ &= [K_{h0}(Z)/K_{z0}(Z)] \exp[\lambda_h(X,Y,Z) - \lambda_z(X,Y,Z)] \end{aligned} \tag{4.1.2}$$

and where:

$$\xi(X,Y,Z) = \lambda_h(X,Y,Z) - \lambda_z(X,Y,Z). \tag{4.1.3}$$

The boundary conditions are:

$$\text{at } Z = 0: \begin{cases} \Phi = F, \quad E_x = -\partial F/\partial X, \quad E_y = -\partial F/\partial Y \\ \partial Q_z/\partial Z = \partial/\partial X(K_h\, \partial F/\partial X) + \\ \qquad + \partial/\partial Y(K_h\, \partial F/\partial Y) \end{cases} \tag{4.1.4a}$$

$$\text{at } Z = 1 : \begin{cases} \partial\Phi/\partial Z = 0, & \partial E_x/\partial Z = 0, \\ \partial E_y/\partial Z = 0, & Q_z = 0. \end{cases} \tag{4.1.4b}$$

A necessary condition for the mathematical equivalence of several different flow problems is that the dimensionless number ϵ should be the same for all the problems (see Silberberg and McKetta, 1953). However, this is not a sufficient condition in itself. As we shall see later, the dimensionless number has yet another important part to play. It will be seen that the condition $\sqrt{\epsilon} \approx 2\pi$ indicates the order of magnitude of the depth of the 'impermeable base,' and this is especially important in numerical model studies where the dimensions must be chosen for the three-dimensional region to be modeled.

4.2. The Dupuit approximation and transmissivity

The solution of the flow problem can be found by means of a series expansion of the dimensionless Φ, E_x, E_y and Q_z in powers of ϵ, as follows:

$$\Phi = \Phi_0 + \epsilon\, \Phi_1 + \epsilon^2\, \Phi_2 + \cdots + \epsilon^n\, \Phi_n + \ldots \tag{4.2.1a}$$

$$E_x = E_{x0} + \epsilon\, E_{x1} + \epsilon^2\, E_{x2} + \cdots + \epsilon^n\, E_{xn} + \ldots \tag{4.2.1b}$$

$$E_y = E_{y0} + \epsilon\, E_{y1} + \epsilon^2\, E_{y2} + \cdots + \epsilon^n\, E_{yn} + \ldots \tag{4.2.1c}$$

$$Q_z = Q_{z0} + \epsilon\, Q_{z1} + \epsilon^2\, Q_{z2} + \cdots + \epsilon^n\, Q_{zn} + \ldots \tag{4.2.1d}$$

These series expansions can be substituted into the dimensionless equations (4.1.1) through (4.1.4), and the equations will then also become power series of ϵ with sum zero. Since ϵ can have any value, all separate terms in the power series must equate to zero. This analytical mathematical method is known as 'perturbation calculus.' For further details see Van Dyke (1975) or Nayfeh (1973).

The terms in the equations belonging to the zeroth power of ϵ result in equations with the well-known Dupuit approximation as solutions. Calculated back to dimensional quantities, the zeroth-order equations are given by:

$$\partial/\partial z(k_z\ \partial\phi_0/\partial z) = 0 \tag{4.2.2a}$$

$$\partial/\partial z(k_z\ \partial e_{x0}/\partial z) = 0 \tag{4.2.2b}$$

$$\partial/\partial z(k_z\ \partial e_{y0}/\partial z) = 0 \tag{4.2.2c}$$

$$\begin{aligned}\partial/\partial z(k_h^{-1}\ \partial q_{z0}/\partial z) + e_{x0}\ \partial^2\lambda_h/\partial z\partial x +\\ + e_{y0}\ \partial^2\lambda_h/\partial z\partial y = 0.\end{aligned} \tag{4.2.2d}$$

From the above equations and the boundary conditions (4.1.4) (calculated back to dimensional form) it follows that the Dupuit approximation is given by the following well-known equations:

$$\Phi_0(x,y,z,t)\ = f(x,y,t) \tag{4.2.3a}$$

$$e_{x0}(x,y,z,t) = -\partial f(x,y,t)/\partial x \tag{4.2.3b}$$

$$e_{y0}(x,y,z,t) = -\partial f(x,y,t)/\partial y \tag{4.2.3c}$$

$$\begin{aligned}q_{z0}(x,y,z,t)\ &= -\,\partial/\partial x[t_0(x,y,z)\ \partial f(x,y,t)/\partial x]\\ &\quad -\,\partial/\partial y[t_0(x,y,z)\ \partial f(x,y,t)/\partial y]\end{aligned} \tag{4.2.3d}$$

where $t_0(x,y,z)$ $[m^3{\cdot}Pa^{-1}{\cdot}s^{-1};\ m^3{\cdot}dbar^{-1}{\cdot}d^{-1} \approx m^3{\cdot}mH_2O^{-1}{\cdot}d^{-1}]$ is the 'three-dimensional' transmissivity:

$$t_0(x,y,z) = \int_z^d k_h(x,y,z')\,\mathrm{d}z'. \tag{4.2.4}$$

This transmissivity is called 'three-dimensional' because it is a function of the two lateral coordinates x and y, and also of the vertical coordinate z. Usually in geohydrology the 'two-dimensional' transmissivity, or briefly the transmissivity $T(x,y)$ or $kD(x,y)$ is equal to $t_0(x,y,0)$, which is a function of the two lateral coordinates x and y only. It should be noted, however, that $t_0(x,y,0)$ is the transmissivity of a basin with thickness d. When the thickness of the water-saturated part of the basin is varying due to variations in the height of the water table, an additional term must be added to $t_0(x, y, 0)$ to obtain the phreatic transmissivity (see later).

It will be observed that the Dupuit approximations for e_x, e_y and q_z satisfy the continuity equation but not Darcy's law.

If the fluxes e_x and e_y were calculated with the aid of $e_x = -\partial\phi/\partial x$, $e_y = -\partial\phi/\partial y$ from the Dupuit approximation for the potential ϕ_0, then the correct zeroth-order approximations e_{x0} and e_{y0} would be obtained. However, if q_z is calculated with the aid of Darcy's law $q_z = -k_z\ \partial\phi/\partial z$, the result would be $q_z = 0$ instead of the q_{z0} given by expression (4.2.3d). Consequently, a warning must be given about the use of approximation methods, which also include numerical approximation methods such as the finite-difference or finite-element methods. In situations where $\epsilon << 1$, the potential can be calculated with a (numerical) approximation method. However, with some approximations (e.g., rounding-off errors due to prematurely stopped iterations needed to solve the system of linear equations obtained with finite-differences or finite-elements) it is possible that the higher-order terms in the potential (terms ϕ_1 etc.) will not be calculated with sufficient accuracy, and, in that case, the vertical velocity component q_z will be calculated as 'noise' around the value zero (see Figure 4). Consequently, this difficulty can be avoided by direct computation of q_z with equation (4.1.1d). Equivalently, the computation of q_z from e_x and e_y with the aid of the continuity equation will also give the correct result (see Section 4.7). Another example of the difference in approximated computations of potential and of flux is given in Section 5.3.

Let us now define the depth-integrated volumetric flow rate $\boldsymbol{J}$ with Cartesian components J_x and J_y $[m^3{\cdot}m^{-1}{\cdot}s^{-1};\ m^3{\cdot}m^{-1}{\cdot}d^{-1}]$:

$$J_x(x, y, t) = \int_{-h_f}^{d} q_x(x, y, z, t)\ \mathrm{d}z \tag{4.2.5a}$$

$$J_y(x, y, t) = \int_{-h_f}^{d} q_y(x, y, z, t)\ \mathrm{d}z \tag{4.2.5b}$$

where the thickness of the basin is equal to $d + h_f(x, y, t)$. From $q_x = k_h\ e_x$, $q_y = k_h\ e_y$ and with the aid of the Dupuit approximations (4.2.3b) and

(4.2.3c) the following well-known expressions are derived:

$$J_{x0}(x,y,t) = -[T(x,y) + k_{h0}(x,y)\, h_f(x,y,t)] \cdot \cdot\, \partial f(x,y,t)/\partial x \tag{4.2.6a}$$

$$J_{y0}(x,y,t) = -[T(x,y) + k_{h0}(x,y)\, h_f(x,y,t)] \cdot \cdot\, \partial f(x,y,t)/\partial y. \tag{4.2.6b}$$

In equations (4.2.6), $k_{h0}(x,y)$ is the horizontal conductivity in the depth interval between $z = 0$ and the water table $z = -h_f$; in the derivation it is assumed that $k_{h0}(x,y)$ is independent of z in that interval. $T(x,y) = = t_0(x,y,0)$ $[m^3 \cdot Pa^{-1} \cdot s^{-1};\ m^3 \cdot dbar^{-1} \cdot d^{-1} \approx m^3 \cdot mH_2O^{-1} \cdot d^{-1}]$ is the conventional ('two-dimensional') transmissivity of the 'projected' basin with thickness d; $T(x,y) + k_{h0}(x,y)\, h_f(x,y,t)$ is the phreatic transmissivity of the basin with thickness $d + h_f(x,y,t)$.

The two equations (4.2.6) are *approximations* that have been derived from Darcy's law under the rather restrictive zeroth-order conditions $e_{x0}(x,y,z,t) = -\partial f(x,y,t)/\partial x$ and $e_{y0}(x,y,z,t) = -\partial f(x,y,t)/\partial y$ (see equations (4.2.3b) and (4.2.3c)). Nevertheless, these equations are often also called 'Darcy's law' in everyday parlance, and this careless use of language can cause misunderstandings. What can one make of the paradox, for instance: 'Darcy's law' does not exactly comply with Darcy's law. It would be better to speak of the Dupuit approximation in this case.

On the other hand, the additional equation (4.2.3d) required for the determination of the vertical volumetric flow rate $q_{z0}(x,y,z,t)$, exactly reflects conservation of mass and it is therefore justified to call this equation the continuity equation. Hence, the equation for the vertical volumetric flow rate on the water table $N(x,y,t) = q_z(x,y,-h_f(x,y,t),t)$ can be obtained by integrating the continuity equation from $z = -h_f$ to $z = d$. Under the zeroth-order Dupuit condition $\phi_0(x,y,z,t) = f(x,y,t)$ (see equation (4.2.3a)) this integration is very simple, and it results in:

$$\begin{aligned} N_0(x,y,t) = & -\partial/\partial x[T(x,y)\, \partial f(x,y,t)/\partial x] \\ & -h_f(x,y,t)\, \partial/\partial x[k_{h0}(x,y)\, \partial f(x,y,t)/\partial x] \\ & -\partial/\partial y[T(x,y)\, \partial f(x,y,t)/\partial y] \\ & -h_f(x,y,t)\, \partial/\partial y[k_{h0}(x,y)\, \partial f(x,y,t)/\partial y] \end{aligned} \tag{4.2.6c}$$

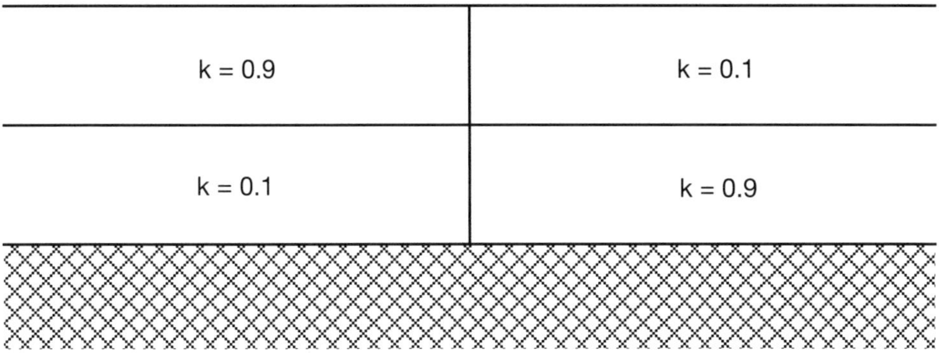

Figure 5. **Laterally heterogeneous conductivity resulting in homogeneous transmissivity.**
Cross-section of a laterally heterogeneous subsurface with thickness of 1 m composed of two layers with thicknesses of 0.5 m. The transmissivity is everywhere 0.5 $m^2{\cdot}dbar^{-1}{\cdot}d^{-1}$ independent of the location x. Nevertheless, a much lower 'effective' transmissivity would be expected near the location $x = 0$ of the lateral heterogeneity. Furthermore, the extension of the transition zone in which the conductivities change from 0.1 to 0.9 $m^2{\cdot}dbar^{-1}{\cdot}d^{-1}$ is also expected to be important. For a discontinuous jump, as in Figure 5, the 'effective' transmissivity is expected to be close to 0.1 $m^3{\cdot}dbar^{-1}{\cdot}d^{-1}$, whereas the smoother the transition zone, the closer this value will be to 0.5 $m^3{\cdot}dbar^{-1}{\cdot}d^{-1}$.

where $N_0(x, y, t) = q_{z0}(x, y, -h_f(x, y, t), t)$ is the zeroth-order vertical volumetric flow rate on the water table $z = -h_f(x, y, t)$.

The zeroth-order accuracy of the Dupuit approximations (4.2.3) and (4.2.6) can give rise to large errors with respect to the exact solutions of Darcy's law. As an extreme but illustrative example, let us consider the following conductivity distribution in a two-dimensional, well-conducting cross-section of the subsurface (see Figure 5):

for $x < 0$:

$$k_h = k_z = 0.1\ m^2 \cdot dbar^{-1} \cdot d^{-1} \quad \text{for } 0\ m \leq z < 0.5\ m \tag{4.2.7a}$$

$$k_h = k_z = 0.9\ m^2 \cdot dbar^{-1} \cdot d^{-1} \quad \text{for } 0.5\ m < z \leq 1\ m \tag{4.2.7b}$$

for $x \geq 0$:

$$k_h = k_z = 0.9\ m^2 \cdot dbar^{-1} \cdot d^{-1} \quad \text{for } 0\ m \leq z < 0.5\ m \tag{4.2.7c}$$

$$k_h = k_z = 0.1\ m^2 \cdot dbar^{-1} \cdot d^{-1} \quad \text{for } 0.5\ m < z \leq 1\ m. \tag{4.2.7d}$$

In this example, the transmissivity $T(x)$ is equal to $T = 0.5\ m^3 \cdot dbar^{-1} \cdot d^{-1}$ for all values of x, and therefore there is continuity (no large jump like in the conductivity) in the transmissivity at $x = 0$. Based on the Dupuit approximation we then find, for a linear potential variation on the top surface with a gradient $\partial f / \partial x$ of, say, $-0.01\ dbar \cdot m^{-1}$, a uniform lateral (depth-integrated) flow rate of $0.005\ m^2 \cdot d^{-1}$ and no vertical flow. However, the above-presented conductivity distribution immediately gives the impression that there exists a flow barrier in the subsurface at $x = 0$, which is totally different from that calculated with the Dupuit approximation. In fact, the Dupuit approximation does not hold because the lateral derivatives of the conductivities are very large at $x = 0$, so that the higher-order terms in the perturbation series expansion (4.2.1) are not negligible. Consequently, the higher-order terms will be examined more closely in Sections 4.4 and 4.7. For the present it will suffice to observe that, with care, it seems possible in some situations to generalize the Dupuit approximation by introducing a generalized transmissivity which takes account of lateral derivatives in the conductivities (see Section 4.5).

4.3. The Dupuit-Forchheimer approximation

The Dupuit approximation (4.2.6c) for $N_0(x, y, t) = q_{z0}(x, y, -h_f(x, y, t), t)$, in combination with the dynamic and kinematic boundary conditions (2.9.1) and (2.10.8), yields the well-known Dupuit-Forchheimer approximation describing the time-evolution of the water table $h_f(x, y, t)$:

$$\begin{aligned}(\theta_0/\rho_0\, g)\, \partial h_f/\partial t = \partial/\partial x[(T + k_{h0}\, h_f)\, \partial h_f/\partial x] +\\ + \partial/\partial y[(T + k_{h0}\, h_f)\, \partial h_f/\partial y] + P_e/(\rho_0\, g).\end{aligned} \tag{4.3.1}$$

For convenience, we have assumed here that we are dealing with an exactly horizontal-vertical system of coordinates (x, y, z). A Dupuit-Forchheimer approximation in a coordinate system (x, y, z) that is sloping in relation to an exactly horizontal plane is also possible. Such an equation is obtained by taking into account that the top potential $f(x, y, t)$ is related to the height of the water table with respect to an *exactly horizontal* plane, whereas $h_f(x, y, t)$ is the height of the water table with respect to the sloping plane $z = 0$. However, geohydrologists generally prefer an exactly horizontal-vertical coordinate system, especially when regional flow systems are under consideration.

The derivation of equation (4.3.1) holds under the condition that in the depth interval between $z = 0$ and $z = -h_f$ the lateral conductivity $k_h(x, y, z) = k_{h0}(x, y)$ is independent of z, and $\theta_0(x, y)$ and $\rho_0(x, y)$ are values at the water table. The result is a nonlinear equation describing the time-evolution of the water table $h_f(x, y, t)$. As can be observed, the Dupuit-Forchheimer equation (4.3.1) is nonlinear since the phreatic transmissivity $T_{ph}(x, y, t) = T(x, y) + k_{h0}(x, y)\, h_f(x, y, t)$ is dependent on the height of the water table $h_f(x, y, t)$.

Nonlinear equations present serious mathematical difficulties with respect to linear equations. However, if the water table is steady (i.e., $\partial h_f/\partial t = 0$) and if, in addition, T and k_{h0} are independent of x and y (i.e., perfect layering), then $\partial T_{ph}(x, y, t)/\partial x = k_{h0}\, \partial h_f(x, y, t)/\partial x$ and $\partial T_{ph}(x, y, t)/\partial y = = k_{h0}\, \partial h_f(x, y, t)/\partial y$, resulting in the following Dupuit-Forchheimer equation for the discharge potential $\Phi_{ph} = \left(k_{h0}^{-1}/2\right)\, T_{ph}^2$ (see Strack, 1989, pp. 37-40):

$$\partial^2 \Phi_{ph}/\partial x^2 + \partial^2 \Phi_{ph}/\partial y^2 + P_e/(\rho_0\, g) = 0. \tag{4.3.2}$$

Under the assumption that $P_e(x, y, t)$ is independent of Φ_{ph} (see below), the steady Dupuit-Forchheimer equation (4.3.2) is linear and is, therefore, frequently applied as a basis for analytical techniques to solve steady groundwater flow problems in perfectly layered basins (see Strack, 1989).

For the sake of understanding the differences between equations (4.2.6c) and (4.3.2) it is repeated here that $N_0(x, y, t)$ in equation (4.2.6c) is the *vertical* component of the volumetric flow rate on the water table. On the other hand, $P_e(x, y, t)$ is the component of the groundwater replenishment perpendicular to the water table multiplied by the term $\sqrt{[1 + (\partial h_f/\partial x)^2 + (\partial h_f/\partial y)^2]}$ (see Section 2.10). Since it is reasonable to assume that the groundwater replenishment from the unsaturated zone to the water table is directed vertically downward, its component normal (i.e., perpendicular) to the water table is obtained by dividing this vertically directed groundwater replenishment by $\sqrt{[1 + (\partial h_f/\partial x)^2 + (\partial h_f/\partial y)^2]}$ (see equation (2.10.4). As a consequence, $P_e(x, y, t)$ is equal to this vertically directed groundwater replenishment, or effective precipitation, which is, in good approximation, independent of Φ_{ph}. However, this does *not* mean that $N_0(x, y, t) = P_e(x, y, t)$ under steady conditions, since the components of the volumetric flow rate $\boldsymbol{q}(\boldsymbol{x}, t)$ and of the groundwater replenishment parallel to the water table will generally differ. This explains the difference between equations (4.2.6c) and (4.3.2). Only if the water table is sufficiently horizontal, $N_0(x, y, t) \approx P_e(x, y, t)$ and equations (4.2.6c) and (4.3.2) are equivalent under steady-state conditions.

The Dupuit-Forchheimer approximation is a second-order partial differential equation in the spatial coordinates, i.e., second derivatives and not higher derivatives of $h_f(x, y, t)$ with respect to x and y occur. As will be shown in Sections 4.4 and 4.7, combination of the exact solution for $q_z(x, y, 0, t)$ with the previously discussed projected dynamic and kinematic boundary conditions will give, in the same way as with the Dupuit-Forchheimer approximation for $q_{z0}(x, y, 0, t)$, an equation which describes the evolution-in-time of $h_f(x, y, t)$. However, in this equation n-th derivatives of $h_f(x, y, t)$ with respect to x and y occur, including terms with $n > 2$. A higher than second-order equation is not practical for handling numerical methods. Both the finite-difference method and the finite-element method are based on second-order derivatives ($n = 2$) as in the Dupuit-Forchheimer approximation.

Apart from that, the result of including the computation of higher-order terms in the calculation of the evolution of $h_f(x, y, t)$ would be pseudo-precision. In the first place, it so happens that the deep groundwater flow which concerns us here is strongly influenced by spatial variations in the water table, but conversely the deep groundwater flow has negligible influence on the water table itself. This means that it would be superfluous to introduce detailed calculations of the deep groundwater flow into the equation for the evolution of the water table.

Moreover, the equation for $h_f(x, y, t)$ also includes the natural groundwater recharge, and the models to define the effective precipitation $P_e(x, y, t)$ are generally global and relatively inexact. So the Dupuit-Forchheimer approximation is generally accurate enough to construct a rational and justifiable interpolation in space and time of a piezometric field measured in observation wells at only a few points in space and time.

The problem-posing discussed in this book is an entirely different one. If modeling (generally with the Dupuit-Forchheimer approximation and a simple model for P_e), calibration, interpolation and informed 'intuition,' is based on piezometric measurements in observation wells, then the construction of the possible time-evolution of the water table will be as credible and rational as possible (also see Van Geer *et al.*, 1991; Zhou *et al.*, 1991). The next step will then be to determine the effects of this water table on flow and transport; this is the natural flow problem defined as the main subject matter of this book.

The Dupuit approximation can be applied here to demonstrate qualitatively an important aspect of the natural flow problem definition. In accordance with this problem definition, equations (4.2.6) are used to determine $J_x(x, y, t)$, $J_y(x, y, t)$ and $N(x, y, t)$ from specified values of the top potential $f(x, y, t)$. Having determined these quantities, also $\phi(x, y, z, t)$, $q_x(x, y, z, t)$, $q_y(x, y, z, t)$ and $q_z(x, y, z, t)$ can be determined by applying equations (4.2.3). It is, however, often objected that such a procedure is 'ill-posed' from a computational point of view, because it involves differentiation of the top potential $f(x, y, t)$ with respect to x and y. Indeed, suppose that $f(x, y, t) = f_t(x, y, t) + f_\delta(x, y, t)$ where $f_t(x, y, t)$ is the 'true' top potential and $f_\delta(x, y, t)$ represents some amount of 'measurement noise.' Even if the amplitude of the noise $f_\delta(x, y, t)$ is small, differentiation of the top potential $f(x, y, t)$ may result in an amplification of the noise $f_\delta(x, y, t)$ with respect to the true top potential $f_t(x, y, t)$. This is especially the case when the noise component $f_\delta(x, y, t)$ is composed of spatial Fourier modes with short wavelengths (large wave numbers w) that do not occur in the true top potential $f_t(x, y, t)$. Of course, the potential $\phi(x, y, z, t)$ does not suffer from low amplitude noise. However, especially the vertical flow component $N(x, y, t)$, and $q_z(x, y, z, t)$ derived from $N(x, y, t)$, are very vulnerable because $N(x, y, t)$ is obtained by differentiating twice (i.e., by multiplying with w^2). To avoid this differentiation, hydrologists sometimes advocate assuming a reasonable 'model' for the effective precipitation $P_e(x,y,t) = P_{et}(x,y,t) + P_{e\delta}(x,y,t)$, and then calculating $f(x,y,t)$ (or, equivalently, $h_f(x,y,t)$) by integration of equation (4.3.1). Indeed, this latter procedure is numerically 'well-posed' in the sense that short-wave (high wave number) noise Fourier modes of the top flux $P_{e\delta}(x,y,t)$ are attenuated with respect to the true top flux $P_{et}(x,y,t)$ without these short-wave modes. However, the draw-back of this procedure is that $P_e(x,y,t)$ has to be 'calibrated' in such a way that the calculated values of $f(x,y,t)$ match the known (measured) values of $f(x,y,t)$. In our opinion such a procedure does only make sense if $P_e(x,y,t)$ is really known. But in general, and especially in regions where surface water is present, the top flux $P_e(x,y,t)$ is largely unknown. This means that the 'ill-posedness' of numerical differentiation has now been shifted to the 'calibration' procedure, which is performed 'manually' and is therefore subjective and irreproducible. Therefore, in our opinion it is preferable to directly perform the numerical differentiations, however with special care to avoid inaccuracies caused by the noise in the top potential $f(x,y,t)$. Elimination of noise can, for instance, be obtained by filtering the noisy short-wave Fourier modes out of the Fourier spectrum of $f(x,y,t)$. This filtering can also be obtained by avoiding too small discretization dimensions Δx and Δy in the Fast Fourier Transform of $f(x,y,t)$. A somewhat related alternative is to apply finite difference approximations to the derivatives of $f(x,y,t)$, avoiding too small discretization dimensions. This means, for instance, the replacement of $\partial^2 f/\partial x^2$ by the central difference approximation $(f_{n+1} - 2f_n + +f_{n-1})/\Delta x^2$ with a not too small value of Δx (see Figure 6). Such filtering procedures are called regularization (see, for instance, Tikhonov and Arsenin, 1977) and in these examples Δx and Δy may each be considered as a kind of regularization parameter. Furthermore, it will be shown in Chapter 5 that short-wave Fourier modes do not penetrate very deeply into the subsurface, which makes the need for small values of Δx and Δy less urgent.

The above discussion on ill-posedness and regularization has been presented for the sake of finding solutions $\phi(x,y,z,t)$, $e_x(x,y,z,t)$, $e_y(x,y,z,t)$ and $q_z(x,y,z,t)$ to the Dupuit approximations (4.2.3). However, similar arguments hold for the determination of the solutions to the exact continuity equation and Darcy's law. Also in that case, the solution $\phi(x,y,z,t)$ depends only on the specified top potential $f(x,y,t)$, whereas the solutions

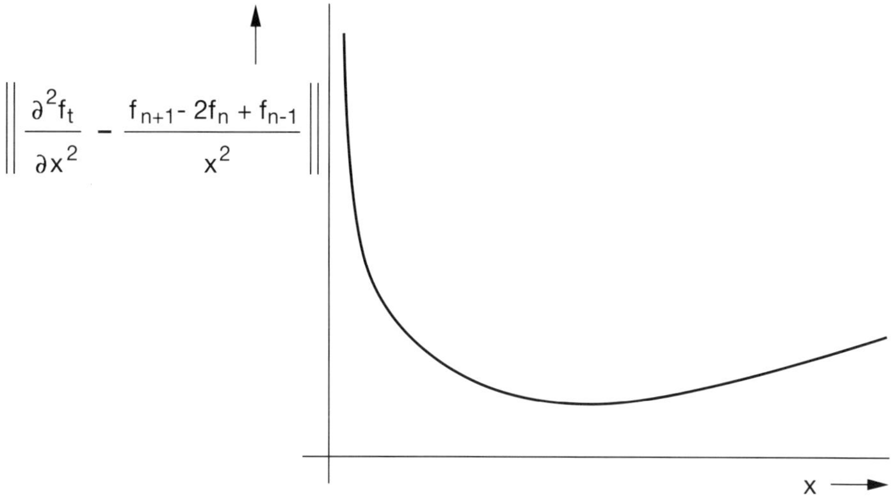

Figure 6. **Error as a function of discretization.**
Measurement (including interpolation of data) of the water table results in an approximation $f(x, y, t)$ to the true water table potential $f_t(x, y, t)$. In general, $\| f_t - f \| / \| f_t \|$ is sufficiently small ($\| \cdot \|$ denotes a suitably chosen norm). But if f contains high wave number measurement noise (of the form $A \sin(w\, x)$, with $|A|$ very small and $|w|$ very large) then $\| \partial^2 f_t/\partial x^2 + - \partial^2 f/\partial x^2 \| / \| \partial^2 f_t/\partial x^2 \|$ is not small at all. Yet, vertical subsurface flow is highly controlled by $\partial^2 f_t/\partial x^2$. In any numerical method the term $\partial^2 f/\partial x^2$, occurring somewhere in the algorithm, is approximated by some form of discretization. As an example, consider the finite difference approximation $\partial^2 f/\partial x^2 \approx \approx (f_{n+1} - 2f_n + f_{n-1})/\Delta x^2$ ($x = n\, \Delta x$, $n = 1, 2, \ldots, N$). The error of this approximation vanishes in the limit $\Delta x \to 0$ and increases if Δx increases. This means that Δx should be sufficiently small. On the other hand, in the limit $\Delta x \to 0$ the high wave number measurement noise gives rise to an intolerably large error in the approximation to $\partial^2 f_t/\partial x^2$, whereas increase of Δx will damp the influence of the high wave number noise. This means that Δx should be sufficiently large. The optimum accuracy in the approximation to $\partial^2 f_t/\partial x^2$ is obtained for some finite Δx, as is shown qualitatively in Figure 6.

$e_x(x,y,z,t)$, $e_y(x,y,z,t)$ and $q_z(x,y,z,t)$ depend on the first and second derivatives of $f(x,y,t)$. That this is the case can be observed from the four top boundary conditions that form the driving force in the four (generally coupled) Laplace-type equations for the four unknowns $\phi(x,y,z,t)$, $e_x(x,y,z,t)$, $e_y(x,y,z,t)$ and $q_z(x,y,z,t)$ (see Section 3.4). In the three top boundary conditions (3.4.5a) for the three equations (3.4.2) for $e_x(x,y,z,t)$, $e_y(x,y,z,t)$ and $q_z(x,y,z,t)$, the aforementioned first and second derivatives of $f(x,y,t)$ occur. The 'ill-posedness' of specifying the top potential explicitly shows up in these three boundary conditions, but this cannot be used as an argument against the application of the three (coupled) Laplace-type equations (3.4.2) for the three unknowns $e_x(x,y,z,t)$, $e_y(x,y,z,t)$ and $q_z(x,y,z,t)$ in numerical models. The ill-posedness reflects a physical phenomenon that also will show up when first solving numerically $\phi(x,y,z,t)$ with its Laplace-type equation (3.1.1), and then solving $q_x(x,y,z,t)$, $q_y(x,y,z,t)$ and $q_z(x,y,z,t)$ with the aid of Darcy's law (2.7.2) which involves a differentiation of $\phi(x,y,z,t)$. Consequently, to determine flow velocities a much higher accuracy is required in the measurements of the water table (or the potential at any position) than is required for the determination of the potentials.

Let us conclude this section with two quotations from the work of the well-known transport physicist Morris Muskat (1937, p. 359):

> *Although the assumptions of the Dupuit-Forchheimer theory of gravity-flow systems now appear to be so questionable as to make the whole theory untrustworthy unless applied with great care, its widespread and indiscriminate use even at present demands at least a brief outline of its fundamental features.*

and (see Muskat, 1937, pp. 364-365):

> *In spite of the practical importance of a knowledge of the variations in groundwater levels in problems of drainage and irrigation, it seems better to await the development of a more satisfactory theory than present in the implications of the linearized Dupuit-Forchheimer equation which will involve errors of unknown magnitude.*

The accuracy of the (linearized) Dupuit approximations, and of the (linearized) Dupuit-Forchheimer approximation based on it, is obviously dependent on the magnitude of the first-order and higher-order corrections. Indeed, negligible potential differences in the vertical z-direction, as is assumed in the Dupuit approximation, are usually justified when dealing with perfectly layered thin basins, but are often of great importance in laterally heterogeneous basins. Consequently, in the following sections a more accurate theory than the Dupuit approximation should be developed.

4.4. First-order corrections: characteristic conductivities

In the following discussions on higher-order corrections to the zeroth-order Dupuit approximations, only *linear* theory will be applied. The reason for linearization is that in higher-order approximations the nonlinear terms are much more complicated than in the zeroth-order approximation discussed before. On the other hand, retaining the nonlinear terms would hardly add to the insight in the flow processes and the applicability of the theory. The linearization introduced here is obtained by projection of the dynamic water table boundary condition on the 'top plane' $z = 0$. Such a linearization results in a valid approximation if the spatial variations in the water table $h_f(x, y, t)$ have sufficiently small amplitudes.

Terms belonging to the powers ϵ, ϵ^2, ϵ^3, etc. can be considered as corrections to the Dupuit approximation. The first-order correct solution in dimensional form for *perfectly layered* basins is given by:

$$\phi = f + r_1(z)\,(\partial^2/\partial x^2 + \partial^2/\partial y^2)\,f + O(\epsilon^2) \tag{4.4.1a}$$

$$e_x = -\partial f/\partial x - r_1(z)\,\partial/\partial x(\partial^2/\partial x^2 + \partial^2/\partial y^2)\,f + O(\epsilon^2) \tag{4.4.1b}$$

$$e_y = -\partial f/\partial y - r_1(z)\,\partial/\partial y(\partial^2/\partial x^2 + \partial^2/\partial y^2)\,f + O(\epsilon^2) \tag{4.4.1c}$$

$$\begin{aligned} q_z = & -t_0(z)\,(\partial^2/\partial x^2 + \partial^2/\partial y^2)\,f \\ & -t_1(z)\,(\partial^2/\partial x^2 + \partial^2/\partial y^2)^2\,f + O(\epsilon^2). \end{aligned} \tag{4.4.1d}$$

In which:

$$r_1(z) = \int_0^z t_0(z')/k_z(z')\,\mathrm{d}z' \quad [m^2] \tag{4.4.2}$$

$$\begin{aligned} t_1(z) &= \int_z^d r_1(z')\,k_h(z')\,\mathrm{d}z' \\ & [m^5{\cdot}Pa^{-1}{\cdot}s^{-1};\ m^5{\cdot}dbar^{-1}{\cdot}d^{-1}]. \end{aligned} \tag{4.4.3}$$

The Dupuit approximation and its first-order correction can be used to obtain a meaningful definition of the lateral and vertical characteristic conductivities for perfectly layered basins (see Appendix C, Section C.6):

$$k_{hc} = t_0(0)/d = T/d \tag{4.4.4a}$$

$$k_{zc} = k_{hc}^2 \, d^3/[3t_1(0)]. \tag{4.4.4b}$$

With this choice for a homogeneous subsurface with conductivities k_h^o and k_z^o, we find that $k_{hc} = k_h^o$ and $k_{zc} = k_z^o$. The expression for the lateral characteristic conductivity k_{hc} correlates with the well-known arithmetic mean value obtained for the parallel connection of conductors. However, the expression for the vertical characteristic conductivity k_{zc} appears to deviate from the usual harmonic mean value obtained for the series connection of conductors. The difference is that, in the definition given by expression (4.4.4b), the part of the subsurface nearer the top has a greater influence on k_{zc} than the part nearer the impermeable base, whereas with a series connection of conductors, each conductor has the same influence.

For basins with nonzero lateral derivatives of the conductivities, an averaged value of the above expressions (4.4.4) for the characteristic conductivities must be taken over the lateral dimensions; an introduction to a suitable averaging procedure is given in Section 4.8. The definition of characteristic conductivities is useful for numerical modeling, where one characteristic value per grid-block or element must be specified for each of the conductivities (see Section 4.8).

Also the first-order correct solutions conform exactly with the continuity equation, but not exactly with Darcy's law. More accurate solutions can be obtained by including higher-order terms in the computations.

First-order approximations can also be given for a basin with *lateral derivatives* of the conductivities as follows:

$$\phi = f + a_j \, \partial f/\partial x_j + r_1 \, \partial^2 f/\partial {x_j}^2 + O(\epsilon^2) \tag{4.4.5a}$$

$$\begin{aligned} e_i = & -(\delta_{ij} + \partial a_j/\partial x_i) \, \partial f/\partial x_j \\ & -(\partial r_1/\partial x_i) \, \partial^2 f/\partial x_j^2 - a_j \, \partial^2 f/\partial x_j \partial x_i \\ & - r_1 \, \partial/\partial x_i(\partial^2 f/\partial {x_j}^2) + O(\epsilon^2) \end{aligned} \tag{4.4.5b}$$

$$q_z = \partial/\partial x_j \left[\int_z^d k_h(x_i, z')\, e_j(x_i, z')\, \mathrm{d}z' \right] + O(\epsilon^2) \qquad (4.4.5c)$$

where $x_i = (x_1, x_2) = (x, y)$, and where the occurrence of the same index twice means summation over that index (the Einstein convention), which means: $a_j b_j = a_1 b_1 + a_2 b_2$; δ_{ij} is the Kronecker delta, $\delta_{ij} = 0$ when $i \neq j$ and $\delta_{ij} = 1$ when $i = j$. The Einstein convention does not hold for x_i $(i = 1, 2)$ occurring as the argument of a function of x_i. The quantities t_0 and r_1 are defined in the same way as for a perfectly layered basin, but are now obviously a function of x_i and z ($t_0 = t_0(x_i, z)$ and $r_1 = r_1(x_i, z)$); the two quantities $a_i(x_j, z)$ $i = 1, 2$, i.e., $a_1(x, y, z)$ and $a_2(x, y, z)$, are defined as:

$$a_i(x_j, z) = \int_0^z [\partial t_0(x_j, z')/\partial x_i]/k_z(x_j, z')\, \mathrm{d}z'. \qquad (4.4.6)$$

If there are very small lateral derivatives of the conductivities, such that $|\partial k_h/\partial x_i| = O(k_{hc}/l_c)$ and $|\partial k_z/\partial x_i| = O(k_{zc}/l_c)$ (the symbol $O(h)$ means having order of magnitude h), then it will be seen from the above formulae (4.4.5) and (4.4.6) that all first-order terms are of the order of magnitude $\epsilon = (d/l_c)^2\ (k_{hc}/k_{zc})$. Thus, for the above-given very small lateral derivatives of the conductivities, all the first-order terms can be neglected for $\epsilon << 1$.

In conclusion, therefore, it can be said that the Dupuit approximation given in Section 4.2 is valid only for such small lateral derivatives of the conductivities. This is a particularly severe limitation of the popular Dupuit approximation, so the question of whether it can be generalized to broaden its validity will be discussed in the next paragraph.

4.5. The generalized transmissivity tensor

Let us start with an example which allows somewhat stronger larger lateral derivatives of the conductivities. Let us suppose that $|\partial k_h/\partial x_i| = O[k_{hc}/\sqrt{(d\, l_c)}]$ and $|\partial k_z/\partial x_i| = O[k_{zc}/\sqrt{(d\, l_c)}]$ (again, the symbol $O(h)$ means having order of magnitude h). It follows then that, after introduction of dimensionless quantities (see Section 4.1), the term $(\partial a_j/\partial x_i)\ (\partial f/\partial x_j)$ has order of magnitude $(d/l_c)\ k_{hc}/k_{zc} = \epsilon\ l_c/d$; the term $(\partial r_1/\partial x_i)\cdot$

$\cdot\partial^2 f/\partial x_j^2 + a_j\ \partial^2 f/\partial x_j \partial x_i$ has order of magnitude $\sqrt{(d/l_c)^3}\cdot(k_{hc}/k_{zc}) =$ $= \epsilon\ \sqrt{(l_c/d)}$, and the term $r_1\ \partial/\partial x_i(\partial^2 f/\partial x_j^2)$ has order of magnitude $(d/l_c)^2\cdot(k_{hc}/k_{zc}) = \epsilon$.

Let us now consider a situation where terms with an order of magnitude ϵ and $\epsilon\ \sqrt{(l_c/d)}$ can be neglected, but where terms with an order of magnitude $\epsilon\ l_c/d$ must be carried on. For instance, $k_{hc}/k_{zc} = 10$ and $d/l_c = 1/32$ gives $\epsilon = 0.01$ (negligible), $\epsilon\ \sqrt{(l_c/d)} = 0.057$ (negligible) and $\epsilon\ l_c/d = 0.32$ (not negligible). Under these conditions, the first-order accurate solutions will reduce to:

$$e_i(x_k, z, t) = -[\delta_{ij} + \partial a_j(x_k, z)/\partial x_i]\ \partial f(x_k, t)/\partial x_j \tag{4.5.1a}$$

$$q_z(x_k, z, t) = -\partial/\partial x_i[t^{\S}_{ij}(x_k, z)\ \partial f(x_k, t)/\partial x_j] \tag{4.5.1b}$$

where:

$$t^{\S}_{ij}(x_k, z) = \int_z^d k_h(x_k, z')\ \Big[\delta_{ij} + \partial a_j(x_k, z')/\partial x_i\Big]\ \mathrm{d}z' \tag{4.5.2}$$

represents the generalized 'three-dimensional' transmissivity, which is a two-dimensional tensor $(i, j = 1, 2)$.

These equations give a generalization of the Dupuit approximation for nonzero, but sufficiently small lateral derivatives of the conductivities. Because of the nonzero derivatives, the generalized 'three-dimensional' transmissivity is a two-dimensional tensor $\underline{\boldsymbol{t}}^{\S}(\boldsymbol{x})$ with four components $t^{\S}_{ij}(x, y, z)$:

$$\underline{\boldsymbol{t}}^{\S}(\boldsymbol{x}) = \begin{bmatrix} t^{\S}_{xx}(x,y,z) & t^{\S}_{xy}(x,y,z) \\ t^{\S}_{yx}(x,y,z) & t^{\S}_{yy}(x,y,z) \end{bmatrix}. \tag{4.5.3}$$

This generalized transmissivity tensor is generally *nonsymmetric*.

We have seen in Section 4.4 that $\epsilon << 1$ is a necessary and sufficient condition for the Dupuit approximation to be valid, provided that the basin is perfectly layered. If the basin is not perfectly layered, the Dupuit approximation may still be applied if the lateral variations in conductivity are sufficiently small. It is now possible to introduce a more practical condition than the order of magnitude estimations given in Section 4.4 for imperfectly layered basins under which the Dupuit approximation is valid. This condition is that the four dimensionless quantities $\left|\mu_{ij}\right| = \left|\partial a_j/\partial x_i\right| \geq 0$, $i, j = 1, 2$, are sufficiently small with respect to 1. It is more practical to introduce one dimensionless number, the lateral heterogeneity number μ as:

$$\mu = \sqrt{(\mu_{11}^2 + \mu_{12}^2 + \mu_{21}^2 + \mu_{22}^2)/4}. \tag{4.5.4}$$

As a consequence, the two necessary and sufficient conditions for the Dupuit approximation, and especially for the Dupuit expression (4.2.4) for the transmissivity (i.e., for the conventional, nongeneralized transmissivity), to be valid, are that the dimensionless lateral heterogeneity number $\mu << 1$ and that the dimensionless number $\epsilon << 1$.

On the basis of the above-presented theory it is possible to propose a classification of the subsurface with respect to its flow properties. A well-known, generally accepted classification is: a basin with exactly constant conductivity components is *homogeneous*, whereas a basin which is not homogeneous is *heterogeneous*. Strictly speaking, the above definition holds on the point scale (the scale of a representative elementary volume, the cm-scale, say). However, the above definition is often generalized to larger scales. For instance, consider a body of porous material on the large scale, say the scale of $1\ km \times 1\ km \times 100\ m$. On the cm-scale the porous medium may be heterogeneous, but the averaged conductivities on, say, the m-scale may be such that the whole body of porous medium is homogeneous.

Now we propose to classify *heterogeneity* (on any spatial scale) in the following way:

1. A basin for which $\mu = 0$ (exactly) is *perfectly layered.*
2. A basin for which $\mu << 1$ (say $\mu \leq 0.1$) is *ideally layered.*
3. A basin for which $\mu < 1$ (say $0.1 < \mu < 1$) is *approximately layered.*
4. A basin for which $\mu > 1$ is *strongly heterogeneous.*

In the above definitions the lateral heterogeneity number μ is based on derivatives of the conductivities $k_h(x, y, z)$ and $k_z(x, y, z)$ in the x- and y-directions and not on derivatives in the z-direction. It is possible to introduce a primed coordinate system with coordinates (x', y', z'), where the z-axis is a 'lateral' coordinate, say $(x', y', z') = (z, y, x)$. If $k_h(x, y, z)/$ $/k_z(x, y, z)$ is constant, the above-presented theory also holds in the primed coordinate system and the values μ' can be calculated. If μ' has approximately the same value as μ, then the terminology will be:

1. A basin for which $\mu = \mu' = 0$ (exactly) is *perfectly homogeneous.*
2. A basin for which $\mu \approx \mu' << 1$ is *ideally homogeneous.*
3. A basin for which $\mu \approx \mu' < 1$ is *approximately homogeneous.*
4. A basin for which $\mu \approx \mu' > 1$ is *strongly heterogeneous.*

Flow in perfectly homogeneous and ideally homogeneous porous media may be described mathematically in the same way. The same holds for perfectly layered and ideally layered porous media. Flow in approximately homogeneous porous media may be described by making use of derivatives of the conductivities in all three directions, whereas for the description of flow in approximately layered porous media derivatives in the two lateral directions only may be applied. Flow in strongly heterogeneous porous media must be described without making use of derivatives of the conductivities. This means that also the character of the flow behavior will differ; for instance, flow properties derived for (ideally) homogeneous porous media cannot simply be transferred to flow in (ideally) layered media, and results for ideally layered porous media do not generally hold for approximately layered porous media. Also the inverse is true, results for strongly heterogeneous porous media do not generally apply to, say, (ideally) layered porous media. To elucidate the variety in flow patterns, it is desirable to apply special mathematical techniques well-suited to the particular flow property class under consideration. For instance, the present emphasis on strongly heterogeneous porous media, e.g., discontinuous shales in small-scale sand bodies (see Haldorsen and Lake, 1984), may obscure phenomena important in, say, approximately layered porous media, e.g., deltaic and eolian formations (see Weber and Van Geuns, 1990). See also Chapters 8 and 9 where a combination of approaches is proposed for numerical modeling.

As we have seen, the generalized transmissivity tensor (4.5.3) is generally nonsymmetric. However, there are conductivity distributions for which the generalized transmissivity is in fact symmetric. For example, if $k_h(x,y)$ and $k_z(x,y)$ are independent of z, and if the anisotropy factor $k_h(x,y)/k_z(x,y) = \alpha$ is a constant, it follows that the generalized transmissivity tensor is symmetric (i.e., $t^{\S}_{ij} = t^{\S}_{ji}$). In the latter case, it also follows for the (two-dimensional) generalized transmissivity $\underline{\boldsymbol{T}}^{\S}(x,y) = \underline{t}^{\S}(x,y,0)$

that its Cartesian components are given by:

$$T^{\S}_{xy}(x,y) = \alpha\ k_h(x,y)\ (d^3/3)\ \partial^2\lambda_h/\partial x\partial y \tag{4.5.5a}$$

$$T^{\S}_{yx}(x,y) = \alpha\ k_h(x,y)\ (d^3/3)\ \partial^2\lambda_h/\partial y\partial x \tag{4.5.5b}$$

$$T^{\S}_{xx}(x,y) = k_h(x,y)\ d\ [1+\alpha\ (d^2/3)\ \partial^2\lambda_h/\partial x^2] \tag{4.5.5c}$$

$$T^{\S}_{yy}(x,y) = k_h(x,y)\ d\ [1+\alpha\ (d^2/3)\ \partial^2\lambda_h/\partial y^2] \tag{4.5.5d}$$

Only if the generalized transmissivity is symmetric can we find two principal axes perpendicular to each other (see Section 4.6).

Under certain flow conditions, we can also find a generalized Dupuit approximation with generalized transmissivity for even stronger lateral derivatives of the conductivities (see Section 4.7).

4.6. Generalized principal axes

In Section 4.5 we have seen that the generalized transmissivity $\underline{\boldsymbol{T}}^{\S}$ is generally a nonsymmetric two-dimensional tensor. For symmetric transmissivities two mutually perpendicular principal axes exist. (Also, for symmetric conductivities three principal axes exist.) Only along these mutually perpendicular axes the flux and the potential gradient have opposite directions. The notion of principal axes has very practical consequences that are almost unconsciously applied in geohydrology. In fact, the principal axes are the mutually perpendicular eigenvectors of the transmissivity tensor, which exist *for symmetric tensors only*. In this section we investigate how to extend the notion of principal axes to nonsymmetric transmissivity tensors by making use of the singular value decomposition (see Stewart, 1973, pp. 317-326), which can be seen as a generalization of the eigenvalue decomposition.

For notational convenience we omit the superscript § in this section. The Cartesian coordinates of the two-dimensional tensor $\underline{\boldsymbol{T}}$ are given in matrix form by:

$$\underline{\boldsymbol{T}} = \begin{pmatrix} T_{11} & T_{12} \\ T_{21} & T_{22} \end{pmatrix} \tag{4.6.1}$$

where, in general, $T_{12} \neq T_{21}$. We will furthermore assume that $\underline{\boldsymbol{T}}$ is non-singular, i.e., that $\underline{\boldsymbol{T}}^{-1}$ exists; a necessary and sufficient condition for non-singularity is that the determinant of $\underline{\boldsymbol{T}}$, det $\underline{\boldsymbol{T}} = T_{11}T_{22} - T_{12}T_{21} \neq 0$ (see Stewart, 1973, pp. 400-401). Furthermore, since flow in porous media dissipates mechanical energy, the transmissivity must be positive definite. For a definition of positive definiteness see Stewart (1973, p. 139). This positive definiteness means that det $\underline{\boldsymbol{T}} = T_{11}T_{22} - T_{12}T_{21} > 0$.

Here, the transpose of $\underline{\boldsymbol{T}}$ will be denoted by $\underline{\boldsymbol{T}}^*$ and is given by:

$$\underline{\boldsymbol{T}}^* = \begin{pmatrix} T_{11} & T_{21} \\ T_{12} & T_{22} \end{pmatrix}. \tag{4.6.2}$$

Now define the following two symmetric tensors $\underline{\boldsymbol{M}}$ and $\underline{\boldsymbol{N}}$:

$$\begin{aligned} \underline{\boldsymbol{M}} = \underline{\boldsymbol{T}}^* \cdot \underline{\boldsymbol{T}} &= \begin{pmatrix} T_{11} & T_{21} \\ T_{12} & T_{22} \end{pmatrix} \cdot \begin{pmatrix} T_{11} & T_{12} \\ T_{21} & T_{22} \end{pmatrix} = \\ &= \begin{pmatrix} T_{11}^2 + T_{21}^2 & T_{11}T_{12} + T_{21}T_{22} \\ T_{11}T_{12} + T_{21}T_{22} & T_{12}^2 + T_{22}^2 \end{pmatrix} \end{aligned} \tag{4.6.3}$$

and:

$$\begin{aligned} \underline{\boldsymbol{N}} = \underline{\boldsymbol{T}} \cdot \underline{\boldsymbol{T}}^* &= \begin{pmatrix} T_{11} & T_{12} \\ T_{21} & T_{22} \end{pmatrix} \cdot \begin{pmatrix} T_{11} & T_{21} \\ T_{12} & T_{22} \end{pmatrix} = \\ &= \begin{pmatrix} T_{11}^2 + T_{12}^2 & T_{11}T_{21} + T_{12}T_{22} \\ T_{11}T_{21} + T_{12}T_{22} & T_{21}^2 + T_{22}^2 \end{pmatrix}. \end{aligned} \tag{4.6.4}$$

From expressions (4.6.3) and (4.6.4) the following determinant Δ can be defined:

$$\Delta = \det(\underline{\boldsymbol{M}} - \lambda\, \underline{\boldsymbol{I}}) = \det(\underline{\boldsymbol{N}} - \lambda\, \underline{\boldsymbol{I}}) = \lambda^2 - \lambda\, S^2 + D^2 \tag{4.6.5a}$$

where $\underline{\boldsymbol{I}} = \delta_{ij}$ $(i, j = 1, 2)$ is the unit tensor with Cartesian components given by:

$$\underline{\boldsymbol{I}} = \delta_{ij} = \begin{pmatrix} 1 & 0 \\ 0 & 1 \end{pmatrix} \tag{4.6.5b}$$

and where S^2 and D^2 are given by:

$$S^2 = T_{11}^2 + T_{12}^2 + T_{21}^2 + T_{22}^2 > 0 \tag{4.6.5c}$$

$$D^2 = (T_{11}T_{22} - T_{12}T_{21})^2 = (\det \underline{\boldsymbol{T}})^2 = (\det \underline{\boldsymbol{T}}^*)^2 > 0. \tag{4.6.5d}$$

The eigenvalues of $\underline{\boldsymbol{M}}$ and $\underline{\boldsymbol{N}}$ are the values of λ for which $\Delta = 0$; this results in the following two eigenvalues:

$$\lambda_1 = \left[S^2 + \sqrt{(S^4 - 4D^2)} \right] /2 \tag{4.6.6a}$$

$$\lambda_2 = \left[S^2 - \sqrt{(S^4 - 4D^2)} \right] /2 \tag{4.6.6b}$$

with

$$\begin{aligned} S^4 - 4D^2 = & \ [(T_{11} + T_{22})^2 + (T_{12} - T_{21})^2] \cdot \\ & \cdot [(T_{11} - T_{22})^2 + (T_{12} + T_{21})^2]. \end{aligned} \tag{4.6.6c}$$

From expression (4.6.6c) it follows that $S^4 - 4D^2 \geq 0$, which means that the square roots in expressions (4.6.6a) and (4.6.6b) exist. Furthermore, since $\sqrt{(S^4 - 4D^2)} < S^2$, it follows that $\lambda_2 > 0$. As a consequence, we find $\lambda_1 \geq \lambda_2 > 0$.

After having determined the eigenvalues λ_1 and λ_2 of $\underline{\boldsymbol{M}}$ and $\underline{\boldsymbol{N}}$, we can determine the eigenvectors $\boldsymbol{a}_1$ and $\boldsymbol{a}_2$ of tensor $\underline{\boldsymbol{M}}$, and the eigenvectors $\boldsymbol{b}_1$ and $\boldsymbol{b}_2$ of tensor $\underline{\boldsymbol{N}}$.

By definition, the eigenvectors of $\underline{\boldsymbol{M}}$ are given by:

$$\underline{\boldsymbol{M}} \cdot \boldsymbol{a}_1 = \lambda_1 \boldsymbol{a}_1 \tag{4.6.7a}$$

$$\underline{\boldsymbol{M}} \cdot \boldsymbol{a}_2 = \lambda_2 \boldsymbol{a}_1. \tag{4.6.7b}$$

Taking the scalar product of equation (4.6.7a) with $\boldsymbol{a}_2$, the scalar product of equation (4.6.7b) with $\boldsymbol{a}_1$ and subtracting results in:

$$\boldsymbol{a}_2 \cdot \underline{\boldsymbol{M}} \cdot \boldsymbol{a}_1 - \boldsymbol{a}_1 \cdot \underline{\boldsymbol{M}} \cdot \boldsymbol{a}_2 = (\lambda_1 - \lambda_2)\ \boldsymbol{a}_1 \cdot \boldsymbol{a}_2. \tag{4.6.8}$$

Since $\underline{\boldsymbol{M}}$ is symmetric $\boldsymbol{a}_2 \cdot \underline{\boldsymbol{M}} \cdot \boldsymbol{a}_1 = \boldsymbol{a}_1 \cdot \underline{\boldsymbol{M}} \cdot \boldsymbol{a}_2$, i.e., the left-hand side of equation (4.6.8) is equal to zero. If $\lambda_1 > \lambda_2$ it follows then that $\boldsymbol{a}_1 \cdot \boldsymbol{a}_2 = 0$, i.e., $\boldsymbol{a}_1$ and $\boldsymbol{a}_2$ are mutually orthogonal. The condition $\lambda_1 = \lambda_2 = \lambda$ occurs if $S^4 - 4D^2 = 0$. This is the case if $T_{11} = T_{22} = \lambda$ and $T_{12} = T_{21} = 0$. (The case $T_{12} = T_{21} = \lambda$ and $T_{11} = T_{22} = 0$ satisfies $S^4 - 4D^2 = 0$, but does not result in a positive definite tensor $\underline{\boldsymbol{T}}$.) In that case any arbitrary set of orthogonal vectors $\boldsymbol{a}_1$ and $\boldsymbol{a}_2$ may be considered as eigenvectors. In a similar way we find that the eigenvectors $\boldsymbol{b}_1$ and $\boldsymbol{b}_2$ of $\underline{\boldsymbol{N}}$ are orthogonal.

Furthermore, the orthogonal eigenvectors may be normed to length 1; i.e., $\boldsymbol{a}_1 \cdot \boldsymbol{a}_1 = \boldsymbol{a}_2 \cdot \boldsymbol{a}_2 = \boldsymbol{b}_1 \cdot \boldsymbol{b}_1 = \boldsymbol{b}_2 \cdot \boldsymbol{b}_2 = 1$. Consequently, both the pair of vectors $\boldsymbol{a}_1$, $\boldsymbol{a}_2$ and the pair of vectors $\boldsymbol{b}_1$, $\boldsymbol{b}_2$ are orthonormal. The orthonormal eigenvectors are not yet uniquely defined; also $-\boldsymbol{a}_1$, $-\boldsymbol{a}_2$, etc. are eigenvectors.

Since the eigenvectors are orthonormal, they may be denoted by the following Cartesian components:

$$\boldsymbol{a}_1 = (a_{11},\ a_{21}) = (\cos\alpha,\ -\sin\alpha) \tag{4.6.9a}$$

$$\boldsymbol{a}_2 = (a_{12},\ a_{22}) = (\sin\alpha,\ \cos\alpha) \tag{4.6.9b}$$

and:

$$\boldsymbol{b}_1 = (b_{11},\ b_{21}) = (\cos\beta,\ -\sin\beta) \tag{4.6.10a}$$

$$\boldsymbol{b}_2 = (b_{12},\ b_{22}) = (\sin\beta,\ \cos\beta). \tag{4.6.10b}$$

We now define the following two tensors $\underline{\boldsymbol{A}}$ and $\underline{\boldsymbol{B}}$:

$$\underline{\boldsymbol{A}} = \begin{pmatrix} a_{11} & a_{12} \\ a_{21} & a_{22} \end{pmatrix} = \begin{pmatrix} \cos\alpha & \sin\alpha \\ -\sin\alpha & \cos\alpha \end{pmatrix} \tag{4.6.11a}$$

and:

$$\underline{\boldsymbol{B}} = \begin{pmatrix} b_{11} & b_{12} \\ b_{21} & b_{22} \end{pmatrix} = \begin{pmatrix} \cos\beta & \sin\beta \\ -\sin\beta & \cos\beta \end{pmatrix} \tag{4.6.11b}$$

The i-th column of the matrix $\underline{\boldsymbol{A}}$ is (the components of) the eigenvector belonging to the i-th eigenvalue λ_i of $\underline{\boldsymbol{M}}$ ($i = 1, 2$). Similarly, the i-th column of the matrix $\underline{\boldsymbol{B}}$ is (the components of) the eigenvector belonging to the i-th eigenvalue λ_i of $\underline{\boldsymbol{M}}$ ($i = 1, 2$). The tensors $\underline{\boldsymbol{A}}$ and $\underline{\boldsymbol{B}}$ are unitary (see Stewart, 1973, p. 275), i.e., $\underline{\boldsymbol{A}}^{-1} = \underline{\boldsymbol{A}}^*$ and $\underline{\boldsymbol{B}}^{-1} = \underline{\boldsymbol{B}}^*$ (see Stewart, 1973, p. 259). Furthermore, $\underline{\boldsymbol{A}}$ and $\underline{\boldsymbol{B}}$ represent a rotation over an angle α and β, respectively.

From equations (4.6.7) and (4.6.11) it follows that:

$$\underline{\boldsymbol{A}}^* \cdot \underline{\boldsymbol{M}} \cdot \underline{\boldsymbol{A}} = \underline{\boldsymbol{B}}^* \cdot \underline{\boldsymbol{N}} \cdot \underline{\boldsymbol{B}} = \begin{pmatrix} \lambda_1 & 0 \\ 0 & \lambda_2 \end{pmatrix} \tag{4.6.12}$$

Since $\underline{\boldsymbol{A}}^* \cdot \underline{\boldsymbol{M}} \cdot \underline{\boldsymbol{A}} = \underline{\boldsymbol{A}}^* \cdot \underline{\boldsymbol{T}}^* \cdot \underline{\boldsymbol{T}} \cdot \underline{\boldsymbol{A}} = \underline{\boldsymbol{A}}^* \cdot \underline{\boldsymbol{T}}^* \cdot \underline{\boldsymbol{B}} \cdot \underline{\boldsymbol{B}}^* \cdot \underline{\boldsymbol{T}} \cdot \underline{\boldsymbol{A}} = (\underline{\boldsymbol{B}}^* \cdot \underline{\boldsymbol{T}} \cdot \underline{\boldsymbol{A}})^* \cdot \cdot (\underline{\boldsymbol{B}}^* \cdot \underline{\boldsymbol{T}} \cdot \underline{\boldsymbol{A}})$ and, similarly, $\underline{\boldsymbol{B}}^* \cdot \underline{\boldsymbol{N}} \cdot \underline{\boldsymbol{B}} = (\underline{\boldsymbol{B}}^* \cdot \underline{\boldsymbol{T}} \cdot \underline{\boldsymbol{A}}) \cdot (\underline{\boldsymbol{B}}^* \cdot \underline{\boldsymbol{T}} \cdot \underline{\boldsymbol{A}})^*$, it follows from equation (4.6.12) that $(\underline{\boldsymbol{B}}^* \cdot \underline{\boldsymbol{T}} \cdot \underline{\boldsymbol{A}})^* \cdot (\underline{\boldsymbol{B}}^* \cdot \underline{\boldsymbol{T}} \cdot \underline{\boldsymbol{A}}) = (\underline{\boldsymbol{B}}^* \cdot \underline{\boldsymbol{T}} \cdot \underline{\boldsymbol{A}}) \cdot \cdot (\underline{\boldsymbol{B}}^* \cdot \underline{\boldsymbol{T}} \cdot \underline{\boldsymbol{A}})^*$. From expressions (4.6.3) and (4.6.4) it can be seen that if $\underline{\boldsymbol{T}}^* \cdot \underline{\boldsymbol{T}} = \underline{\boldsymbol{T}} \cdot \underline{\boldsymbol{T}}^*$ then $\underline{\boldsymbol{T}}$ is symmetric. Hence, $\underline{\boldsymbol{B}}^* \cdot \underline{\boldsymbol{T}} \cdot \underline{\boldsymbol{A}}$ is symmetric, i.e., $(\underline{\boldsymbol{B}}^* \cdot \underline{\boldsymbol{T}} \cdot \underline{\boldsymbol{A}})^* = \underline{\boldsymbol{B}}^* \cdot \underline{\boldsymbol{T}} \cdot \underline{\boldsymbol{A}}$. As a consequence, the following important theorem holds (see Stewart, 1973, pp. 318-319):

$$\underline{\boldsymbol{B}}^* \cdot \underline{\boldsymbol{T}} \cdot \underline{\boldsymbol{A}} = \begin{pmatrix} T_1 & 0 \\ 0 & T_2 \end{pmatrix} \tag{4.6.13a}$$

where $T_1 = \pm\sqrt{\lambda_1}$ and $T_2 = \pm\sqrt{\lambda_2}$. The directions of the eigenvectors can be chosen in such a way that the plus signs hold, i.e.:

$$T_1 = \sqrt{\lambda_1} \tag{4.6.13b}$$

$$T_2 = \sqrt{\lambda_2}. \tag{4.6.13c}$$

The above expressions (4.6.13) represent the singular value decomposition of tensor $\underline{\boldsymbol{T}}$. The values $T_1 \geq T_2$ are called the singular values of the tensor $\underline{\boldsymbol{T}}$ and they have the physical dimension of a transmissivity $[m^3 \cdot Pa^{-1} \cdot s^{-1};\ m^3 \cdot dbar^{-1} \cdot d^{-1} \approx m^3 \cdot mH_2O^{-1} \cdot d^{-1}]$. From equations (4.6.13) it follows that:

$$\underline{\boldsymbol{T}} \cdot \boldsymbol{a}_1 = T_1 \underline{\boldsymbol{b}}_1 \tag{4.6.14a}$$

$$\underline{\boldsymbol{T}} \cdot \boldsymbol{a}_2 = T_2 \underline{\boldsymbol{b}}_2. \tag{4.6.14b}$$

The expressions (4.6.14) will give insight in the essential difference between symmetric and nonsymmetric tensors. This will be explained as follows.

Let us consider a curvilinear orthogonal coordinate system with unit directions $\boldsymbol{a}_1$ and $\boldsymbol{a}_2$ and with coordinates ξ_1 and ξ_2. In this coordinate system a potential gradient **grad** f has the following components:

$$\mathbf{grad}\, f = (h_1^{-1}\ \partial f/\partial\xi_1,\ h_2^{-1}\ \partial f/\partial\xi_2) \tag{4.6.15a}$$

where h_1 and h_2 are the scale factors belonging to the curvilinear orthogonal coordinate system (see Morse and Feshbach, 1953, part I, pp. 21-31). Equation (4.6.15a) can also be written as:

$$\mathbf{grad}\, f = h_1^{-1}\ \partial f/\partial\xi_1\ \boldsymbol{a}_1 + h_2^{-1}\ \partial f/\partial\xi_2\ \boldsymbol{a}_2. \tag{4.6.15b}$$

Then, according to (4.6.14), $\underline{\boldsymbol{T}}\cdot$ **grad** f is given by:

$$\underline{\boldsymbol{T}}\cdot \mathbf{grad}\, f = T_1\ h_1^{-1}\ \partial f/\partial\xi_1\ \boldsymbol{b}_1 + T_2\ h_2^{-1}\ \partial f/\partial\xi_2\ \boldsymbol{b}_2. \tag{4.6.16}$$

Substitution of equation (4.6.16) into the Dupuit equations $\boldsymbol{J} = -\boldsymbol{T}\cdot \mathbf{grad}\, f$ for the depth-integrated flux $\boldsymbol{J}$ (see equations (4.2.5)) yields:

$$\boldsymbol{J} = -T_1\ h_1^{-1}\ \partial f/\partial\xi_1\ \boldsymbol{b}_1 - T_2\ h_2^{-1}\ \partial f/\partial\xi_2\ \boldsymbol{b}_2 \tag{4.6.17}$$

or, equivalently:

$$(J_1, J_2) = -(T_1\ h_1^{-1}\ \partial f/\partial\xi_1,\ T_2\ h_2^{-1}\ \partial f/\partial\xi_2) \tag{4.6.18}$$

where the components J_1 and J_2 of the vector $\boldsymbol{J}$ are given in a coordinate system with unit vectors $\boldsymbol{b}_1$ and $\boldsymbol{b}_2$! Consequently, it is very important to realize that expression (4.6.18) expresses the Dupuit relationship between the two components of $\boldsymbol{J}$ and **grad** f *in two different coordinate systems* which are rotated over an angle $\beta - \alpha$ with respect to each other (see Figure 7). Of course, the vectors $\boldsymbol{b}_1$ and $\boldsymbol{b}_2$ in equation (4.6.17) may be replaced by the vectors $\underline{\boldsymbol{B}}\cdot\underline{\boldsymbol{A}}^*\cdot\boldsymbol{a}_1$ and $\underline{\boldsymbol{B}}\cdot\underline{\boldsymbol{A}}^*\cdot\boldsymbol{a}_2$, respectively, which results in a final equation in only one coordinate system, but then the two insight-giving scalar relationships between J_1 and $(\mathbf{grad}\, f)_1 = h_1^{-1}\ \partial f/\partial\xi_1$, and between J_2 and $(\mathbf{grad}\, f)_2 = h_2^{-1}\ \partial f/\partial\xi_2$ are lost.

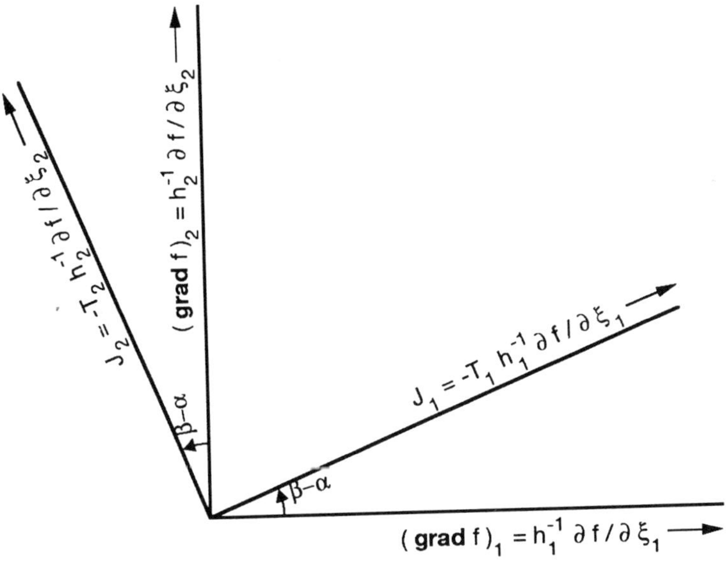

Figure 7. **Generalized principal axes of a nonsymmetric transmissivity tensor.**
Figure 7 shows two orthogonal coordinate systems which are rotated over an angle $\beta - \alpha$ with respect to each other. In one coordinate system the two components of the gradient in the top potential $\mathbf{grad}\, f = ((\mathbf{grad}\, f)_1, (\mathbf{grad}\, f)_2) = (h_1^{-1}\ \partial f/\partial\xi_1, h_2^{-1}\ \partial f/\partial\xi_2)$ *are given. In the other coordinate system the two components of the depth-integrated volumetric flow rate* $\boldsymbol{J} = (J_1, J_2)$ *are given. Since the two coordinate systems represent generalized principal axes, the Dupuit equation* $\boldsymbol{J} = -\underline{\boldsymbol{T}}\cdot \mathbf{grad}\, f$ *yields* $J_1 = -T_1\ h_1^{-1}\ \partial f/\partial\xi_1$ *and* $J_2 = -T_2\ h_2^{-1}\ \partial f/\partial\xi_2$*. Only if the transmissivity tensor* $\underline{\boldsymbol{T}}$ *is symmetric,* $\beta = \alpha$ *and the two coordinate systems coincide.*

If the transmissivity $\underline{\boldsymbol{T}}$ is a symmetric tensor, then $\underline{\boldsymbol{M}} = \underline{\boldsymbol{N}}$ holds, which makes it possible to choose the eigenvectors in such a way that $\boldsymbol{a}_1 = \boldsymbol{b}_1$ and $\boldsymbol{a}_2 = \boldsymbol{b}_2$. Then, since a transmissivity is always positive definite, the plus-sign expressions (4.6.13b) and (4.6.13c) hold and T_1 and T_2 are the eigenvalues of $\underline{\boldsymbol{T}}$. In this symmetric case the two different coordinate systems coincide, which considerably enhances the practical applicability of equation (4.6.18).

In cases where the transmissivity $\underline{\boldsymbol{T}}$ is symmetric, the directions

$\boldsymbol{a}_1 = \boldsymbol{b}_1$ and $\boldsymbol{a}_2 = \boldsymbol{b}_2$ are generally called the principal axes of $\underline{\boldsymbol{T}}$. Therefore, for a nonsymmetric transmissivity $\underline{\boldsymbol{T}}$, we propose to call $\boldsymbol{a}_1$ and $\boldsymbol{a}_2$ the generalized principal axes of $\underline{\boldsymbol{T}}$ with respect to **grad** f, and $\boldsymbol{b}_1$ and $\boldsymbol{b}_2$ the generalized principal axes of $\underline{\boldsymbol{T}}$ with respect to $\boldsymbol{J}$.

4.7. Second-order and higher-order corrections

The exact solutions for a *perfectly layered* subsurface are given by the following expansions in infinite series:

$$e_x = -\sum_{n=0}^{\infty} r_n(z)\ \partial/\partial x(\partial^2/\partial x^2 + \partial^2/\partial y^2)^n\ f \tag{4.7.1a}$$

$$e_y = -\sum_{n=0}^{\infty} r_n(z)\ \partial/\partial y(\partial^2/\partial x^2 + \partial^2/\partial y^2)^n\ f \tag{4.7.1b}$$

$$q_z = -\sum_{n=0}^{\infty} t_n(z)\ (\partial^2/\partial x^2 + \partial^2/\partial y^2)^{(n+1)}\ f. \tag{4.7.1c}$$

Wherein:

$$r_n(z) = \int_0^z t_{n-1}(z')/k_z(z')\ \mathrm{d}z' \quad [m^{2n}] \tag{4.7.2}$$

$$t_n(z) = \int_z^d r_n(z')\ k_h(z')\ \mathrm{d}z' \quad [m^{2n+3}.Pa^{-1}.s^{-1}] \tag{4.7.3}$$

with

$$r_0(z) = 1 \tag{4.7.4}$$

$$t_0(z) = \int_z^d k_h(z')\ \mathrm{d}z'. \tag{4.7.5}$$

If the higher-order corrections are determined for basins with *lateral heterogeneities*, then we find more extra terms in which lateral derivatives of integrals over the conductivities appear.

For sufficiently small values of the dimensionless number ϵ, the above series (4.7.1) will converge to the exact solutions to the flow equations. In the literature, methods are described to find solutions even for divergent series, the nonlinear sequence-to-sequence transformations (see Shanks, 1955). However, the above series should in the first place be considered as tools to give insight in the fluid mechanical processes under consideration, especially in the ranges of validity of the highly popular Dupuit approximation. To find full solutions in situations where the Dupuit approximation does not hold, the following more robust technique exists.

For piecewise constant perfectly layered porous media it can be proved that one unique solution exists for any value of the dimensionless number $0 \leq \epsilon < \infty$ (see Van Veldhuizen *et al.*, 1992a). For that case, a much more reliable solution method than the nonlinear sequence-to-sequence method can be given for any spatial Fourier component of the top potential $f(x, y, t)$ (see Bervoets, 1991; and Bervoets *et al.*, 1991). In this way, the main task is to decompose the top potential $f(x, y, t)$ into Fourier components as reliably as possible.

From the series (4.7.1) it can be seen that the n-th order correct solutions, i.e., the approximate solutions obtained by omitting all the $(n+1)$-th terms, will satisfy the continuity equation $\operatorname{div} \boldsymbol{q} = 0$ exactly. However, Darcy's law will not be satisfied exactly. We find, as residuals in Darcy's law, that:

$$\begin{aligned} &\partial e_x/\partial z - \partial/\partial x(q_z/k_z) = \\ &\quad = t_n(z)\ \partial/\partial x(\partial^2/\partial x^2 + \partial^2/\partial y^2)^{(n+1)}\ f \end{aligned} \tag{4.7.6a}$$

$$\begin{aligned} &\partial e_y/\partial z - \partial/\partial y(q_z/k_z) = \\ &\quad = t_n(z)\ \partial/\partial y(\partial^2/\partial x^2 + \partial^2/\partial y^2)^{(n+1)}\ f. \end{aligned} \tag{4.7.6b}$$

These expressions become equal to zero if the residuals (the right-hand terms) become small enough for $n \to \infty$.

This last statement can be elucidated as follows. In general, i.e., for all types of heterogeneous subsurfaces, it holds that:

$$(\mathbf{curl\, e})_x = \partial/\partial y(q_z/k_z) - \partial e_y/\partial z = 0 \tag{4.7.7a}$$

$$(\mathbf{curl\, e})_y = \partial e_x/\partial z - \partial/\partial x(q_z/k_z) = 0 \tag{4.7.7b}$$

$$(\mathbf{curl}\, \boldsymbol{e})_z = \partial e_y/\partial x - \partial e_x/\partial y = 0. \qquad (4.7.7c)$$

$$\operatorname{div} \boldsymbol{q} = \partial/\partial x(k_h e_x) + \partial/\partial y(k_h e_y) + \partial q_z/\partial z = 0 \qquad (4.7.7d)$$

Expressions (4.7.7a), (4.7.7b) and (4.7.7c) for the three components of the curl of $\boldsymbol{e}$ are equivalent to Darcy's law. Indeed, each vector $\boldsymbol{e}$ which is irrotational (i.e., $\mathbf{curl}\, \boldsymbol{e} = \mathbf{0}$) can be expressed as the gradient of a scalar ϕ (i.e., $\boldsymbol{e} = -\,\mathbf{grad}\,\phi$; see Section 2.5). Equation (4.7.7c) is automatically satisfied by the choice of the top boundary conditions (3.4.5a); $e_x(x, y, 0) = -\partial f(x, y)/\partial x$, $e_y(x, y, 0) = -\partial f(x, y)/\partial y$. The two equations (4.7.7a) and (4.7.7b) are written in dimensionless form as:

$$\epsilon\, \partial/\partial Y(Q_z/K_z) - \partial E_y/\partial Z = 0 \qquad (4.7.8a)$$

$$\partial E_x/\partial Z - \epsilon\, \partial/\partial X(Q_z/K_z) = 0. \qquad (4.7.8b)$$

Note that the dimensionless number ϵ appears in these two dimensionless equations, which means that terms of the order $n + 1$ (E_{xn+1}, E_{yn+1}) in these equations are coupled with terms of the order n (Q_{zn}). The two equations (4.7.8) must be completed with the continuity equation (4.7.7d) in dimensionless form:

$$\partial/\partial X(K_h\; E_x) + \partial/\partial Y(K_h\; E_y) + \partial Q_z/\partial Z = 0. \qquad (4.7.9)$$

The dimensionless number ϵ does not appear in this last equation, which means that terms of the order n are related to each other in this equation only.

The three dimensionless equations (4.7.8a), (4.7.8b) and (4.7.9) for E_x, E_y and Q_z are equivalent to the three dimensionless Laplace-type equations (4.1.1b), (4.1.1c) and (4.1.1d) presented in Section 4.1, with boundary conditions (4.1.4a) and (4.1.4b) for E_x, E_y and Q_z. All the solutions obtained from the latter equations could also be obtained from the three former dimensionless equations given above (see also Appendix D). This is sometimes an easier mathematical procedure, provided the calculations are purely analytical; the Laplace-type equations are more manageable for numerical calculations. Consequently, the solutions obtained from the Laplace-type equations can now be checked with the aid of the three equations (4.7.8a), (4.7.8b) and (4.7.9) given above. Sometimes direct calculation of e_x and e_y

is not required. In that case e_x and e_y can be derived from Laplace-type equation (3.1.1) for ϕ and by $e_x = -\partial\phi/\partial x$, $e_y = -\partial\phi/\partial y$. On the other hand, however, calculation of q_z by $q_z = -k_z\ \partial\phi/\partial z$ might be very inaccurate (see Sections 3.5 and 4.2). In that case equation (4.7.7d) (or equation (4.7.9)) can be used to calculate q_z from e_x and e_y. This latter alternative can rather simply be implemented in conventional finite difference and finite element models based on Laplace-type equation (3.1.1).

As already shown in Section 4.5, the Dupuit approximation of Section 4.2 is valid only if the higher-order corrections are negligibly small in relation to the zeroth-order terms. This is the case for a perfectly layered subsurface if $\epsilon << 1$, but for a lateral heterogeneous subsurface there is an additional requirement that the lateral derivatives of the conductivities must be sufficiently small ($\mu << 1$). Thus, the Dupuit approximation is also an approximation for ideally layered subsurfaces.

Section 4.5 also gives a generalization of the Dupuit approximation given for a somewhat less weak lateral heterogeneous subsurface, namely for an approximately layered subsurface. This last generalization holds under the condition that $|\partial k_h/\partial x_i| = O\left[k_{hc}/\sqrt{(d\ l_c)}\right]$ and $|\partial k_z/\partial x_i| = = O\left[k_{zc}/\sqrt{(d\ l_c)}\right]$, where terms in the order of magnitude ϵ and $\epsilon\ \sqrt{(l_c/d)}$ are negligibly small, but where terms in the order of magnitude $\epsilon\ l_c/d$ must be carried on. The example $k_{hc}/k_{zc} = 10$, $d/l_c = 1/32$, $\epsilon = 0.01$ (negligible), $\epsilon\ \sqrt{(l_c/d)} = 0.057$ (negligible) and $\epsilon\ l_c/d = 0.32$ (not negligible) has already been given. The terms in the perturbation series with an order higher than first-order are then negligible, so that the generalized Dupuit approximation of Section 4.5 will hold for $\epsilon\ \sqrt{(l_c/d)} << 1$.

We can, however, examine still stronger lateral heterogeneities. For instance, if $|\partial k_h/\partial x_i| = O(k_{hc}/d)$ and $|\partial k_z/\partial x_i| = O(k_{zc}/d)$, then all the terms in the series expansion with derivatives of the conductivities will be of the order of magnitude $(k_{hc}/k_{zc}) = \epsilon\,(l_c/d)^2$. Higher derivatives of $\partial f/\partial x_i$ will have an order of magnitude $\epsilon\ l_c/d$ at most. Thus, if terms of the order of magnitude $\epsilon\ l_c/d = (d/l_c)\ (k_{hc}/k_{zc})$ are neglected, we again find a generalized Dupuit approximation, but with a generalized transmissivity in which the derivatives of the conductivities also appear in higher order than in the generalized transmissivity of Section 4.5. So the conclusion is that the Dupuit approximation will always hold for sufficiently shallow basins, but the transmissivity tensor in this approximation is related in a very complex manner to the conductivities and their higher-order lateral derivatives.

The equations (4.1.1b), (4.1.1c), (4.1.1d), (4.1.4a) and (4.1.4b) or, equivalently, the equations (4.7.8a), (4.7.8b) and (4.7.9) for E_x, E_y and Q_z, which also require smoothness in the lateral heterogeneities, may in this respect also be viewed as 'generalized Dupuit equations,' valid for much greater lateral heterogeneity of the conductivities.

4.8. The large-scale conductivity tensor

In general, we are dealing with groundwater flow in large regional areas where it is impossible to give the local values of the conductivities at every point $\boldsymbol{x}$ of the three-dimensional subsurface, so it is a practical necessity to work with large-scale values of the conductivities.

An impression of the local conductivity distribution can be obtained by simulating the geology of sedimentary basins. This can be achieved in a stochastic way, and also in a deterministic way. For the stochastic way see, for instance, Dagan (1989), and see also Section 3.5. where Ababou's improvement of Gelhar's theory for random conductivity distributions has briefly been introduced. For the deterministic way see, for example, Tetzlaff and Harbaugh (1989), and Stam *et al.* (1989), where the various geological processes which led to the development of sedimentary basins are described and properly modeled in a computer code.

The values of local conductivities obtained from the two approaches can then be used to obtain large-scale values by 'averaging.' The production of local conductivity distributions will not be discussed further, and attention will now be directed to the averaging from small-scale to large-scale values.

In a sense, large-scale values have already been encountered with the four generalized (two-dimensional) transmissivity components $T^{\S}_{xx}(x,y)$, $T^{\S}_{xy}(x,y)$, $T^{\S}_{yx}(x,y)$ and $T^{\S}_{yy}(x,y)$ $[m^3 \cdot Pa^{-1} \cdot s^{-1};\ m^3 \cdot dbar^{-1} \cdot d^{-1} \approx m^3 \cdot \cdot mH_2O^{-1} \cdot d^{-1}]$. The quantity $T^{\S}_{ij}(x_k)/d$ $(i,j,k = 1,2)$ $[m^2 \cdot Pa^{-1} \cdot s^{-1};$ $m^2 \cdot dbar^{-1} \cdot d^{-1} \approx m^2 \cdot mH_2O^{-1} \cdot d^{-1}]$ can be viewed as a depth-averaged conductivity tensor, which is a quantity on the larger *vertical* scale. But what is also needed is a large-scale conductivity on the lateral scale.

To derive such a large-scale value, we shall look next at two-dimensional flow in a vertical cross-section of the basin. The relevant Dupuit equations for a basin with thickness d at lateral locations x' are:

$$J_x(x',t) = -T^{\S}(x')\ \partial f(x',t)/\partial x' \tag{4.8.1a}$$

$$N(x',t) = -\partial/\partial x'[T^{\S}(x')\ \partial f(x',t)/\partial x'] \tag{4.8.1b}$$

where $T^{\S}(x')$ is the generalized transmissivity and $N(x',t) = q_z(x',0,t)$.

Of course, with the aid of equation (4.8.1b) we can always choose a course of $f(x',t)$ so that $q_z(x',0,t) = 0$; the choice of such a course will mean that the characteristic lateral length scale l_C of spatial variations in $f(x',t)$ becomes equal to the characteristic length scale l_L of the lateral heterogeneities in $T^{\S}(x')$. If $N(x',t) = 0$, then the continuity equation (4.8.1b) requires that $J_x(x',t) = J_x(x,t)$, where x is a fixed value independent of x', and we can write:

$$J_x(x,t)/T^{\S}(x') = -\partial f(x',t)/\partial x'. \tag{4.8.2}$$

Integrating over x' from $x' = a$ to $x' = b$, $(b > a)$, i.e., integrating over a lateral distance $b - a$, gives:

$$J_x(x,t)/d = -[(b-a)/d]\left[\int_a^b T^{\S}(x')^{-1}\ \mathrm{d}x'\right]^{-1} \cdot \cdot [f(b,t) - f(a,t)]/(b-a) \tag{4.8.3a}$$

where $x = (a+b)/2$ has been chosen.

According to Taylor's theorem, the following expression holds: $[f(b,t) - f(a,t)]/(b-a) = [\partial f(x',t)/\partial x']_{x'=x} + O[(b-a)^2\ \partial^3 f(x',t)/ /\partial x'^3]_{x'=x}$. If we choose the order of magnitude for the lateral extent $b-a$ equal to $\sqrt{(d\ l_C)}$, i.e., $b-a = O[\sqrt{(d\ l_C)}]$, then the term $O[(b-a)^2\ \partial^3 f(x')/ /\partial x'^3]_{x'=x}$ is the order of magnitude $O(d/l_C)$ with respect to the term $[\partial f(x',t)/\partial x']_{x'=x}$. In the context of the generalized Dupuit approximation, terms of the order of magnitude d/l_C may be neglected, so we can then posit that $[f(b,t) - f(a,t)]/(b-a) = \partial f(x,t)/\partial x$. In this way we find:

$$J_x(x,t)/d = -[(b-a)/d]\left[\int_a^b T^{\S}(x')^{-1}\ \mathrm{d}x'\right]^{-1} \partial f(x,t)/\partial x \tag{4.8.3b}$$

where $-[J_x(x,t)/d]/[\partial f(x,t)/\partial x]$ derived from equation (4.8.3b) may be considered as an expression for the laterally averaged transmissivity. It is noted that the assumption $J_x(x',t) = J_x(x,t)$, independent of x', is a basic assumption in this derivation; therefore, this assumption will be discussed in some more detail below.

Until now we considered depth-averaged values over the whole depth d of the basin. It would, however, be desirable to extend the above-presented theory to a thinner block within the basin; for instance, to a grid-block for a numerical model. We can then write equation (4.8.1a) as a kind of block-scale equation averaged over the thickness d of this block:

$$<q_x>'(x,t) = -<k_h>'(x)\,\partial f'(x,t)/\partial x \tag{4.8.4}$$

where:

$$<q_x>'(x,t) = J'_x(x,t)/d \tag{4.8.5}$$

is the depth-averaged volumetric flow rate at point x, and where:

$$<k_h>'(x) = T'^{\S}(x)/d \tag{4.8.6}$$

is a depth-averaged lateral conductivity component. (All symbols are primed, the reason for this prime ($'$) will be explained later on.) The notation $<\bullet>$ denotes depth-averaged, where the algorithm how to depth-average may differ depending on the quantity that is depth-averaged.

In the large-scale equation (4.8.4) with depth-averaged lateral conductivity component (4.8.6), $f'(x,t)$ is the potential on the *upper* boundary of the rectangular block, whereas a zero vertical volumetric flow rate is specified on the *lower* boundary. Of course, it follows from the continuity equation that the vertical volumetric flow rate on the upper boundary, denoted by $<q_z>'(x,t)$, is generally nonzero. In general, the potential $\phi'(x,z,t)$ is not equal to $f'(x,t)$, but the zeroth-order solution for the potential, $\phi'_0(x,z,t)$ is equal to $f'(x,t)$.

In general, we assume that the first-order correction to the potential is relatively small, so that it makes sense to define the depth-averaged vertical potential difference $\Delta<\phi>'(x,t)$ as the zeroth-order potential difference between the upper and lower boundary, i.e., $\Delta<\phi>'(x,t) = \phi'_0(x,d,t) + - f'(x,t)$ (for a more rigorous justification, see Appendix D). From this definition it follows that for zero vertical volumetric flow rate on the lower boundary $\Delta\langle\phi\rangle'(x,t) = 0$. If we had specified a nonzero vertical volumetric flow rate on the lower boundary, we would have found expressions for the depth-averaged volumetric flow rate in the block, caused by a nonzero depth-averaged vertical potential difference $\Delta\langle\phi\rangle'(x,t) \neq 0$. However, in this section we want to consider the result of a zero depth-averaged vertical potential difference $\Delta\langle\phi\rangle'(x,t) = 0$, which justifies the specification of a zero vertical volumetric flow rate at the bottom.

There is, however, an arbitrary asymmetry in the boundary conditions. In the same way, we could equally well have derived depth-averaged equations like equations (4.8.4), (4.8.5) and (4.8.6) for the problem definition where a potential $f''(x,t)$ is specified on the lower boundary and where the zero vertical volumetric flow rate is specified on the upper boundary. In this way a similar equation is found to that given by equation (4.8.4), but with a lateral conductivity $\langle k_h\rangle''(x)$ which is generally not equal to $\langle k_h\rangle'(x)$. Of course, also here the vertical volumetric flow rate $\langle q_z\rangle''(x,t)$ on the lower boundary will generally be nonzero.

In both problem definitions the vertical volumetric flow rate on the upper or lower boundary has been chosen equal to zero. This means that in both problems the depth-averaged vertical potential differences $\Delta\langle\phi\rangle'(x,t)$ and $\Delta\langle\phi\rangle''(x,t)$ have the same value, namely, zero. Now we choose for both problems also the same horizontal potential gradient, i.e., we choose $f'(x,t) = f''(x,t) = f(x,t)$ and we denote $f(x,t)$ as the depth-averaged potential $\langle\phi\rangle(x,t)$. From this choice we can calculate the two mean volumetric flow rates $\langle q_x\rangle(x,t) = (\langle q_x\rangle' + \langle q_x\rangle'')/2$ and $\langle q_z\rangle(x,t) = (\langle q_z\rangle' + \langle q_z\rangle'')/2$, which will be called the depth-averaged volumetric flow rates.

Now the asymmetry has been removed. The expression for the depth-averaged lateral volumetric flow rate $\langle q_x\rangle(x,t)$ leads to a component of the depth-averaged Darcy's law $\langle q_x\rangle(x,t) = -\langle k_{xx}\rangle(x)\, \partial\langle\phi\rangle(x,t)/\partial x$. This depth-averaged Darcy's law is an expression similar to equation (4.8.4), but with a depth-averaged conductivity component $\langle k_{xx}\rangle(x) = = [\langle k_h\rangle'(x) + \langle k_h\rangle''(x)]/2$. The expression for the depth-averaged vertical volumetric flow rate $\langle q_z\rangle(x,t)$ leads to another component of the depth-averaged Darcy's law; a component with a depth-averaged conductivity component $\langle k_{zx}\rangle(x)$ expressing net flow in the vertical direction caused by a potential gradient in the horizontal direction. For further details, see Appendix D. The depth-averaged Darcy's law can also be extended to flow caused by a nonzero depth-averaged vertical potential difference (see Appendix D).

Under the assumption of a constant depth-averaged volumetric flow rate over the lateral extension of the block, this depth-averaged Darcy's law can be integrated over the lateral interval $b - a$ to obtain a large-scale Darcy's law for the block-averaged quantities $\boldsymbol{<q>}$ and $\boldsymbol{<\phi>}$ with block-averaged conductivity $\boldsymbol{<\underline{k}>}$ (see Appendix D). However, the assumption of constant depth-averaged volumetric flow rate in the block needs some justification. Let us first consider almost one-dimensional flow. For instance, almost one-dimensional flow is encountered in laboratory experiments, where flow through a porous medium contained in a cylindrical column with impervious walls is studied (see Figure 8). Because of the impervious cylinder walls, the flow will be almost one-dimensional in the direction along the axis of the column. Furthermore, we consider the situation where a particular block in the cylinder is surrounded in the axial direction by a very large number of other blocks, where all blocks together constitute the total flow domain. The particular block under consideration (the test block) may have a locally inhomogeneous conductivity distribution, but all other blocks are assumed to have the same locally homogeneous conductivity distribution. In addition we exclude situations where the conductivity of the test block is extremely small with respect to the conductivity of the other blocks. Under that condition, the axial potential differences over the test block, which will generally differ in the radial and tangential directions because of the local heterogeneities in the block, will generally be negligibly small with respect to the total axial potential difference over the whole column. It follows then that the axial volumetric flow rate through the test block will be almost completely determined by the conductivity of all the neighboring blocks, *independent* from the conductivity distribution in the test block under consideration. Therefore, in this situation it is reasonable to determine the block-scale averaged conductivity with the aid of a constant depth-averaged axial flow rate in the block. Furthermore, since the depth-averaged radial flow rate is equal to zero, the local radial flow rate will be relatively small, which means that it is allowed to determine the block-scale averaged conductivity $\boldsymbol{<\underline{k}>}$ under the approximation that all three components of the depth-averaged flow rate vector are constant in the test block, i.e., $<\boldsymbol{q}> = \boldsymbol{<q>}$ is constant in the block. Of course, the above approximations do not exclude local radial potential differences. As we expect, these local radial potential differences result in a flow pat-

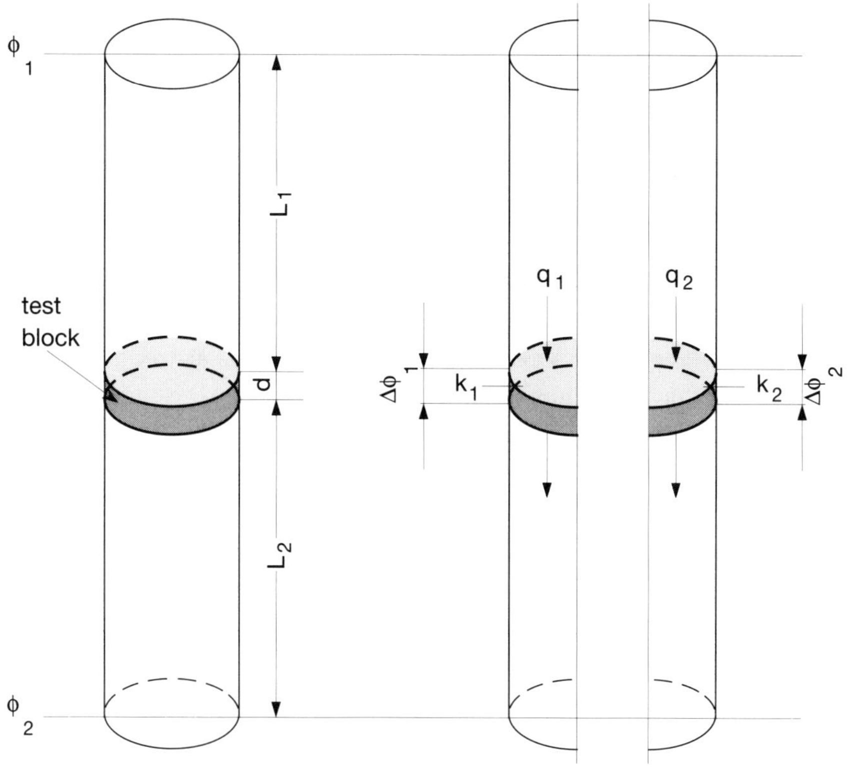

Figure 8. **Flow through a locally heterogeneous block.**

Figure 8a shows a cylinder filled with a homogeneous porous medium with conductivity k, except for the thin test block which is heterogeneous. The cylinder wall is impervious, which forces the flow to be essentially one-dimensional in the axial direction. Figure 8b shows two halves of the same cylinder; the left half of the test block has conductivity k_1 and the right half of the test block has conductivity k_2. The volumetric flow rate through the left half of the cylinder equals $q_1 = (\phi_1 - \phi_2)/[(L_1 + L_2)/k + d/k_1]$; the volumetric flow rate through the right half of the cylinder equals $q_2 = (\phi_1 - \phi_2)/[(L_1 + L_2)/k + d/k_2]$. The thickness of the thin test block is such that $d/k_1 << (L_1 + L_2)/k$ and $d/k_2 << (L_1 + L_2)/k$; it follows then that $q_1 = q_2 = q$. However, the potential difference over the left half of the test block equals $\Delta\phi_1 = q/k_1$ and the potential difference over the right half of the test block equals $\Delta\phi_2 = q/k_2$, i.e., the potential differences may differ considerably. This means that a block-scale conductivity value obtained by averaging over the cylinder cross-section should be based upon a volumetric flow rate which is constant over that cross-section.

tern where flow lines are bent around the poorly conducting parts in the heterogeneous test block.

The validity of the above argument to assume constant depth-averaged volumetric flow rate $<\boldsymbol{q}>$ in the block, is not generally warranted for real three-dimensional flow in all types of heterogeneous domains. On the other hand, for practical reasons we are bound to a definition of block-scale averaged conductivity that is *independent of the surrounding domain-scale conductivity distribution*, and under that practical constraint the definition using constant depth-averaged volumetric flow rate $<\boldsymbol{q}>$ within the block is proved here to be a natural choice. This brings us to a more general point. In Section 2.1 it has been mentioned that, in the 'organic model' of nature, an event at the lower level cannot be explained fully without reference to the event at the higher level. Here the event is groundwater flow, and when we consider the block-scale level as the lower level, and the domain-scale level as the higher level, the above discussion may be considered as an illustration of the organic model.

Further information on these aspects will be found in Zijl and Stam (1992), and Stam and Zijl (1992), and in Appendix D.

A correct choice of the grid-block dimensions, or the integration rectangle $d\,(b - a)$ as discussed above, is far from trivial. The first criterion is that the size of the rectangle must be small enough to ensure that the scale on which the flow processes play a role does not get lost. For example, if we are studying a local problem where the velocity variations are on a lateral scale of, say, 10 m, where smaller-scale variations may be neglected, then in effect we may not define the large-scale conductivities on a scale much larger than 10 m. However, the choice of too large scales happens quite often in practice, particularly in combination with analytical mathematical solution techniques in which one value is usually given for the transmissivity over a very large region. The consequence of such an approach is that the values given for the macro-dispersion lengths are then much too large (see Section 6.3).

The second criterion for the correct choice of the grid-blocks, for instance the n integration rectangles with areas $d_i\,(b_i - a_i)$, $i = 1, 2, \ldots, n$ (no summation over the index i) as considered above, of which the whole flow domain is composed, is as follows. Let us suppose that in the above n rectangles the block-scale conductivity obtained from the point-scale conductivity distribution $\underline{\boldsymbol{k}}(\boldsymbol{x})$ is denoted by $<\underline{\boldsymbol{k}}>_1(\boldsymbol{x})$. The changeover from these n rectangles with block-scale conductivity distribution $<\underline{\boldsymbol{k}}>_1(\boldsymbol{x})$ to one larger rectangle $d_T\ (b_T - a_T) = \sum d_i\ (b_i - a_i)$ (summation over the index i) with large-scale conductivity $<\underline{\boldsymbol{k}}>_2(\boldsymbol{x})$ must give practically the same result as direct computation of this large-scale conductivity $<\underline{\boldsymbol{k}}>_2(\boldsymbol{x})$ from the local-scale conductivity distribution $\underline{\boldsymbol{k}}(\boldsymbol{x})$, without the calculation of the smaller integration rectangles as intermediate step. This criterion is met exactly if the porous medium has a periodic structure and if the blocks have dimensions equal to integer values of the periodic dimensions (see also Bensoussan *et al.*, 1978). Practically speaking, a 'more or less periodic' structure like a random distribution with constant mean value suffices. Anyhow, inclusion in one integration block of different formation types like, for instance, a part of an aquifer and a part of an aquitard, will lead to misleading results.

From the theory presented earlier, it follows that the concept of generalized transmissivity only holds if d/l_c is sufficiently small, i.e., if the second and third terms in the right-hand side of equation (4.4.5b) are negligibly small with respect to the first term in the right-hand side. This means that the concept of generalized transmissivity only holds for the Fourier components of $f(x, y)$ with sufficiently long wavelengths with respect to the thickness of the block. If the block is chosen too thick with respect to the minimum wavelength under consideration, a large-scale Darcy's law does not exist, which means that higher than first derivatives of f must also be taken into account. However, for practical reasons, regional modeling studies are always performed with a large-scale Darcy's law (i.e., with an expression without higher derivatives of f).

A completely general derivation of the large-scale Darcy's law, without using the Dupuit approximation and the need for differentiability of the lateral heterogeneities, has been presented by Quintard and Whitaker (1987); see also Quintard and Whitaker (1988; 1990a; and 1990b). However, though this general theory gives much valuable insight, it has the drawback that a system of partial differential equations must be solved to obtain the final large-scale conductivity tensor, and this obstructs its practical applications. By contrast, the theory based on the Dupuit approximation is more practicable in application, and can also be extended to all the components of the conductivity tensor; this has been worked out in detail in Appendix D. (Also see Zijl and Stam (1991) and Stam and Zijl (1991).) It turns out that the large-scale conductivity tensor is generally nonsymmetric. Since the theory on two-dimensional tensors presented in Section 4.6 can be extended to three-dimensional tensors, similar conclusions about principal axes for symmetric tensors and generalized principal axes for nonsymmetric tensors hold.

Chapter 5

Introduction to Flow Systems Analysis

5.1. The homogeneous subsurface as a simple example

In this chapter an introduction to flow systems analysis will be presented. In the previous chapter we have seen that the subsurface is generally heterogeneous. We have also seen that, under certain conditions, homogeneous large-scale conductivities may be applied in blocks subdividing the subsurface to replace the heterogeneous subsurface. Such a procedure is required by numerical approximation methods like the finite difference or finite element method. Still a further simplification is to conceive these blocks as layers constituting a perfectly layered subsurface. A perfectly layered subsurface with piecewise constant conductivities can be fully treated by analytical mathematical tools; see Van Veldhuizen *et al.* (1992a,b) for the mathematical aspects, and Bervoets (1991), Bervoets *et al.* (1991) and Nieuwenhuizen (1992) for the more applied aspects. However, to obtain a qualitative insight into the thoughts underlying flow systems analysis, it is most useful to replace the heterogeneous subsurface by an 'equivalent' homogeneous subsurface. Therefore, to simplify the following discussion, let us consider a homogeneous subsurface.

All flow properties already derived for a perfectly layered basin also hold for a homogeneous subsurface, of course. As explained in Section 3.6, each Fourier component $f_w(x,y,t) = \rho g\ h_{f_w}(x,y,t)$ of the potential on the top plane $\phi(x,y,0,t) = \rho g\ h_f(x,y,t) = f(x,y,t)$ is considered separately. For each arbitrary Fourier component $f_w(x,y,t)$ the derivatives with respect to x and y, viz. $\partial f_w(x,y,t)/\partial x$ and $\partial f_w(x,y,t)/\partial y$, can be determined. As a consequence, the lateral flux components on the top plane, viz. $q_{wx}(x,y,0,t)$ and $q_{wy}(x,y,0,t)$, can also be simply determined by Darcy's law for each Fourier component $f_w(x,y,t)$ of the Fourier series expansion of $f(x,y,t)$.

As has been shown in Section 3.6, the general expression for each Fourier mode of the potential on the top plane is given by:

$$f_w(x,y,t) = a(t)\ \cos(w_x x + w_y y) + \\ + b(t)\ \sin(w_x x + w_y y) \tag{5.1.1}$$

where $a(t) = \rho g\ a_f(w_x, w_y, t)$ and $b(t) = \rho g\ b_f(w_x, w_y, t)$ (see equation (3.6.3)), and where w_x and w_y are the wave numbers $[m^{-1}]$ in the x- and y-direction, respectively. In general, the wave numbers may be positive and negative $(-\infty < w_x,\ w_y < +\infty)$, but for a finite region the wave numbers are positive $(0 \leq w_x,\ w_y < +\infty)$. Sometimes it is more convenient to use the inverses of the two wave numbers: $l_x = 1/w_x$ and $l_y = 1/w_y$.

To obtain a clearer picture of one particular Fourier mode, it is preferable to define the following lateral x'-y' coordinate system in the x-y plane:

$$x' = (w_x x + w_y y) \Big/ \sqrt{(w_x^2 + w_y^2)} \tag{5.1.2a}$$

$$y' = (-w_y x + w_x y) \Big/ \sqrt{(w_x^2 + w_y^2)}. \tag{5.1.2b}$$

These x'-y' axes are obtained by rotating the original x-y axes over an angle α with $\cos\alpha = w_x \Big/ \sqrt{(w_x^2 + w_y^2)}$ and $\sin\alpha = w_y \Big/ \sqrt{(w_x^2 + w_y^2)}$. After this coordinate transformation, $f_w(x,y,t)$ can be written as a function $\tilde{f}_w(x',t)$ of only the coordinate x' and the time t:

$$f_w(x,y,t) = \tilde{f}_w(x',t) = \\ = a(t)\ \cos(x'/l_c) + b(t)\ \sin(x'/l_c) \tag{5.1.3}$$

where $l_c = 1 \Big/ \sqrt{(w_x^2 + w_y^2)} = 1 \Big/ \sqrt{(1/l_x^2 + 1/l_y^2)}$, and $2\pi l_c$ is the characteristic wavelength of the Fourier mode under consideration. This wavelength characterizes the *spatial scale* of the flow subsystem.

From the above expressions we find, by differentiation and by applying Darcy's law, that on the top plane $z = z' = 0$ the two lateral flux components are given by:

$$q_{wx}(x,y,0,t) = k_h^o\ g_w(x,y,t)/l_x \tag{5.1.4a}$$

$$q_{wy}(x,y,0,t) = k_h^o\ g_w(x,y,t)/l_y \tag{5.1.4b}$$

where:

$$\begin{aligned} g_w(x,y,t) &= a(t)\ \sin(w_x x + w_y y) \\ &\quad - b(t)\ \cos(w_x x + w_y y). \end{aligned} \tag{5.1.5}$$

Equivalently, in the x'-y' coordinate system this becomes:

$$q_{wx'}(x',0,t) = k_h^o\ \tilde{g}_w(x',t)/l_c \tag{5.1.6}$$

$$q_{wy'}(x',0,t) = 0 \tag{5.1.7}$$

where:

$$\begin{aligned} \tilde{g}_w(x',t) &= g_w(x,y,t) = \\ &= a(t)\ \sin(x'/l_c) - b(t)\ \cos(x'/l_c). \end{aligned} \tag{5.1.8}$$

In Chapter 4, an infinite series in the vertical coordinate z and the dimensionless number ϵ (or in the coordinate z and the depth of the impervious base d) was obtained for each Fourier mode as the solution of the flow field in the basin below the top. Unfortunately, this series converges only for sufficiently small values of ϵ (or d). However, for a homogeneous basin this series solution turns out to be equivalent to a closed-form solution in cos-, sin-, and exp-functions. Of course, these closed-form solutions can also be continued to all values of $0 < \epsilon < \infty$ (or all values of $0 < d < \infty$). For each Fourier mode we then find a solution in the three space coordinates x, y, z, or, equivalently, in the two space coordinates x', z' (see Appendix C). In a completely different way (namely by separation of variables), Tóth (1963) found the same solution in the x'-z' cross-section. From this general solution for one Fourier mode, the solution for an infinitely deep basin $\epsilon \to \infty$ (or $d \to \infty$) will be presented here as an example. For that solution we find that the vertical flux component on the top plane, $q_{wz}(x,y,0,t) = q_{wz'}(x',0,t)$, is equal to:

$$q_{wz}(x,y,0,t) = \sqrt{(k_h^o\ k_z^o)}\ f_w(x,y,t)/l_c \tag{5.1.9}$$

or, equivalently:

$$q_{wz'}(x',0,t) = \sqrt{(k_h^o\ k_z^o)}\ \tilde{f}_w(x',t)/l_c. \tag{5.1.10}$$

Furthermore, we find that the magnitudes of the flux components decay exponentially with increasing depth according to:

$$\begin{aligned} q_{wx}(x,y,z,t) &= q_{wx}(x,y,0,t)\ \cdot \\ &\quad \cdot\ \exp[-(z/l_c)\ \sqrt{(k_h^o/k_z^o)}] \end{aligned} \tag{5.1.11a}$$

$$\begin{aligned} q_{wy}(x,y,z,t) &= q_{wy}(x,y,0,t)\ \cdot \\ &\quad \cdot\ \exp[-(z/l_c)\ \sqrt{(k_h^o/k_z^o)}] \end{aligned} \tag{5.1.11b}$$

$$\begin{aligned} q_{wz}(x,y,z,t) &= q_{wz}(x,y,0,t)\ \cdot \\ &\quad \cdot\ \exp[-(z/l_c)\ \sqrt{(k_h^o/k_z^o)}] \end{aligned} \tag{5.1.11c}$$

or, equivalently:

$$\begin{aligned} q_{wx'}(x',z',t) &= q_{wx'}(x',y',0,t)\ \cdot \\ &\quad \cdot\ \exp[-(z'/l_c)\ \sqrt{(k_h^o/k_z^o)}] \end{aligned} \tag{5.1.12a}$$

$$q_{wy'}(x',z',t) = 0 \tag{5.1.12b}$$

$$\begin{aligned} q_{wz'}(x',z',t) &= q_{wz}(x',y',0,t)\ \cdot \\ &\quad \cdot\ \exp[-(z'/l_c)\ \sqrt{(k_h^o/k_z^o)}] \end{aligned} \tag{5.1.12c}$$

where $z' = z$.

As is clear from the above expressions, the flow pattern caused by one specific Fourier mode is two dimensional in the x'-z' plane for any value of y', $-\infty < y' < +\infty$; this fact facilitates making pictures of the resulting flow pattern.

From the above expressions it can also be seen that Fourier modes with relatively short wavelengths $2\pi l_c$ decay faster with increasing depth than Fourier modes with relatively long wavelengths. As a consequence, we may expect that the short-wave Fourier modes cause shallow flow with recharge and discharge points close together, whereas long-wave Fourier modes cause deep flow with recharge and discharge points far apart. Each Fourier mode not only extends infinitely for $z = z' \to \infty$, but is also periodic in its x'-direction. This means that the wavy pattern repeats itself infinitely from $x' \to -\infty$ to $x' \to +\infty$ with a wavelength of $2\pi l_C$. This spatial periodicity makes it possible to distinguish, for every Fourier component denoted by (w_x, w_y), an infinite number of two-dimensional 'Fourier cells.' These Fourier cells are defined in the (x', z') plane with a lateral dimension of half the wavelength πl_C and, theoretically speaking, with infinite depth. The inter-cell boundaries $x' = x'_{ic}$ parallel to the z'-axis will be chosen at the places where the wave has its maximal or its minimal value as a function of x'. If there were only one Fourier mode, the inter-cell boundaries would be water divides.

Until now we have not yet defined the meaning of the word 'flow system.' According to Laszlo (1972, p. 30), a 'natural system' is defined as a 'nonrandom accumulation of matter-energy, in a region of space-time, which is nonrandomly organized into coacting interrelated subsystems or components.' (See also Laszlo (1983, pp. 18-20)). Accordingly, in geohydrology the notion of 'flow systems' is used to describe the fact that there are shallow flow paths with their recharge points relatively close to their discharge points, and that there are also deeper flow paths, flowing below the shallow flow paths, with their recharge points farther away from their discharge points. This hierarchy of nested flow line patterns may repeat itself on every spatial scale and can thus be considered as a hierarchy of systems and subsystems (see Engelen and Jones, 1986, p. 125).

In agreement with Laszlo's above-given definition of a 'natural system,' but with a slight shift in terminology, we propose here to define a *flow system* as one specific cell of a Fourier mode with wave numbers w_x and w_y and amplitudes $a(t)$ and $b(t)$. In Laszlo's terminology such a flow system would be a 'subsystem' or 'component.' The sum of all flow systems, i.e., the 'natural system' in Laszlo's terminology, will then be called the *flow field*. By this definition, the groundwater flow field, as it is observable in nature, consists of a combination of many flow systems, viz. the superposition of all occurring Fourier modes, all oriented in their own, generally different, two-dimensional x'-z' planes. The flow paths of the fluid particles can be obtained only after having combined all flow systems. In this way, a flow system is a concept that, generally speaking, cannot be observed in the field in isolation from the other flow systems at the same place. However, at certain lateral locations and at certain depths below the top of the basin, there will sometimes be only one dominant flow system determining the main features of the groundwater flow at these places. That this can be the case will be explained later. If this is the case, i.e., if only one flow system is dominant at certain points in the basin, these points will be called the *location* of the flow system. In the same spirit, an artificial flow system induced by a man-made well may be defined as the flow caused by the well in the absence of spatial variations in the water table. Also this well-flow system must be superimposed on the natural flow systems, and it will turn out that, in the neighborhood of the well, the well-flow system is the dominant system, i.e., the well-flow system is located in the vicinity around the well. The above definition of 'flow system' is given to assess the relation between the horizontal spatial scales and the vertical spatial scales (and the time scales). In Section 7.2 another definition of 'flow system' will be given in relation to the transport of dissolved mass. For the description of the transport of contaminants, concepts like transport velocity and flow path are the key concepts. However, for a better understanding of the scale dependency of these processes, a decomposition of the flow field into flow systems with different spatial and temporal scales will be of great help.

At the depth $z = d^*$, the three flux components are given by $q_{wi}(x, y, 0, t)\ \exp(-\sqrt{\epsilon^*})$, $i = x, y, z$, in which $\sqrt{\epsilon^*} = (d^*/l_c)\ \sqrt{(k_h^o/k_z^o)}$. If we choose the depth d^* so that $\sqrt{\epsilon^*} = \pi/2$, we find a decrease in magnitude by a factor $\exp(-\pi/2) \approx 0.21$; if we choose d^* so that $\sqrt{\epsilon^*} = \pi$, we find a decrease in magnitude by a factor $\exp(-\pi) \approx 0.043$; for $\sqrt{\epsilon^*} = 3\pi/2$ we find a decrease in magnitude of ca. 0.01, and for $\sqrt{\epsilon^*} = 2\pi$ we find a decrease in magnitude of ca. 0.002. Somewhat dependent on the situation, we have a negligibly small vertical flux component with respect to $q_z(x, y, 0, t)$ for one of these values of $\sqrt{\epsilon^*}$. We do not need to choose an infinite depth d^*; we already get a good approximation of the solution when the depth d of the 'impervious' base is chosen so that $\sqrt{\epsilon} \approx 2\pi$ ($\approx 2\pi$ means varyingly dependent on the details of the flow situation), i.e., $d \approx 2\pi l_c\ \sqrt{(k_z^o/k_h^o)}$. Indeed, the solution for the flow problem with impervious base $d \to \infty$ is little different from the solution for the flow problem with impervious base $d \approx 2\pi l_c\ \sqrt{(k_z^o/k_h^o)}$. The depth d defined in this way is therefore dependent on the wavelength $2\pi l_c$ of the spatial variation in the water table.

5.2. The 'impervious' base in a layered subsurface

The idea of an 'impervious' base is in fact a more meaningful idea in a layered subsurface where aquifers and aquitards alternate. Indeed, in that case the damping factor jumps abruptly in value (see Section 3.3) and it becomes clear that the 'impervious' base must be at the bottom of an aquifer and at the top of an aquitard. In that case, it is also possible to arrive at an exact solution for perfectly layered subsurfaces with piecewise constant conductivities (see Van Veldhuizen *et al.*, 1992a; and Bervoets *et al.*, 1991).

Notwithstanding the effectiveness of such exact solutions in a computer program, the simple rule of thumb $d \approx 2\pi l_c \sqrt{(k_{zc}/k_{hc})}$ (or $\sqrt{\epsilon} \approx 2\pi$) remains valuable for estimating the order of magnitude of the characteristic depth of the flow problem. The formula $d \approx 2\pi l_c \sqrt{(k_{zc}/k_{hc})}$ (or $\sqrt{\epsilon} \approx 2\pi$) plays an extremely important part in the qualitative comprehension of the nested character of flow paths (see Section 5.3), an important aspect of the 'flow systems analysis' of Tóth and Engelen c.s. (see Engelen and Jones, 1986). On the basis of empirical field observations obtained from many kinds of maps and cross-sections, combined with speculative thought to obtain a hypothesis, Engelen arrived at the following expression (see Engelen and Jones, 1986, p. 125):

$$`Q_s' = `k_s' \cdot `i_s' \cdot `D_s' \tag{5.2.1}$$

where: $`Q_s' = \sqrt{(a^2+b^2)}\ [J \cdot m^{-3}]$ is the potential energy per unit volume of Fourier mode $w$ available for flow; $`k_s' = \sqrt{(k_{hc}/k_{zc})}/(2\pi)\ [-]$ is the 'energy conversion capacity'; $`i_s' = \sqrt{(a^2+b^2)}/l_c\ [J \cdot m^{-4}]$ is the lateral 'gradient' of the potential energy per unit volume; and $`D_s' = d$ is the depth of the flow system caused by Fourier mode w. Indeed, from $d \approx 2\pi l_c \sqrt{(k_{zc}/k_{hc})}$ and the above-presented definitions it is found that: $`Q_s' = `k_s' \cdot `i_s' \cdot `D_s'$.

The rule of thumb $d \approx 2\pi l_c \sqrt{(k_{zc}/k_{hc})}$ (or, equivalently, $`Q_s' = = `k_s' \cdot `i_s' \cdot `D_s'$) is especially important for the preparation of more detailed numerical modeling where an impervious base must be chosen at a certain depth. Also, the decision whether a thin layer with low conductivity may be neglected within an aquifer, or must be treated as a semipervious layer between two aquifers, can be answered roughly by this rule of thumb.

5.3. Nesting of flow paths

Based on the above theory and concepts, the nested character of groundwater flow and the meaning of the different spatial scales can be described very broadly. We have already seen in Section 3.6 that the smallest wavelength occurring in the Fourier series is equal to two times the pixel dimensions of the map, and the largest wavelength is equal to the linear dimensions of the region presented on the map. In other words, there is a relationship between the lateral extension of a spatial variation in water table height and the wavelength of a Fourier component. For instance, in a large-scale super-regional model a dominant lateral extension of variation in the water table gives rise to a wavelength $2\pi l_c$ of 20 km (so $l_c = 20/(2\pi)\ km = 3.18\ km$) and, because of the subsurface, the characteristic anisotropy factor is $k_{hc}/k_{zc} \approx 400$. Then it follows from $d \approx 2\pi l_c \sqrt{(k_{zc}/k_{hc})}$ that, for this super-regional wave, the characteristic depth of the impervious base $d \approx 1000\ m$. The regional variations in the water table superimposed on the super-regional wave will have a characteristic extension such that the wavelength $2\pi l_c$ is 2 km, for instance, which implies for this regional extension an impervious base with a characteristic depth $d \approx 100\ m$. So below 100 m depth we find flow caused by the super-regional wave only. Above 100 m depth the flow pattern is determined by superposition of the flow due to the super-regional and the regional extension, where the regional component is usually dominant. In this way, we find that deep flow paths take a much longer route in the lateral direction than do shallow flow paths; moreover, the deep and shallow flow paths are not necessarily in the same direction.

In reality, the situation is more complicated than suggested above. The rule of thumb $d \approx 2\pi l_c \sqrt{(k_{zc}/k_{hc})}$ (or, equivalently, 'Q_s' = 'k_s'$\cdot$'i_s'$\cdot$'D_s') associated with a damping factor of ca. 0.002, is related to the damping of a Fourier component with increasing depth with respect to its value at the top of the basin. Since the rule of thumb criterion does not recognize the mutual comparison of two different Fourier components, the reasoning given earlier will apply only for the potential resulting from a water table where all the occurring Fourier components have the same amplitude. Let us assume that this is the case, then, in the above example with wavelengths of 20 km and 2 km; the top flux of the Fourier component with wavelength

20 km is a factor 0.1 of the top flux with wavelength 2 km. The Fourier component of 20 km will hardly be damped at $d = 100\ m$, whereas the Fourier component of 2 km will be damped to only 0.002 of its top value. This means therefore that, at $d = 100\ m$, the flux amplitude ratio between the 20-km wave and the 2-km wave is $1 : 0.02 = 50$. In this case the criterion $d \approx 2\pi l_c \sqrt{(k_{zc}/k_{hc})}$ would be a safe choice, but it need not always be so. For instance, the criterion $d \approx \pi l_c \sqrt{(k_{zc}/k_{hc})}$ would have led to incorrect solutions. So $d \approx 2\pi l_c \sqrt{(k_{zc}/k_{hc})}$ is only a rough guide and, after a first estimation, it would be advisable to take account of the different amplitude ratios between the Fourier components of the water table. This can be done with a relatively simple computer program based upon the exact solution in a perfectly layered basin (see Bervoets, 1991; and Bervoets *et al.*, 1991).

In the above example, we can also see that flux components with short wavelengths exert influence at much greater depth than potential components with the same wavelengths. This is due to the fact that the amplitude of the top flux of a Fourier component is inversely proportional to the wavelength of that Fourier component, which results in a different flux amplitude ratio between short waves and long waves with respect to the potential amplitude ratio. In other words, if the potential differences of Fourier components with short wavelength are negligible by comparison with those with long wavelengths at a certain depth, then that need not necessarily be the case for the flux components. As with the Dupuit approximation (see Section 4.2), this phenomenon indicates the importance of obtaining approximate (often numerical) solutions of the flux components from their own equations, instead of determining the flux from the (numerically) approximated potential. Obviously, this will not apply if exact, not approximated, solutions of the potential are used.

We should always bear in mind that the deep flow is very slow by comparison with the shallow flow, and therefore makes little contribution to the transport of solutes from infiltration to exfiltration regions. Nevertheless, research into the behavior of deep groundwater flow is certainly important for both science and technology, and it is reassuring that research is continuing these studies in depth. Knowledge of deep groundwater flow is especially important for understanding the accumulation of poisonous or radioactive wastes in the deep subsurface, for instance, and to explain the rise and fall of salt content in the course of centuries, or to explain the presence of oil and gas and minerals in certain areas. It is now generally accepted that the transportation of CO_2 by deep groundwater flow is an important factor in the metamorphosis of rocks (re-crystallization due to changes in the physicochemical environment). It is also possible in principle for relatively large quantities of deep groundwater carrying bicarbonate ions and calcium ions to rise to the surface in some places and affect the microbiological activity and vegetation there.

A further consideration to be borne in mind is that the deep groundwater flow is often influenced by factors other than spatial variations in the water table, and particularly by the compaction and decompaction which occurs in consolidated rocks with changes in the rock pressure. Presumably the geopressurization caused by electro-osmosis in clay layers may also play a role (see Mitchell, 1976).

5.4. Time-dependence of flow systems; temporal scales

We can speculate that the super-regional spatial course of the water table is generally associated with the topography of the landscape. This idea was first expressed by Tóth (1963). Since the topography changes only slowly with time, the deep groundwater flow will be almost steady and the results can therefore be recorded on a map. This can be made more plausible by recognizing that the time dependence of the amplitude of a Fourier component depends on the characteristic length l_c of that component. As long as the linearized and projected kinematic boundary condition (2.11.3) on the 'top' plane $z = 0$ may be considered as a sufficiently accurate approximation to the nonlinear kinematic boundary condition (2.10.8) on the water table $z = -h_f(x, y, t)$, the superposition principle will hold. Consequently, the linearized and projected kinematic boundary condition (2.11.3) will also hold for each Fourier mode:

$$\theta \, \partial h_{fw}(x, y, t)/\partial t = P_{ew}(x, y, t) - q_{wz}(x, y, 0, t) \tag{5.4.1}$$

where P_{ew} is the Fourier component of P_e with wave numbers (w_x, w_y).

The fact that the effective precipitation P_e has spatial variability may appear strange at first sight. One might think that rain intensity is almost equally distributed all over the area and that evaporation by vegetation does not differ very much from place to place, with, as a consequence, an almost equally distributed effective precipitation. However, P_e also includes the effective infiltration at places where surface water occurs (therefore, a better name for P_e would be groundwater replenishment). Now, surface waters like rivers, canals and ditches often have a value of P_e that is controlled by surface water flow in such a way that the water level remains approximately constant in time. Because of this overland flow, the effective infiltration, i.e., the values of P_e at the locations where surface water occurs, has completely different values from those of the precipitation minus evaporation found elsewhere. Therefore, P_e has spatial variations which can be decomposed in Fourier modes in which the wavelengths are determined by the orientation of, and the distances between, the surface waters.

Further, we have seen that, in a homogeneous porous medium, the following expression holds for $q_{wz}(x, y, 0, t)$ (see equation (5.1.9)):

$$q_{wz}(x, y, 0, t) = \sqrt{(k_h^o\ k_z^o)}\ f_w(x, y, t)/l_c \tag{5.4.2}$$

where $f_w(x, y, t) = \rho g\ h_{fw}(x, y, t)$. The combination of equations (5.4.1) and (5.4.2) gives:

$$\theta\ \partial h_{fw}/\partial t = P_{ew} - \rho g\ \sqrt{(k_h^o\ k_z^o)}\ h_{fw}/l_c. \tag{5.4.3}$$

Let us assume for the moment that $P_{ew} = 0$ (no effective precipitation and no contact with surface water) and that the Fourier component of the groundwater table $h_{fw}(x, y, 0)$ is known at a time-point $t = 0$. The Fourier components of the spatial variations in the water table flatten out with increasing time (are decaying with time) in accordance with the solution:

$$\begin{aligned} h_{fw}(x, y, t) &= h_{fw}(x, y, 0) \cdot \\ &\cdot \exp[-\left\{\rho g\ \sqrt{(k_h^o\ k_z^o)}/(\theta\, l_c)\right\}\ t]. \end{aligned} \tag{5.4.4}$$

The characteristic time t_c for time variability is thus equal to:

$$t_c = \theta\ l_c/[\rho g\ \sqrt{(k_h^o\ k_z^o)}]. \tag{5.4.5}$$

This formula (5.4.5) can also give an order of magnitude estimation of the characteristic time t_c for layered subsurfaces; for this purpose, the previously defined (Section 4.4) characteristic conductivities k_{hc} and k_{zc} must be used instead of k_h^o and k_z^o.

From the above expression we see that the characteristic time t_c becomes greater as the spatial extent of the variation in the water table becomes greater (and we also see that gravity plays a part). For instance, with a regional wave with wavelength $2\pi l_c = 6.3\ km$, i.e., with characteristic length $l_c = 2\ km$, and a subsurface with $k_{hc} = 1\ m^2 \cdot dbar^{-1} \cdot d^{-1}$, $k_{zc} = 0.0025\ m^2 \cdot dbar^{-1} \cdot d^{-1}$, $\theta = 0.25$ and with $\rho g = 1\ dbar \cdot m^{-1}$, the characteristic time $t_c = 10\ 000\ d \approx 27$ years. So flow at a depth greater than ca. 315 m (see Section 5.2) will react to changes in natural groundwater recharge with a lag of more than 27 years. In practical terms this means that this flow component is independent of the effective precipitation, but is related to the topography and particularly to the surface water within it. Since this flow component changes so slowly, it can be charted on maps with a life span of say, 20 years.

On the other hand, from formula (5.4.5) for t_c, it follows that a relationship with the topography is much less likely for the local spatial course superimposed on the super-regional and regional courses; these 'short-wave' water tables fluctuate in time with the seasons (with a lag due to flow in the unsaturated zone), and mapping local results would then have little meaning since the maps would quickly be out of date. For instance, with a local wave with $l_c = 10\ m$ (wavelength of ca. 63 m) and a subsurface with $k_{hc} = 1\ m^2 \cdot dbar^{-1} \cdot d^{-1}$, $k_{zc} = 0.1\ m^2 \cdot dbar^{-1} \cdot d^{-1}$ and $\theta = 0.35$, then the characteristic time $t_c \approx 10\ d$ (bearing in mind that the relation k_{hc}/k_{zc} is generally reduced with shorter, thus less deep, waves). So these short waves (with a maximum depth of ca. 20 m; see Section 5.3) will respond almost immediately to the seasonal fluctuations in P_e.

These examples are very general, however, and more detailed analysis would be needed for any real or actual problems.

5.5. Net exfiltration/infiltration maps; spatial scales

As already noted, in many cases it can be assumed that the large-scale spatial variations in the water table (the long waves) will have a reasonably steady character. On this basis, we can make maps of $q_z(x, y, 0, t_0)$, where t_0 is a representative point in time. Such maps will then show the exfiltration/infiltration (discharge/recharge) patterns associated with wavelengths greater than a particular value ($2\pi l_x$, $2\pi l_y > 2 l_{pix}$). In a Fourier series, averaging over a specific lateral spatial dimension l_{pix} amounts to eliminating terms with a wavelength less than $2 l_{pix}$. Consequently, the exfiltration/infiltration patterns for wavelengths less than that particular value $2l_{pix}$ should be added to ascertain the actual exfiltration/infiltration occurring at a particular point, but not represented on the map. The patterns represented on the map give the net exfiltration/infiltration and not the actual exfiltration/infiltration occurring at a particular point. This net exfiltration/infiltration pattern is interpreted to be approximately equal to the average of the exfiltration/infiltration actually occurring over a pixel area with surface dimension l^2_{pix}, the size of the pixel. From this it may be clear that the presentation of maps with locally varying dimensions of pixel size could lead to entirely wrong interpretations, especially if the maps also show the associated flow paths (the pseudo-flow paths). Since the effects of small-scale variations are damped out (decay) more quickly with increasing depth than the effects of large-scale variations, the net exfiltration/infiltration map provides information on the exfiltration/infiltration and the flow paths that are actually occurring at greater depths, namely, at depths with an order of magnitude greater than ca. $2l_{pix}\ \sqrt{(k_{zc}/k_{hc})}$.

The presence of exfiltration (discharge) implies a relatively constant influx of water into the biosphere, and in most cases it also means a regular influx of bicarbonate and calcium ions. These ions have an important influence on microbiological activity, and the consequent mineralization of nitrogen and phosphate, as well as the oxygen balance in the soil and the measure of phosphate available in plant-extractable form. There are indications that phosphate availability is particularly influenced by a reduction in exfiltration, and can result in great changes of species in the plant bed. Continuing research is advisable for proper regulation of the extent and chemical composition of discharge flows since their effects on the cycle of nutrients and protons can be highly significant for the preservation of our natural habitat.

5.6. The pre-modeling phase of flow systems analysis

One thing is clear from these examples: the correct way to begin a regional flow systems analysis is to start with a general analysis of the problem itself, along the lines suggested above, before going ahead with numerical modeling (or mapping). With the help of mathematical analysis, including the rules of thumb given here and more-elaborated computer programs, the scales (both the lateral scales and the associated depth-scales and time-scales) of the problem should be examined before choosing dimensions of the grid-blocks or elements, the relevant time-points, and the depth of the 'impervious' base. Based on the reasoning given in this chapter, all kinds of refinements can still be introduced when appropriate. For instance, the amplitude relationships in the water table belonging to the various Fourier waves occurring simultaneously may also be important, and the amplitude-ratios of the velocities, and the relationship between the different wavelengths will also play a role (to be analyzed with a computer program based on the analytical solution for flow in perfectly layered basins (see Bervoets, 1991; and Bervoets *et al.*, 1991). Much further research still remains to be done with regard to the influence of lateral heterogeneities and the time-dependency, especially for the deeper parts of the subsurface where compressibility, compaction-decompaction and geopressurization may also play an important role.

Chapter 6

Transport of Dissolved Matter

6.1. Convection, diffusion and mechanical dispersion

Due to limitations on expanding the land use for cattle and crop production, intensive cultivation of land has led to an increased production and use of manure as a fertilizer, and to an increased use of agrochemicals like fertilizers, pesticides and herbicides. This excessive use has induced a concern about the associated negative side effects. One of the most serious problems is the resulting groundwater pollution due to movement of chemical agents into the surface waters and groundwaters. Especially since the groundwater is moving slowly, the subsurface pollution is advancing as well, while gradually occupying larger and larger parts of the subsurface. In this way presently unpolluted regions are expected to be polluted in the future. In order to guide responsible decision-making and management for the prevention of further deterioration of the natural environment, the transport processes governing the movement of the contaminants have to be understood in more detail.

When we look at the groundwater flow on the pore scale (see Figure 9), the pressure field and velocity field are described by the nonlinear Navier-Stokes equations and the no-flow boundary conditions on the fluid-solid interfaces. In most cases the fluid flow velocity is low and the temporal changes in flow occur slowly, i.e., the Reynolds number and the Strouhal number are much smaller than one, so the Navier-Stokes equations may reasonably be simplified to the steady Stokes equations (see, for instance, Tritton, 1977). The Stokes equations are linear in pressure and velocity, which explains the linearity of Darcy's law (see also Appendix A). The transport of solutes on the pore scale is described by convection and molecular diffusion only. Convection represents a reversible process, i.e., if $-t$ is substituted for t and $-\boldsymbol{x}$ for $\boldsymbol{x}$, the equation describing convection remains the same. However, molecular diffusion represents an irreversible process.

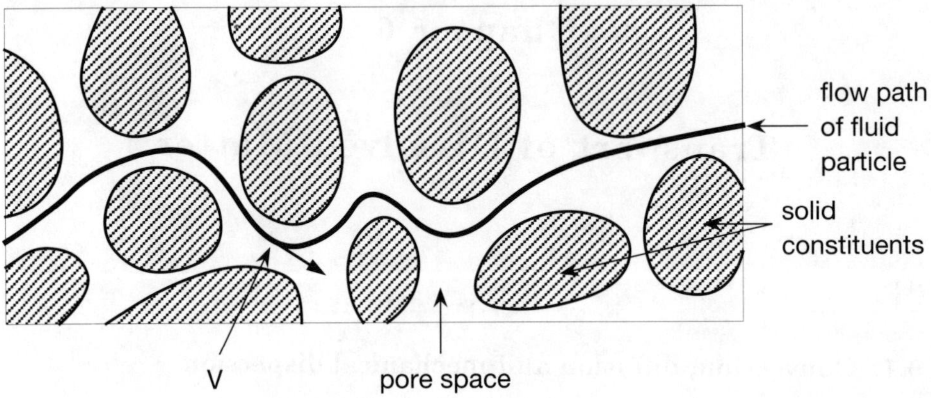

Figure 9. **Flow in the pore space channels.**
On the microscopic scale, groundwater is flowing through the channels of the pore space between the solid constituents composing the porous medium. In the theory of porous media we deal with averaged values over a representative elementary volume (REV) composed of a sufficiently large number of flow channels and solid constituents.

(Note also that Darcy's law combined with the continuity equation for compressible flow describes an irreversible process; however, combined with the continuity equation for incompressible flow ($\operatorname{div} \boldsymbol{q} = 0$), it describes a reversible process.)

On the scale of Darcy's law (the scale of a representative elementary volume, say the scale of ca. 1 cm), the transport velocity $\boldsymbol{v}$ has the meaning of an averaged value of the pore scale Navier-Stokes velocity over many flow-effective pores. Around this average value $\boldsymbol{v}$ there is obviously a certain spread in the Navier-Stokes velocity, which indicates that reversible convective transport takes place at a mean velocity with a velocity spread around it. Combining these velocity spreads with a small amount of irreversible molecular diffusion ultimately results in an irreversible process. In the limit of vanishingly small molecular diffusion, this process is called mechanical microdispersion. To this mechanical microdispersion the finite, nonvanishingly small diffusion must be added. The combination of finite diffusion and mechanical microdispersion is called microdispersion.

Microdispersion is a theoretically well-understood process (see, for instance, Dagan, 1989).

We usually deal, however, with large-scale studies, where for practical reasons we cannot solve the local Darcy's law (i.e., Darcy's law on the scale of ca. 1 cm). We must then make use of a large-scale Darcy's law, an example of which is given in Section 4.8. In deriving the generalized Dupuit approximation, and the large-scale Darcy's law which is based upon it, we have seen that the flow in regional basins seems to follow a much more tortuous local course than the large-scale velocity $<\boldsymbol{v}>$ would suggest. A combination of velocity spread around this large-scale value $<\boldsymbol{v}>$ and microdispersion gives an irreversible process called macrodispersion.

Macrodispersion is much more difficult to describe theoretically, and is consequently much discussed in the professional literature. Even so, an 'on-the-job' or 'practician' formulation is nearly always adopted in practical studies. This practice-formulation is based on the same formulation as for microdispersion, but a much greater dispersion length is adopted. In general, this macrodispersion length depends not only on the considered spatial scales, as is the case for the conductivity, but also on the temporal scales. Considerations of these scale dependencies are mostly outside the scope of this book. Instead, it is the microdispersion, with an introduction to the scale-theory, and the 'practice-formulation' for macrodispersion which will be developed briefly in the following sections. For more details, see Dagan (1989); Bear and Verruijt (1987), and for more practical applications, see Kinzelbach (1987). However, here an alternative approach based on 'stream-line analysis' will be presented as an approximate, but practical alternative.

6.2. Conservation of mass for dissolved matter

Naturally, the first fundamental premise, the conservation of mass, also applies to the matter dissolved in water. Expressed mathematically, the equation for the conservation of dissolved matter on a local scale appears as follows:

$$\partial/\partial t(\omega\,\rho\,\theta) + \mathrm{div}[\omega\,\rho\,\theta(\boldsymbol{v} + \boldsymbol{u})] = \rho\theta b. \tag{6.2.1}$$

Here ω is the mass fraction of the dissolved matter, ρ is the density of the water-solute mixture, θ is the effective porosity of the porous medium,

$\boldsymbol{v} = (v_x, v_y, v_z)$ is the transport velocity of the flowing mixture (a vector) and $\boldsymbol{u} = (u_x, u_y, u_z)$ is the velocity of the dissolved matter relative to the velocity of the flowing mixture (a vector), so that $\boldsymbol{v} + \boldsymbol{u}$ is the total velocity of the dissolved matter, b is the source term describing the process of taking matter from or into the flowing solution. The contribution to b coming from the stagnant fluid in the pores is given by $-(\theta^* - \theta)/\theta \; \partial\omega/\partial t$, where θ^* is the porosity including flowing *and stagnant* fluid. This term can be treated in the same way as the sorption term; see Section 6, equation (6.6.2a). The more popular quantity concentration is defined by $c = \rho\omega$; however, the mass fraction is the more fundamental quantity; see Section 6.3. For other quantities expressing the amount of dissolved matter see Section 6.7.

unit of mass fraction [−]; SI unit of density [$kg{\cdot}m^{-3}$]; SI unit of concentration [$kg{\cdot}m^{-3}$]; conventional unit of concentration [$mg{\cdot}L^{-1} = g.m^{-3}$]; unit of porosity [−]; SI unit of velocity [$m{\cdot}s^{-1}$]; conventional unit of velocity [$m{\cdot}d^{-1}$]; SI unit of source term [s^{-1}]; conventional unit of source term [d^{-1}].

After some mathematical manipulation, it follows that the equation for the conservation of mass of the dissolved matter can also be expressed as:

$$\begin{aligned} &\rho\theta\,(\partial\omega/\partial t \;+ \boldsymbol{v}\cdot\mathbf{grad}\,\omega) + \mathrm{div}(\omega\,\rho\,\theta\,\boldsymbol{u}) + \\ &\qquad + \omega\,[\partial(\rho\,\theta)/\partial t + \mathrm{div}(\rho\,\theta\,\boldsymbol{v})] = \rho\theta b. \end{aligned} \tag{6.2.2}$$

Substitution of the law of conservation of mass for the mixture (the continuity equation (2.2.1)) $\partial/\partial t(\rho\,\theta) + \mathrm{div}(\rho\,\theta\,\boldsymbol{v}) = 0$ in the above equation, and some further mathematical manipulation gives:

$$\begin{aligned} &\partial\omega/\partial t \;+\; \boldsymbol{v}\cdot\mathbf{grad}\,\omega + (\omega/\rho)\;\boldsymbol{u}\cdot\mathbf{grad}\,\rho\,+ \\ &\qquad + \mathrm{div}(\omega\,\theta\,\boldsymbol{u})/\theta = b. \end{aligned} \tag{6.2.3}$$

6.3. The dispersive flux

As has already been described in Section 6.1, the transport velocity $\boldsymbol{v}$ should not be confused with the local Navier-Stokes velocity of the fluid inside the pore space. On a pore scale level, the fluid velocity varies in both magnitude and direction in any representative elementary volume (1 cm^3, say). The flow paths in the pore system are very intricate and cannot be described in full detail. Local variations in grain and void sizes increase the mixing and spreading of the fluid. This spreading is in both longitudinal and transverse directions of average flow. Combined with diffusion, this phenomenon is called microdispersion. In the literature, the usual expression for the dispersive flux $\omega\,\boldsymbol{u}$ is:

$$\begin{aligned} -\omega\,\boldsymbol{u} &= D_m\,\mathbf{grad}\,\omega + \\ &+ (\alpha_L/v)\,(\boldsymbol{v}\cdot\mathbf{grad}\,\omega)\,\boldsymbol{v} + \\ &+ (\alpha_T/v)\,(\boldsymbol{v}\times\mathbf{grad}\,\omega)\times\boldsymbol{v}. \end{aligned} \tag{6.3.1}$$

Wherein D_m $[m^2{\cdot}s^{-1};\ m^2{\cdot}d^{-1}]$ is the molecular diffusion coefficient; α_L $[m]$ is the longitudinal dispersion length; α_T $[m]$ is the transverse dispersion length, $v = \sqrt{(\boldsymbol{v}\cdot\boldsymbol{v})}$ is the absolute value of the transport velocity; the vector product $\boldsymbol{a}\times\boldsymbol{b}$ of two vectors $\boldsymbol{a} = (a_x, a_y, a_z)$ and $\boldsymbol{b} = (b_x, b_y, b_z)$ is defined as the vector: $\boldsymbol{a}\times\boldsymbol{b} = (a_y b_z - a_z b_y,\ a_z b_x - a_x b_z,\ a_x b_y - a_y b_x)$; this vector stands perpendicular to the vectors $\boldsymbol{a}$ and $\boldsymbol{b}$.

The term $D_m\ \mathbf{grad}\,\omega$ represents the molecular diffusion which always occurs in mixtures, not just in porous media. However, in this case, D_m is smaller than the normal molecular diffusion coefficient D because of the 'tortuousness' of the flow paths around the solids of the porous media. Sometimes the Millington-Quirk relationship $D_m = \theta^{4/3}\,D$ is used (see Millington and Quirk, 1961). Only for fluid mixtures with constant density ρ_0 this term may be written in its well-known form $(D_m/\rho_0)\,\mathbf{grad}\,c$; see Hassanizadeh (1988).

The term $(\alpha_L/v)\ (\boldsymbol{v}\cdot\mathbf{grad}\,\omega)\ \boldsymbol{v}$ has the same direction as the transport velocity $\boldsymbol{v}$; therefore this term describes the longitudinal mechanical dispersion.

The term $(\alpha_T/v)\ (\boldsymbol{v}\times\mathbf{grad}\,\omega)\times\boldsymbol{v}$ has a direction perpendicular to the transport velocity $\boldsymbol{v}$, and therefore describes the transverse mechanical dispersion.

The above expression (6.3.1) is justifiable for microdispersion, but is often used as a 'practice-formulation' for the description of macrodispersion.

In the literature, the above expression (6.3.1) for the dispersive flux is conventionally presented in the following more complicated, but mathematically equivalent way (see Dagan, 1989; Bear and Verruijt, 1987):

$$\omega\,\boldsymbol{u} = -\underline{\boldsymbol{D}}\cdot\mathbf{grad}\,\omega \tag{6.3.2}$$

where $\underline{\boldsymbol{D}}$ $[m^2 \cdot s^{-1};\ m^2 \cdot d^{-1}]$ is the dispersion tensor given by:

$$\begin{aligned} \underline{\boldsymbol{D}} &= (D_m + \alpha_T\, v)\, \underline{\boldsymbol{I}} + [(\alpha_L - \alpha_T)/v]\ \boldsymbol{v}\,\boldsymbol{v} = \\ D_{ij} &= (D_m + \alpha_T\, v)\ \delta_{ij} + [(\alpha_L - \alpha_T)/v]\ v_i\, v_j. \end{aligned} \tag{6.3.3}$$

In which $\underline{\boldsymbol{I}} = \delta_{ij}$ is the unit tensor, and $\boldsymbol{v}\,\boldsymbol{v} = v_i\, v_j\ (i,j = 1,2,3)$ is a tensor (not to be confused with the scalar product $\boldsymbol{v} \cdot \boldsymbol{v} = v_i v_i = v_x^2 + v_y^2 + v_z^2$).

Substitution of expression (6.3.1) for the dispersive flux $\omega\,\boldsymbol{u}$ in the term $(\omega/\rho)\, \boldsymbol{u} \cdot \mathbf{grad}\, \rho$ of the law of mass conservation (6.2.3) shows that the latter term is almost always negligibly small in relation to the term $\boldsymbol{v} \cdot \mathbf{grad}\, \omega$. The law of mass conservation (6.2.3) then reduces to:

$$\partial\omega/\partial t + \boldsymbol{v} \cdot \mathbf{grad}\, \omega = -\,\mathrm{div}(\omega\, \theta\, \boldsymbol{u})/\theta + b. \tag{6.3.4}$$

With the aid of some vector calculus, expression (6.3.1) for the dispersive flux can be rewritten as:

$$\begin{aligned} -\omega\, \boldsymbol{u} = {} & [(\alpha_L - \alpha_T)/v]\ (\boldsymbol{v} \cdot \mathbf{grad}\, \omega)\ \boldsymbol{v} + \\ & + (D_m + \alpha_T\, v)\ \mathbf{grad}\, \omega. \end{aligned} \tag{6.3.5}$$

From substitution of the above expression (6.3.5) in the term $\mathrm{div}(\omega\, \theta\, \boldsymbol{u})/\theta$ we find that:

$$\begin{aligned} -\,\mathrm{div}(\omega\, \theta\, \boldsymbol{u})/\theta = {} & [(\alpha_L - \alpha_T)/q]\ (\boldsymbol{v} \cdot \mathbf{grad}\, \omega)\ \mathrm{div}\, \boldsymbol{q} + \\ & + \boldsymbol{v} \cdot \mathbf{grad}[\{(\alpha_L - \alpha_T)/v\}\ \boldsymbol{v} \cdot \mathbf{grad}\, \omega] + \\ & + \mathrm{div}[(D_m + \alpha_T\, v)\ \theta\ \mathbf{grad}\, \omega]/\theta \end{aligned} \tag{6.3.6}$$

where $\boldsymbol{q} = \theta\, \boldsymbol{v}$.

As we have seen in Section 2.3, $\mathrm{div}\, \boldsymbol{q} = 0$ (equation (2.3.4)) is a good approximation to the continuity equation (2.2.1), so that by substitution of equation (6.3.6) the law of mass conservation (6.3.4) then appears as follows:

$$\begin{aligned} \partial\omega/\partial t\ + \boldsymbol{v} \cdot \mathbf{grad}\, \omega = {} & \boldsymbol{v} \cdot \mathbf{grad}[\{(\alpha_L - \alpha_T)/v\}\ \boldsymbol{v} \cdot \mathbf{grad}\, \omega] + \\ & + \mathrm{div}[(D_m + \alpha_T\, v)\ \theta\ \mathbf{grad}\, \omega]/\theta + b. \end{aligned} \tag{6.3.7}$$

Equation (6.3.7) is usually known as a convection-dispersion equation. The potential ϕ does not appear in this equation, but the groundwater velocity

$\boldsymbol{v} = \boldsymbol{q}/\theta$ does; this fact justifies the considerable attention given in previous chapters to the description of the vector fields $\boldsymbol{e} = \underline{\boldsymbol{k}}^{-1} \cdot \boldsymbol{q}$ and $\boldsymbol{q}$.

The molecular diffusion coefficient has generally the order of magnitude of $D_m \approx 10^{-9}\ m^2 \cdot s^{-1}$. In a locally homogeneous porous medium, the longitudinal microdispersion length is in the order of magnitude of the dimensions of the pores. According to Fried and Combarnous (see Dagan, 1989), it holds that $1.4\, l_p < \alpha_L < 2.2\, l_p$, where l_p is the dimension of the pore scale. The transverse microdispersion length is still one or two orders of magnitude smaller ($0.01\, \alpha_L < \alpha_T < 0.1\, \alpha_L$) (see Dagan, 1989). This means that in the term $\boldsymbol{v} \cdot \mathbf{grad}[\{(\alpha_L - \alpha_T)/v\}\ \boldsymbol{v} \cdot \mathbf{grad}\, \omega]$, α_T may be neglected with respect to α_L. In this way, convection-dispersion equation (6.3.7) simplifies to:

$$\begin{aligned} \partial\omega/\partial t\ + \boldsymbol{v} \cdot \mathbf{grad}\, \omega &= \boldsymbol{v} \cdot \mathbf{grad}\, [(\alpha_L/v)\ \boldsymbol{v} \cdot \mathbf{grad}\, \omega] + \\ &+ \operatorname{div} [(D_m + \alpha_T\ v)\ \theta\ \ \mathbf{grad}\, \omega]\, /\theta + b \end{aligned} \qquad (6.3.8)$$

6.4. Large-scale effective porosity

As pointed out earlier, the above convection-dispersion equation (6.3.7) (or mathematically equivalent versions of it) is often used also on the larger scale and within it a smaller-scale heterogeneous conductivity distribution, but with much larger time-independent values of the dispersion lengths. Theoretically, however, it is not always justifiable to bring the large-scale dispersion into computation in this way; it is only at very long (unrealistically long) time scales that the macrodispersion lengths become time independent. Dispersion on the large scale is called macrodispersion. Remember that mechanical macrodispersion is a description which corrects for the large-scale averaged transport velocity $\langle\boldsymbol{v}\rangle$ being unequal to the local fluid velocity $\boldsymbol{v}$. Therefore, the macrodispersion lengths in a heterogeneous porous medium will depend strongly on the size and form of the heterogeneities in the averaging volume, and thus on the averaging scales.

As has been said before, in reality there is also a great dependency on the temporal scale. Within short time scales after the release of a solute, the transport will take place mainly via the well-conducting 'channels' of the heterogeneous porous medium. Since diffusion and transverse mechanical microdispersion are relatively slow processes in the well-conducting routes, the dominant transport mechanisms are convection and longitudinal mechanical microdispersion. With longer time scales this will change, and the much slower transport to and from the poorly conducting regions of the porous medium becomes more influential on the mass fraction, which changes appreciably by this long time-scale effect. In this context, the idea of effective porosity as described in Section 2.2 can be extended to a large-scale porosity effective for flow. We can define the large-scale flow-effective porosity as the volume percentage of the porous medium with high conductivity, where virtually all water flow takes place. To do so, we first define a 'macroporosity' Θ_{mac}. For example, a relatively simple definition just showing the basic mode of thought could be:

$$\Theta_{mac} = \int\!\!\int\!\!\int k(x,y,z)\,\mathrm{d}x\mathrm{d}y\mathrm{d}z \Big/ \left[k_{max}\int\!\!\int\!\!\int \mathrm{d}x\mathrm{d}y\mathrm{d}z\right] \qquad (6.4.1)$$

whereby the integration is performed over a volume-element which contains a sufficiently large number of through-flowing channels (see Figure 10). In formula (6.4.1) the scalar k equals $k = \sqrt{(\underline{\boldsymbol{k}} : \underline{\boldsymbol{k}})} = \sqrt{(k_{ij}\,k_{ij})}$ (summation over $i = 1,2,3$ and $j = 1,2,3$), and k_{max} is the maximum value of k that occurs in the volume-element. The large-scale flow-effective porosity $\boldsymbol{<}\theta\boldsymbol{>}$ is then equal to the product $\boldsymbol{<}\theta\boldsymbol{>} = \theta\,\Theta_{mac}$. The above model is applicable, for instance, to a sand body containing many clay-lenses. Obviously, the above-presented simple, perhaps oversimplified, formula for Θ_{mac} could be extended to describe more complex conductivity distributions. For instance, an extension is needed for situations where the high conductivities do not coincide with through-flowing channels but are completely surrounded by poor conductivity. If necessary, the direction of flow must also be taken into account for layered substrata. However, the development of models for the macroporosity will not be discussed further.

On the short time scale then, we can describe the flow with the large-scale flow-effective porosity $\boldsymbol{<}\theta\boldsymbol{>} = \theta\,\Theta_{mac}$, i.e., $\boldsymbol{<v>} = \boldsymbol{<q>}/\boldsymbol{<}\theta\boldsymbol{>}$, where the large-scale Darcy's law with large-scale conductivity $\boldsymbol{<}\underline{\boldsymbol{k}}\boldsymbol{>}$ is used to calculate $\boldsymbol{<q>}$. Then the transport may be described with convection and longitudinal mechanical dispersion, neglecting diffusion and transverse mechanical dispersion. Consequently, for this short time-scale transport the governing equation is given by the following simplified form of equation (6.3.8):

$$\begin{aligned}\partial\omega/\partial t\ +\ &\boldsymbol{<v>}\cdot\mathbf{grad}\,\omega = \\ &= \boldsymbol{<v>}\cdot\mathbf{grad}[(\alpha_L/\boldsymbol{<v>})\ \boldsymbol{<v>}\cdot\mathbf{grad}\,\omega] + b.\end{aligned} \qquad (6.4.2)$$

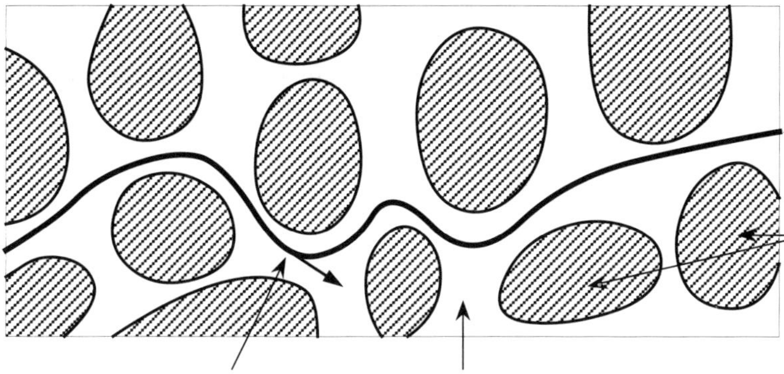

Figure 10. **Flow in well-conducting channels.**
Groundwater is predominantly flowing through the well-conducting channels of a heterogeneous porous medium. In the poorly conducting lenses the flow is almost stagnant. In the theory of large-scale flow-effective porosity we deal with averaged values over a volume composed of a sufficiently large number of well-conducting flow channels and poorly conducting lenses. However, even if there is hardly any flow in the poorly conducting lenses, there will be transport to them, from them, and in them, due to molecular diffusion and transverse dispersion. As a consequence, on a short time scale, large-scale transport-effective porosity will be equal to large-scale flow-effective porosity. But on a long time scale, large-scale transport-effective porosity will be equal to small-scale effective porosity, which is much larger than large-scale flow-effective porosity.

where $\langle v \rangle = |\langle \boldsymbol{v} \rangle|$.

On a somewhat longer time scale, a transition must be made to the general dispersion including diffusion and transverse mechanical dispersion. In that case, the flow pattern has also to be calculated more accurately, taking into account full heterogeneity in the conductivity (i.e., applying $\underline{\boldsymbol{k}}$ and $\boldsymbol{q}$ instead of $\langle \underline{\boldsymbol{k}} \rangle$ and $\langle \boldsymbol{q} \rangle$ in equation (6.4.2)) and applying the small-scale effective porosity θ to obtain $\boldsymbol{v}$ (i.e., $\boldsymbol{v} = \boldsymbol{q}/\theta$). Then, to obtain the full equation (6.3.8), the term $\mathrm{div}[(D_m + \alpha_T\, v)\ \theta\ \mathbf{grad}\,\omega]/\theta$ must be added to the right-hand side of the short time-scale transport equation (6.4.2). This term takes into account the diffusion and transverse mechanical dispersion, especially to and from the impervious or poorly conducting 'lenses.'

However, the full transport equation (6.3.8) is very hard to solve for general three-dimensional flow. Therefore, an approximation is sought. Especially when the flow in the poorly conducting 'lenses' is practically stagnant, an approximation can be introduced. This approximation is to replace the diffusion-transverse dispersion term $\text{div}\,[(D_m + \alpha_T\ v)\ \theta\ \mathbf{grad}\,\omega]\,/\theta$ in convection-dispersion equation (6.3.8) by a mass transfer term $(h_m/\boldsymbol{<}\theta\boldsymbol{>})$ $(\omega_p - \omega)$, where ω is the mass fraction in the well-conducting 'channels,' ω_p is the mass fraction in the poorly conducting 'lenses,' and h_m $[s^{-1};\ d^{-1}]$ is the mass transfer coefficient. This term is added to the right-hand side of the short-time convection-dispersion equation (6.4.2), resulting in:

$$\begin{aligned} &\partial\omega/\partial t + \boldsymbol{<}\boldsymbol{v}\boldsymbol{>}\cdot\mathbf{grad}\,\omega = \\ &\quad = \boldsymbol{<}\boldsymbol{v}\boldsymbol{>}\cdot\mathbf{grad}[(\alpha_L/\boldsymbol{<}v\boldsymbol{>})\ \boldsymbol{<}\boldsymbol{v}\boldsymbol{>}\cdot\mathbf{grad}\,\omega] + \\ &\quad + (h_m/\boldsymbol{<}\theta\boldsymbol{>})\ (\omega_p - \omega) + b. \end{aligned} \tag{6.4.3}$$

In the above simplified transport equation (6.4.3), the transport velocity $\boldsymbol{<}v\boldsymbol{>} = \boldsymbol{<}q\boldsymbol{>}/\boldsymbol{<}\theta\boldsymbol{>}$ is the value in the well-conducting channels. This 'mass transfer' approximation results in a considerable simplification of the flow problem, because the concepts of large-scale conductivity and large-scale flow-effective porosity can still be applied. An additional equation is also needed to describe transport (mainly diffusion) in the poorly conducting 'lenses.' Taking only mass transfer, and not convection, diffusion and dispersion, into account within the lenses, the mass transport equation for the poorly conducting lenses becomes:

$$\partial\omega_p/\partial t = [h_m/(\theta_p - \boldsymbol{<}\theta_p\boldsymbol{>})]\ (\omega - \omega_p) + b_p \tag{6.4.4}$$

where θ_p is the small-scale effective porosity of the poorly conducting lenses, $\boldsymbol{<}\theta_p\boldsymbol{>} = \theta_p\,\Theta_{mac}$ is the large-scale effective porosity of the poorly conducting lenses, and b_p is the source term for the poorly conducting lenses describing the process of taking new matter into solution.

Such a 'two-mass fractions-mixture-approximation' is often applied in the theory of heat transfer in packed beds and heat exchangers (see, for instance, Vafai and Sozen, 1990; or Lage and Bejan, 1990). For a more general overview of continuum mixture theories, see Atkin and Craine (1976). The main problem is now replaced to the determination of the mass transfer coefficient h_m (for heat transfer coefficients see, for instance, Wong, 1977). The usual approach is to relate h_m to the dimensions of the solid particles in the packed bed and to the Péclet number $v\ d_p/D$, where d_p is the linear dimension of the solid particle, and D is the 'normal' diffusion coefficient. As an example for such a relation, applied to heat and mass transfer between a slowly flowing fluid and solids in a packed bed, see Pfeffer (1964). Analogously, in the present context of transport through heterogeneous porous media, a generalized Péclet number $v\ d_p/(D_m + \alpha_T\ v_p)$ should be used, where d_p is the linear dimension of the poorly conducting lenses.

On time scales $t << h_m^{-1}$, mass transfer to the poorly conducting lenses is negligible and transport is described by equation (6.4.2). On time scales $t >> h_m^{-1}$, mass transfer from the well-conducting channels to the poorly conducting lenses is completed. Thanks to the special form of the source terms b and b_p (see Section 6.6), this means that $\omega_p \rightarrow \omega$.

From equations (6.4.3) and (6.4.4) it follows then that:

$$\begin{aligned} \partial\omega/\partial t + \boldsymbol{<v>'}\boldsymbol{\cdot}\,\mathbf{grad}\,\omega &= \\ = \boldsymbol{<v>'}\boldsymbol{\cdot}\,\mathbf{grad}\left[(\alpha_L/\boldsymbol{<v>'})\ \boldsymbol{<v>'}\boldsymbol{\cdot}\,\mathbf{grad}\,\omega\right] + b' \end{aligned} \qquad (6.4.5a)$$

where

$$\boldsymbol{<v>'} = \boldsymbol{<\theta><v>}/(\theta_p + \boldsymbol{<\theta>} - \boldsymbol{<\theta_p>}) \qquad (6.4.5b)$$

and

$$b' = \left[\boldsymbol{<\theta>}\ b + (\theta_p - \boldsymbol{<\theta_p>})\ b_p\right]/(\theta_p + \boldsymbol{<\theta>} - \boldsymbol{<\theta_p>}). \qquad (6.4.5c)$$

In general, $|\boldsymbol{<v>'}| << |\boldsymbol{<v>}|$; for instance, if $\theta = \theta_p$ then $\boldsymbol{<v>'} = = \Theta_{mac}\boldsymbol{<v>} = \boldsymbol{<q>}/\theta$. Comparison of equation (6.4.2) with equations (6.4.5) shows that the large-scale porosity effective for *transport* gradually changes from $\boldsymbol{<\theta>}$ for $t << h_m^{-1}$ to θ for $t >> h_m^{-1}$. In this book, however, we do not elaborate further on this approach.

In most 'practical' applications it is not the above 'two-mass fractions-mixture-approximation,' but the approximation with 'macrodispersion' that is adopted, and relatively large time-independent dispersion lengths (known as the 'macrodispersion lengths') are applied. As a very general rule of thumb, it can be said that the longitudinal macrodispersion length is an order of magnitude smaller (ca. a factor of 10) than the length scale of averaging, and, in turn, the transverse macrodispersion length is an order of magnitude smaller than the longitudinal dispersion length. With large differences in lateral and vertical averaging scales, then the macrodispersion lengths will also vary considerably, depending on the prevailing direction of flow.

6.5. Stream-line analysis

Assuming that the transport velocity field $\boldsymbol{v}(\boldsymbol{x}, t)$ is steady during the time interval $t_0 < t < t_1$, then $\partial \boldsymbol{v}(\boldsymbol{x}, t)/\partial t = \mathbf{0}$. Thus, $\boldsymbol{v} = \boldsymbol{v}(\boldsymbol{x})$ where $\boldsymbol{x} = (x, y, z)$ is the position vector. In the three-dimensional flow field, a stream-line is given by the vector $\boldsymbol{x} = \boldsymbol{X}(\tau) = (X(\tau), Y(\tau), Z(\tau))$, where τ is a scalar variable which represents the position on the stream-line. By definition of the stream-line, the direction $d\boldsymbol{X}(\tau)/d\tau = (dX(\tau)/d\tau,\ dY(\tau)/d\tau,\ dZ(\tau)/d\tau)$ is in the same direction as $\boldsymbol{v}(\boldsymbol{x})$, so that:

$$d\boldsymbol{X}(\tau)/d\tau = (dX(\tau)/d\tau,\ dY(\tau)/d\tau,\ dZ(\tau)/d\tau) = \boldsymbol{v}/s \qquad (6.5.1)$$

where s is an arbitrary scalar function. For example, if we choose s to be $s = v = |\boldsymbol{v}| = \sqrt{(\boldsymbol{v} \cdot \boldsymbol{v})}$, then τ is the actual length coordinate (expressed in meters); if we choose s to be $s = 1$, then τ is the coordinate which represents the time in which the water element has travelled along the stream-line with velocity v.

The stream-line concept is not defined at locations $\boldsymbol{x}$ where $\boldsymbol{v} = \mathbf{0}$, the stagnation points. It is true that the flow field $\boldsymbol{v}(\boldsymbol{x})$ does not have to be a single-valued function in any point $\boldsymbol{x}$, but bifurcation of stream-lines must not occur; that is, there may be a maximum of two values of $\boldsymbol{v}$ at one point $\boldsymbol{x}$, and the two directions must make an angle smaller than $\pi/2$. We are then dealing with the type of stream-lines with refraction at interfaces where discontinuities in the conductivities occur, as described in Section 3.3.

For each scalar function $\omega(\boldsymbol{x}, t)$ with a value on the stream-line, the following expressions hold: $\omega(\boldsymbol{x}, t) = \omega(\boldsymbol{X}(\tau), t) = \omega(x, y, z, t) = \omega(X(\tau), Y(\tau), Z(\tau), t) = \Omega(\tau, t)$ and $\partial\Omega/\partial\tau = (\partial\omega/\partial x)\ (dX/d\tau) + (\partial\omega/\partial y) \cdot (dY/d\tau) + (\partial\omega/\partial z)\ (dZ/d\tau) = (d\boldsymbol{X}/d\tau) \cdot \mathbf{grad}\, \omega$, so that, with expression (6.5.1):

$$\boldsymbol{v} \cdot \mathbf{grad}\,\omega = s\ \partial\Omega/\partial\tau. \tag{6.5.2}$$

This expression says that a term of the form $\boldsymbol{v} \cdot \mathbf{grad}\,\omega$ means differentiation to the stream-line coordinate τ; it thus concerns differentiation along a stream-line without having to consider the directions perpendicular to it.

Therefore, the convection-dispersion equation (6.3.8) can be written as:

$$\begin{aligned}\partial\Omega/\partial t\ + s\ \partial\Omega/\partial\tau &= s\ \partial/\partial\tau[(\alpha_L/v)\ s\ \partial\Omega/\partial\tau] + \\ &+ \operatorname{div}[(D_m + \alpha_T\, v)\ \theta\ \mathbf{grad}\,\omega]/\theta + b.\end{aligned} \tag{6.5.3}$$

If application of the simplified convection-dispersion equation (6.4.3) is justified, we find transport exclusively along the stream-line in accordance with:

$$\begin{aligned}\partial\Omega/\partial t\ + s\ \partial\Omega/\partial\tau &= s\ \partial/\partial\tau[(\alpha_L/v)\ s\ \partial\Omega/\partial\tau] + \\ &+ (h_m/\theta)(\Omega_p - \Omega) + b.\end{aligned} \tag{6.5.4}$$

where the symbols $<\cdot>$ to denote large-scale quantities have been omitted.

The above equation (6.5.4), with derivatives solely along the stream-lines, is a great simplification from a mathematical viewpoint, so it is tempting to use this simplified equation in model codes, etc. Such a simplification does not seem unreasonable in view of the great uncertainties involved in transport through heterogeneous porous media. The 'stream-line analysis' based on this assumption can also be extended to time-dependent flow, where the stream-line pattern changes at time-points t_n and $t_{n+1} (n = 1, 2, \dots)$ with steady stream-lines in between these two time-points.

6.6. Sorption and decay

Heavy metals and many organic substances are chemical components which are absorbed by the sediment; these components may also decay when time proceeds. Whenever there is only one component dissolved in the groundwater it is customary to include sorption in the form of the term $\rho\theta b = -\rho_B\ \partial\Omega_B/\partial t$. A possible decay term can also be inserted in the source term b according to $b = -k_d\,\Omega^n$, or else:

$$b = -[\rho_B/(\rho\,\theta)]\;\partial\Omega_B/\partial t - k_d\;\Omega^n \tag{6.6.1a}$$

where ρ_B $[kg{\cdot}m^{-3}]$ is the bulk density of the solid matter of porous medium, Ω_B $[-]$ is the mass fraction absorbed on the grains in the sediment (i.e., $\rho_B\Omega_B = c_B$ $[kg{\cdot}m^{-3}]$ is the bulk concentration of mass absorbed on the grains in the sediment), k_d $[s^{-1};\ d^{-1}]$ is the decay coefficient and n $[-]$ is the exponent in the decay term. If application of the simplified convection-dispersion equation (6.5.4) is justified, expression (6.6.1a) holds for the well-conducting channels, and a similar term should be used for the poorly conducting lenses:

$$b_p = -[\rho_{pB}/(\rho\theta_p)]\;\partial\Omega_{pB}/\partial t - k_{pd}\Omega_p^n \tag{6.6.1b}$$

Often, in the literature, it is further assumed that Ω_B at a constant temperature is only a function of the mass fraction Ω of the dissolved matter in the groundwater. This can, however, only be the case when the component is in chemical equilibrium with the sediment and the water. If there is no chemical equilibrium, as if often in field situations, an additional term describing the reaction kinetics must be added (see, for example, Van Duijn and Knabner, 1991). Here we will neglect reaction kinetics and consider equilibrium chemistry as a simple example.

In an isothermal flow field we then find:

$$b = -[\rho_B/(\rho\,\theta)]\;(d\Omega_B/d\Omega)\;\partial\Omega/\partial t - k_d\;\Omega^n. \tag{6.6.2a}$$

and

$$b_p = -[\rho_{pB}/(\rho\theta_p)]\;(d\Omega_{pB}/d\Omega)\;\partial\Omega_p/\partial t - k_{pd}\Omega_p^n. \tag{6.6.2b}$$

Inserting the above expressions in equations (6.5.4) and (6.4.4) yields:

$$\begin{aligned}\partial\Omega/\partial t\; + (s/R)\;\partial\Omega/\partial\tau\; &= \\ &= (s/R)\;\partial/\partial\tau[(\alpha_L/v)\;s\;\partial\Omega/\partial\tau] + \\ &\quad + [h_m/(\theta R)]\;(\Omega_p - \Omega) - (k_d/R)\;\Omega^n\end{aligned} \tag{6.6.3a}$$

and

$$\begin{aligned}\partial\Omega_p/\partial t &= [h_m/\{(\theta_p - <\theta_p>)\ R_p\}]\ (\Omega - \Omega_p) \\ &\quad - (k_{pd}/R_p)\ \Omega_p^n\end{aligned} \tag{6.6.3b}$$

where:

$$R = 1 + [\rho_B/(\rho\,\theta)]\ (d\Omega_B/d\Omega) \tag{6.6.4a}$$

and

$$R_p = 1 + [\rho_{pB}/(\rho\theta_p)]\ (d\Omega_{pB}/d\Omega_p) \tag{6.6.4b}$$

are the retardation factors.

From the above equation (6.6.3a) we can see that the convective term $(s/R)\ \partial\Omega/\partial\tau\ =\ (\boldsymbol{v}/R)\boldsymbol{\cdot}\,\mathbf{grad}\,\omega$ is a factor R smaller than it would be without sorption. Apparently there is a slowing-down (retardation) of the convection velocity due to sorption. Further, we see that the dispersion term, the mass transfer term, and the decay term are also a factor R smaller than if there were no sorption. It is now obvious that the retardation factor is extremely important for computations of the transport of dissolved matter. It is also clear that a convenient choice of s is given by $s = R$; the resulting form of the convection-dispersion equation (6.6.3a) then becomes:

$$\begin{aligned}\partial\Omega/\partial t\ + \partial\Omega/\partial\tau &= \partial/\partial\tau[(\alpha_L/(v/R))\ \partial\Omega/\partial\tau] + \\ &\quad + [h_m/(\theta R)]\ (\Omega_p - \Omega) - (k_d/R)\ \Omega^n.\end{aligned} \tag{6.6.5}$$

The above equation (6.6.5) can be solved along a previously defined streamline. It is also possible to extend the above theory to a coupled system of such equations to describe ion exchange for different chemicals with mutually interactive components (see, for instance, Appelo, 1992). Attempts to obtain large-scale values for the sorption parameter are presented by Van Duijn and Van der Zee (1986).

6.7. Quantities and units for amount of dissolved matter

This section attempts to contribute a little bit in bridging the gap between the modes of thought prevailing in the discipline of geohydrochemistry on the one hand, and the discipline of transport physics on the other, as described in the previous sections of this chapter. In geohydrochemistry there are many good reasons to apply a variety of different quantities and accompanying units to express the amount of matter dissolved in water; see, for instance, Appelo (1992). In transport physics, on the other hand, the mass fraction ω, and the concentration $c = \rho\omega$, are the only natural quantities, as has been shown in the previous sections of this chapter. In this section the different geohydrochemical quantities and their accompanying units will be presented briefly in relation to each other. This presentation will be given from the point of view of transport physics, with a strong emphasis on the difference between (dimensional) quantities and (dimensionless) numerical values. In that sense, the following presentation also complements the geohydrochemical literature, in which this difference is hardly made.

For that purpose, let us first recall the basic equation (1.4.1) giving the relation between (dimensional) physicochemical quantities, (dimensionless) numerical values and (dimensional) units:

$$\text{physicochemical quantity} = \text{numerical value} \times \text{unit} \qquad (6.7.1a)$$

or, in symbolic mathematical language:

$$q = \{q\}\ [q] \qquad (6.7.1b)$$

where q is the quantity, $\{q\}$ is the numerical value, and $[q]$ is the unit in which the quantity is expressed.

For instance:

mass = numerical value × unit of mass,
or $M = \{M\}\ [M]$

amount of substance = numerical value × unit of amount of substance,
or $N = \{N\}\ [N]$

atomic mass = numerical value × unit of atomic mass,
or $a = \{a\}\ [a]$

etc.

In the following discussion we present the International System of Units (the SI system of units). In this way our basic equations (6.7.1) are rewritten as:

$$\text{quantity} = \text{numerical value} \times \text{SI unit} \quad (6.7.2a)$$

$$q = \{q\} \text{ SI unit.} \quad (6.7.2b)$$

The following examples are given:

mass = numerical value × kilogram,
or $M = \{M\}\ kg$

mass fraction = numerical value × kilogram per kilogram,
or $\omega = \{\omega\}\ kg{\cdot}kg^{-1} = \{\omega\}$

volume = numerical value × cubic meter,
or $V = \{V\}\ m^3$

concentration = numerical value × kilogram per cubic meter,
or $c = \{c\}\ kg{\cdot}m^{-3}$

amount of substance = numerical value × mole,
or $N = \{N\}\ mol$

atomic mass = numerical value × kilogram per mole,
or $a = \{a\}\ kg{\cdot}mol^{-1}$

molarity = numerical value × mole per cubic meter,
or $n = \{n\}\ mol{\cdot}m^{-3}$

molality = numerical value × mole per kilogram,
or $m = \{m\}\ mol{\cdot}kg^{-1}$

valence = numerical value × Faraday,
or $v = \{v\}\ F$

normality = numerical value × equivalent per cubic meter,
or $r = \{r\}\ eq{\cdot}m^{-3}$

It is important to keep in mind that in the SI system of units the unit of atomic mass (and the unit of molecular mass) is the kilogram per mole [$kg{\cdot}mol^{-1}$]. For instance, the atomic mass of H (atomic hydrogen) is equal to $a{<}H{>} = 0.00100797\ kg{\cdot}mol^{-1}$.

In geohydrochemistry, however, the unit $g{\cdot}mol^{-1} = 10^{-3}\ kg{\cdot}mol^{-1}$ is almost exclusively used for presentation of data. The reason for presenting data in the unit $g{\cdot}mol^{-1}$ is that this results in more convenient numerical values; for instance, the atomic mass of hydrogen is equal to $a<H> = 1.00797\ g{\cdot}mol^{-1}$ resulting in the more convenient numerical value $\{a<H>\} = 1.00797 \approx 1$, and all other numerical values of atomic and molecular mass $\{a\}$ are larger than 1. Of course, in the SI system of units applying decimal multiples of SI units for presentational purposes is allowed. But in virtually all geohydrochemical publications the unit $g{\cdot}mol^{-1}$ is even omitted; its use is supposed to be known. Unfortunately, this practice then leads to application of the numerical value $\{a\}$ with respect to the unit $g{\cdot}mol^{-1}$ as a base unit in mathematical equations.

To show the consequences of what has been stated above, it is important to understand the difference between equations expressing the relation between physicochemical *quantities*, and equations expressing the relation between the *numerical values* of physicochemical quantities expressed in an *a priori specified set of units*. The distinction between the two types of equations is best illustrated by the following examples.

As a first example, let us turn to the following algebraic equation:

$$m_i = N_i/(a_s\ N_s). \tag{6.7.3}$$

In the above expression, N_i is the amount of substance of solute i dissolved in an amount of solvent N_s, where a_s is the atomic or molecular mass of the solvent. The above expression defines the molality m_i of solute i. Application of the basic equation (6.7.1) ($q = \{q\}\ [q]$) to the above expression (6.7.3) yields:

$$\{m_i\} = \{N_i\}/(\{a_s\}\ \{N_s\}) \times [N_i]/([a_s]\,[N_s]\,[m_i]). \tag{6.7.4}$$

When we use coherent units, for instance the SI system of units, the factor $[N_i]/([a_s]\ [N_s]\ [m_i]) = 1$, and in that case the equation for the numerical values is given by:

$$\{m_i\} = \{N_i\}/(\{a_s\}\ \{N_s\}). \tag{6.7.5}$$

However, in geohydrochemistry the usual units are: $[m_i] = mol{\cdot}kg^{-1}$, $[N_i] = [N_s] = mmol = 10^{-3}\ mol$, and $[a_s] = g{\cdot}mol^{-1} = 10^{-3}\ kg{\cdot}mol^{-1}$.

Using these units we yield:

$$[N_i]/([a_s]\ [N_s]\ [m_i]) = mmol/(g{\cdot}mol^{-1}\ mmol\ mol{\cdot}kg^{-1}) = \\ = kg{\cdot}g^{-1} = 10^3. \qquad (6.7.6a)$$

Consequently, the expression for the numerical value of the molality expressed in the above units yields:

$$\{m_i\} = 10^3\ \{N_i\}/(\{a_s\}\ \{N_s\}). \qquad (6.7.6b)$$

This equation can also be written as:

$$m_i = 10^3\ \{N_i\}/(\{a_s\}\ \{N_s\})\ mol{\cdot}kg^{-1}. \qquad (6.7.6c)$$

The latter two equations (6.7.6b) and (6.7.6c) are well known in hydrochemistry (see, for instance, Stumm and Morgan, 1981, p. 45), and the curled parentheses $\{\cdot\}$ denoting the fact that equation (6.7.6b) and the right-hand side of equation (6.7.6c) deal with dimensionless numerical values with respect to an a *priori* given set of units are always omitted in the literature.

From the above example it is observed that numerical factors in equations (the factor 10^3 in equations (6.7.6)) are factors needed for conversion of units, and that equations in which numerical factors occur are equations giving relations between numerical values of quantities expressed in an a priori given set of units. In coherent sets of units such factors do not occur.

Another example is given by the following expression:

$$pH = -\log(n{<}H^{+}{>})/n_0), \quad n_0 = 10^3\ mol{\cdot}m^{-3} \qquad (6.7.7)$$

where pH (dimensionless) is the pH-scale, and $n{<}H^{+}{>}$ is the molarity of hydrogen ions dissolved in water. Application of the basic equation (6.7.1) ($q = \{q\}\ [q]$) to the above expression (6.7.7) yields:

$$pH = -\log\left\{n{<}H^{+}{>}\right\} + \log\{n_0\} - \log([n{<}H^{+}{>}]/[n_0]). \qquad (6.7.8a)$$

Expressing the molarity unit $[n{<}H^{+}{>}]$ in $mol{\cdot}L^{-1} = 10^3\ mol{\cdot}m^{-3}$

($1\ L = 10^{-3}\ m^3 = 1\ liter$) we find:

$$pH = -\log\left\{n{<}H^+{>}\right\}. \tag{6.7.8b}$$

Equation (6.7.8b) is well known and the curled parentheses are almost always omitted (see, for instance, Stumm and Morgan, 1981, p. 131). Nevertheless, we must keep in mind that this equation deals with the numerical value of the hydrogen ion molarity expressed in the unit $mol{\bullet}L^{-1}$. When we use as a unit of molarity $[n{<}H^+{>}] = mmol{\bullet}L^{-1} = mol{\bullet}m^{-3}$ = SI unit of molarity, which is also a popular unit, we find:

$$pH = -\log\left\{n{<}H^+{>}\right\} + 3. \tag{6.7.8c}$$

Now a number of useful physicochemical quantities are presented together with their SI units and 'popular' units.

Mass fraction. The mass fraction ω_i of a solute i of which a mass M_i is dissolved in a solute-solvent mixture with mass M is defined as:

$$\omega_i = M_i/M \tag{6.7.9}$$

(note that M is the mass of solute and solvent together).

The SI unit of mass fraction is $[\omega] = kg{\bullet}kg^{-1}$. The more popular unit among geohydrochemists is $mg{\bullet}kg^{-1} = 10^{-6}\ kg{\bullet}kg^{-1} = ppm$.

Concentration. The concentration c_i of a solute i of which a mass M_i is dissolved in a solute-solvent mixture with volume V is defined as:

$$c_i = M_i/V \tag{6.7.10}$$

(note that V is the volume of solute and solvent together).

The SI unit of concentration is $[c] = kg{\bullet}m^{-3}$. The more popular unit among geohydrochemists is $mg{\bullet}L^{-1} = 10^{-3}\ kg{\bullet}m^{-3}$.

The concentration is related to the mass fraction by the expression:

$$c_i = \rho\,\omega_i \tag{6.7.11}$$

where $\rho = M/V$ is the density of the solute-solvent mixture.

Molarity. The molarity n_i of a solute i of which an amount of substance N_i is dissolved in a solute-solvent mixture with volume V is defined as:

$$n_i = N_i/V. \qquad (6.7.12)$$

Since 1 mol = 6.02205 x 10^{23} particles (atoms, molecules, ions), the molarity is a measure for the number of particles dissolved in a unit volume of mixture (note that V is the volume of solute and solvent together). The SI unit of molarity is $[n_i] = mol{\bullet}m^{-3}$, and the popular unit among geohydrochemists is $mmol{\bullet}L^{-1}$, which is equal to the SI unit.

The relation between molarity and concentration is given by:

$$c_i = n_i\ a_i \qquad (6.7.13)$$

where a_i is the molecular (or atomic) mass of solute i. Equation (6.7.13) can be written as:

$$\{c_i\} = \{n_i\}\ \{a_i\}\ [n_i]\ [a_i]/[c_i]. \qquad (6.7.14a)$$

In the customary units $[c_i] = mg{\bullet}L^{-1}$, $[n_i] = mmol{\bullet}L^{-1}$, and $[a_i] =$ $= g{\bullet}mol^{-1}$ we find that:

$$\{c_i\} = \{n_i\}\ \{a_i\}. \qquad (6.7.14b)$$

Molality. As has already been mentioned, the following expression defines the molality m_i of a solute i:

$$m_i = N_i/(a_s\ N_s) \qquad (6.7.3)$$

where N_i is the amount of substance of solute i dissolved in an amount of substance of solvent N_s, a_s is the molecular mass of the *solvent* (not of the mixture). Note that $M_s = a_s\ N_s$ is the mass of solute and *not* the mass of solute and solvent together.

Valence. Ions are electrically charged in a quantized way with integer multiples of the electrical charge of a proton. For instance, one Cl^- ion has the electrical charge of minus one proton (= charge of one electron) and one Ca^{2+} ion has the electrical charge of two protons. Due to the quantum character of ionic charge, it is natural to express the quantity 'charge per ion' in the unit proton-charge•ion^{-1} (or, equivalently minus-electron-charge•ion^{-1}).

However, the SI unit of amount of substance, which is a unit for a certain (very large) number of particles (e.g., ions), is the mole [mol], where 1 mol = 6.02205 x 10^{23} particles, or 1 particle = 0.166056 x 10^{-23} mol (6.02205 x 10^{23} $particles{\cdot}mol^{-1}$ is called the Avogadro number). The SI unit of electrical charge is the coulomb [C], and the charge of a proton equals 1.6021892 x 10^{-19} C. As a consequence, the unit 1.6021892 x x10^{-19} x 6.02205 x 10^{23} $C{\cdot}mol^{-1}$ = 96 484.6 $C{\cdot}mol^{-1}$ is the SI unit for the quantity 'charge per ion,' and only integer multiples of this value can occur.

However, since the 'charge per ion' expressed in the unit $C{\cdot}mol^{-1}$ does not simply reflect the quantum character of this quantity, it is a very unnatural unit for expressing the quantity 'charge per ion.' To be able to represent the quantum character of the quantity 'charge per ion' exactly, and to be able to use the unit mole [mol] as the unit of amount of substance (number of particles), it is customary in geohydrochemistry to use the unit Faraday [F]. The Faraday is defined in such a way that it is exactly equal to the electrical charge in proton units per ion, i.e., 1 F = 1 proton-charge•ion^{-1} = 96 484.6 $C{\cdot}mol^{-1}$. From this definition it follows that 1 $F{\cdot}mol$ = 96 484.6 C. In this way, the quantity 'charge per ion' expressed in the unit F can only have integer numerical values like −3, −2, −1, 0, 1, 2, 3, etc.

The valence v of a charged ion is now defined as the absolute value of the electrical charge per ion. For instance, the valence of the Ca^{2+} ion is equal to $v = 2$ F, and the valence of the HCO^{3-} ion is equal to $v = 3$ F.

Normality. The normality r_i of an ionized solute i is defined as the molarity of the dissolved ions times the valence of the ions, or in symbolic notation:

$$r_i = n_i \, v_i. \tag{6.7.15}$$

In fact, the normality is a measure for the number of protons, or electrons, dissolved in a unit volume of mixture. In SI units the normality r_i is expressed in $F{\cdot}mol{\cdot}m^{-3}$. It is customary to call 1 $F{\cdot}mol$ = 1 eq = 1 equivalent. In that way the SI unit of normality is expressed in $eq{\cdot}m^{-3}$ = = $meq{\cdot}L^{-1}$, which is also the most popular unit.

Equivalent* $\mathbf{CaCO_3}$ *concentration. Especially in the U.S. it is customary to express ion concentrations in equivalent $CaCO_3$ concentrations. The molecular mass of $CaCO_3$ is equal to 0.1 $kg.mol^{-1}$, which means that a mass of 1 kg $CaCO_3$ contains an amount of 10 mol $CaCO_3$. In solution the amount of 10 $mol{\bullet}kg^{-1}$ $CaCO_3$ is separated in ions resulting in an amount of 10 $mol{\bullet}kg^{-1}$ Ca^{2+} ions. This amount has an electrical charge of 20 $eq{\bullet}kg^{-1}$. In other words, for $CaCO_3$ an electrical charge of 1 eq has a mass of 0.05 kg. Electro-neutrality requires that an ionic solute, say Cl^-, with a normality of $r = \{r\}\ eq{\bullet}m^{-3}$ balances an equal normality of ionized $CaCO_3$. Consequently, the equivalent $CaCO_3$ concentration amounts to 0.05 $\{r\}\ kg{\bullet}m^{-3}$. From this explanation, the following definition of equivalent $CaCO_3$ concentration e_i of ionized solute i will be clear:

$$e_i = \beta\ r_i, \quad \beta = 0.05\ kg{\bullet}eq^{-1} \tag{6.7.16}$$

or:

$$\{e_i\} = \{r_i\}\ \beta\ [r_i]/[e_i]. \tag{6.7.17a}$$

When $[e_i] = mg{\bullet}L^{-1} = 10^{-3}\ kg{\bullet}m^{-3}$ and $[r_i] = meq{\bullet}L^{-1} = eq{\bullet}m^{-3}$, we find $\beta\ [r_i]/[e_i] = 50$, resulting in the well-known equation:

$$\{e_i\} = 50\ \{r_i\}. \tag{6.7.17b}$$

Chapter 7

The System-Analytical View of Groundwater Flow and Transport

7.1. The goal-seeking approach to the system's analysis

When solving real life problems, especially the more complex ones, man has found that the descending approach which starts from abstract generalizations of known facts and ends in detailed decisions has proved to be successful in most of the situations experienced. The descending approach, in particular, applies to all groundwater flow-related problems. Here the complexity stems from the complicated geology, from the physical processes that govern groundwater flow, and from the way we humans change, or intend to change, the course of subsoil phenomena. The reasons for our intervention into the subsoil are almost always the same – to use water (groundwater) for our biological needs with the least possible cost or risk. In particular, the natural tendency to avoid damaging the present or the future water resources has led to developing and then using all kinds of simulation/prediction tools which would allow us to foresee the results of man's intervention into nature. One of the abstract tools which is commonly used for practical problem-solving with the descending approach is the notion of a 'system.' In the following chapters this notion is applied in a way which differs slightly from the manner used until now. It stresses 'the problem decomposition' rather than the 'physical reality decomposition,' though both are strongly related. In the case of a 'weak' human intervention into the subsoil a 'physical reality decomposition' is considered primarily. It means that the flow domain geometry, natural distribution of hydraulic parameters, intensity and distribution of natural boundary conditions are decisive when analyzing (i.e., making a decomposition of) groundwater flow.

If, however, man's activity cannot be considered a slight perturbation of the subsoil, and the exploitation of groundwater resources is, or is intended to be, intensive, it is the purpose and the technical means of the activity that impose and define the structure of the 'system.' With the increasing scarcity of water resources the latter mode is becoming more and more prevalent. Even so, the 'natural decomposition' of groundwater flow described in the previous chapters still plays and will continue to play a basic role in understanding how the complex groundwater system operates. In the following sections of this chapter, a goal-seeking approach to groundwater flow is presented. It shows how the groundwater flow description developed in the previous chapters can be linked to the real-life applications through the notion of 'system.'

Definition 1: (see Nawalany, 1986c). A *system* is a purposely chosen set of objects having attributes specified together with the relations between the attributes.

This definition is very close to the definition given by Gibson (see Klir, 1969). The words 'object,' 'attribute' and 'relation' are primitive concepts. When the concept of a system is applied to natural processes the attributes can be interpreted as the physical variables, the objects as the subsets of the variables, and the relations between the attributes can be formulated by means of mathematical expressions (e.g., formulae, equations, logical relations, etc.). Having this in mind, one can simplify Definition 1 by saying:

Definition 1': A *system* is a purposely chosen set of variables together with the relations between them.

Both definitions contain the expression 'purposely chosen' which is the heart of the goal-seeking approach in the systems theory. The expression implies an existence of two things: the purpose (or the goal) and the criterion of choice. In other words, *the procedure of choosing the variables* (and the relations between them) *which are essential to achieve the goal is what defines the system.* There exists also a dual concept to that of the system – the concept of the system's environment. Its definition states:

Definition 2: *The environment of the given system* is a system consisting of the variables which do not belong to the system.

If not specified, the relations between these variables are assumed to be irrelevant to the problem under consideration. However, there is one more possible set of relations, namely the relations between the variables of the system and those of the environment. Consequently, one can write the following:

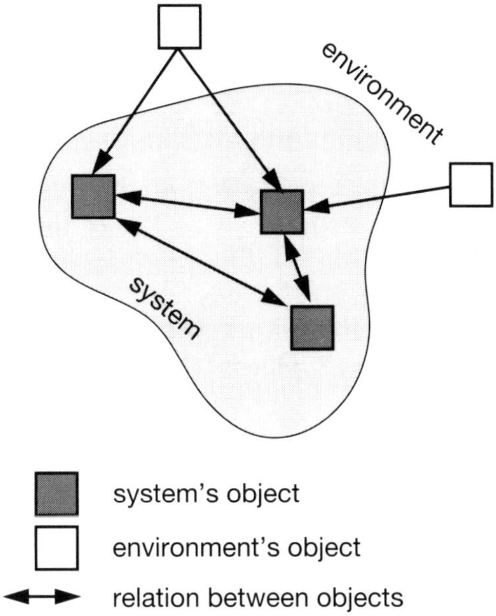

Figure 11. Abstract representation of the system.

Definition 3: The relations between the system variables and the environment variables are called *the interactions between the system and its environment.*

In general, interactions are not symmetric. Of special interest are those interactions for which the variables of the environment can be considered independent of the related system variables. Actually, the art of distinguishing the system is defining the system variables in such a way that either (some of) their interactions with the environment can be considered to be nonexistent (negligible, not relevant to the problem) or, if existing, the related variables of the environment can be assumed to be independent of the system variables – see Figure 11. Consequently, instead of speaking about the interactions it is more practical to speak about *the input variables* which are defined as follows:

Definition 4: *The input variables* for a given system are those variables of the environment which interact with the system and can be considered independent of the system variables. In conclusion for this part of the general considerations, *besides the system variables, the input variables are the only environmental variables that must be specified when the concept* *of the system is used.*

There are three groups of system variables that can be distinguished: parameters, output variables and state variables.

Definition 5: *Parameters* are those variables of the system which are independent of the input variables as well as of the other variables of the system.

Definition 6: *Output variables* are the system variables which have an operational meaning for the problem (the goal) being considered.

In many instances uniqueness of the input-output relations is essential for solving the problem (achieving the goal). Therefore, in such cases, a special set of variables must be defined additionally in the system.

Definition 7: *State variables* are the system variables which satisfy the following conditions:

(i) they are not parameters;

(ii) they form a minimal indexing set for the following set: $\{(x, y) : x\text{-input variable}, y\text{-output variable}\}$.

The expression 'indexing set' in condition (ii) means that the state variables are used for distinguishing between the pairs (input variable, output variable) in which the values of the input variables are the same, and the values of the corresponding output variables are different. Without the indexing the existence of such pairs in the system could make the concept of the system useless. The expression 'minimal ... set' merely means that the set of the state variables is the least numerous among all the indexing sets that make the relation between the given set of the input variables and the given set of the output variables unique. It can also be observed that Definition 7 does not specify what the state variables are; they may be specified in many different ways. Sometimes it is convenient to use (some of) the state variables as output variables or even not to distinguish between the parameters and the state variables. Obviously, in the latter case condition (i) of Definition 7 must be omitted.

One more important concept will be discussed in this section, namely, the subsystem.

Definition 8: *A subsystem of a given system* is a purposely chosen subset of the system variables with their relations restricted to the given subset.

Here the expression 'purposely chosen' means that the given system is divided into subsystems to achieve the goal (to solve the problem) with less effort (because operating on smaller sets is always easier). Since the subsystem is a system for itself, all the variables and relations involved must be redefined accordingly. In particular, all the original system variables which have not been chosen as the subsystem variables now become the subsystem environment variables, and their relations with the subsystem variables become interactions. In general, the interactions between the subsystem and its environment must be consistent with the interactions and relations that existed in the original system. Nevertheless, in many cases simplifications are made when defining the input variables for the subsystem. Some of the interactions are considered negligible for the given subsystem, so they are omitted. This is in fact the second reason for defining subsystems. By eliminating some of the relations the sets involved are made even smaller, and, therefore, easier to operate upon.

Definition 9: *The parameters of the subsystem* are the original system parameters that are restricted to the subsystem.

Definition 10: *The output variables of the subsystem* are the former output variables and/or new ones created from the original system state variables.

Definition 11: *The state variables of the subsystem* are those original system state variables which satisfy conditions (i) and (ii) of Definition 7, and are restricted to the given subsystem.

Definition 11 implies, in particular, that when defining the subsystems, one has to discriminate some autonomous (independent) classes of the state variables which can be used as the indexing sets for the subsystems. It also indicates that not every non-overlapping partition of the system state variables serves the purpose. By making a partition which is not reasonably (purposely) chosen, one can either lose the operational meaning of some output variables (because of possible non-uniqueness of the subsystem's input-output relationship), or make the systems analysis less effective (because of the introduction of superfluous variables and/or relations).

When applying the concepts of the system and the subsystems to the *systems analysis* of existing objects and/or processes (e.g., natural phenomena) one defines *the system first* and tries to distinguish *the subsystems afterwards*. This order is usually reversed for *the systems synthesis*, that is, when the (sub)systems are known (or defined) and the system is assembled from them. The synthesis approach is typical for designing or planning activities in which one is interested in overall behavior of the whole system knowing or assuming the behavior of its subsystems.

7.2. Natural and man-influenced groundwater flow systems

Two classes of groundwater flow systems can readily be distinguished depending on whether one is interested in

> knowing how groundwater flows in the subsoil if man's intervention can be considered only a slight perturbation of the natural state,

or in

> assessing how efficient the groundwater exploitation operations are, or can be, depending on the 'nature' of the groundwater system and on modifications of the system induced by the human activity.

Accordingly, two classes of groundwater systems can be named:

(i) natural systems;
(ii) man-influenced systems.

The two classes are discussed below in terms of the goal-seeking approach to the system's analysis.

Natural systems

Different goals can be formulated when considering groundwater as an object of scientific investigation. For instance, the most general objective is *to describe the flow of water in the subsoil* as a special case of flow in porous, fissured or fractured media. By the description of groundwater flow is then understood the presentation of some mathematical equations that allow us to calculate how much water is flowing from one place to another. Technically it means calculating the space and time evolution of the potential or velocity fields that obey general laws of physics and fulfill conditions specific for the given flow domain. Another objective can be formulated when *describing transport phenomena in the subsoil.* In this case also, a mathematical formulation of physical, chemical or biological processes is indispensible. Naturally, the system then becomes considerably more complicated and also the meaning of 'the system' changes, as will be demonstrated below. There are a number of other goals that can be scientifically interesting and for which different groundwater systems need to be defined; for instance, the surface water-groundwater interaction, the groundwater uptake by plants, the groundwater evaporation process, rock matrix-groundwater chemical interactions, etc. The specific systems are formulated when trying *to set a monitoring system for groundwater* with the objective of checking the theoretical formulations of flow and transport phenomena.

It seems very attractive to show that what is understood by the system clearly depends on the goal set. This will be demonstrated in the following discussion using the three examples of: evaluating the recharge/discharge regions for a subsurface (first example), the time-space evolution of a pollution plume in the subsurface (second example), and making decisions on the abstraction scheme (third example).

Example one

Goal: Knowing the spatial distribution of the water table in a given natural subsurface, find the recharge and the discharge regions.

Remark. A number of assumptions are hidden under the notion 'given subsurface.' It may mean that the three-dimensional geometry of the subsurface, its geological stratification, hydraulic characteristics, etc. are known or can be reasonably estimated. The knowledge of the subsurface may originate from (measurements in) some real subsurface, or it can be a hypothetical knowledge generated for the sake of theoretical investigation. Whatever is the source of information on the subsurface being considered, it is assumed here that all notions necessary for the systems analysis can be identified. Additionally, the subsurface is considered natural, i.e., human interference is negligible.

System objects:

There is only *one object* in the system which is a *flow domain*. It is characterized by the following *attributes*:

(i) velocity field;
(ii) potential field;
(iii) conductivity distribution.

The *relations* between the attributes are:

(i) continuity equation;
(ii) Darcy's law.

The relations are summarized either into one classical flow equation – (8.2.1) for the potential, or into three coupled equations for the velocity field components (the velocity-oriented approach) – (8.2.12).

Environmental objects:

There are two kinds of *objects* which are *not chosen* to be the objects of the system still interacting with the groundwater within the flow domain. The first are all the *impervious strata* that surround the given subsurface. The second are the *surface waters* (rivers, lakes, etc.) which can exchange water with the subsurface. It is tacitly assumed that interactions of all other possible objects like an unsaturated zone, plant roots, small abstraction wells, etc. with the flow domain are summarized in the known position of the water table.

For the two kinds of environment's objects the following *attributes* are specified:

(i) normal component of flow (for the strata);
(ii) resistivity of bottom sediments to flow (for the surface waters);
(iii) water table position (in the surface waters).

The *relations between the system and the environment* are represented by:

(i) a zero normal component of the flow between the flow domain and the impervious strata;
(ii) a simple formula that allows calculation of fluxes of water from/into the surface waters into/from the subsurface out of the known resistivity of sediments, position of water table in surface waters, and water table in the subsurface.

A subdivision of the system's attributes (variables) into the four basic categories can be summarized as follows:

(i) ***state variables:***
- potential field;
- three components of the velocity field.

Depending on the choice of the state variable, one may use one of the two equivalent mathematical descriptions to groundwater flow: the classical description or the velocity-oriented approach.

(ii) ***input variables:***

- normal components of the water flux along the impervious boundaries;
- normal components of the water flux along the common boundaries between the surface waters and the subsurface;
- position of the water table in the subsurface.

The input variables come into the definition of the boundary conditions that are (together with the flow equations) necessary for the existence and the uniqueness of the solution to the groundwater flow problem. For the goal stated above, it is assumed that there is some method of calculating (approximating) the solution of the flow equations, i.e., calculating the potential field and/or the velocity field within the whole flow domain.

(iii) ***output variables:***

- normal component of water velocity on the system's top boundary (i.e., on the water table boundary or on a projection of the water table on a horizontal plane);
- set of groundwater particle paths within the flow domain.

The first output variable enables us to discriminate recharge and discharge regions along the flow domain top boundary. Recharge regions (for which the groundwater velocity vector is pointing inwards) may become starting places for flow paths within the flow domain.

(iv) ***parameters:***

- conductivity distribution within the system;
- resistivity of the bottom sediments in the surface waters.

Subsystems:

By solving the state equations and calculating water trajectories one can make a further subdivision of the flow domain. Out of all paths (trajectories) that can be drawn for the flow domain, specific subsets can be discriminated *in space* and named *flow subsystems*. They are defined as follows:

> A flow subsystem is any family of groundwater paths that start at the same recharge region and leave the flow domain at the same discharge region.

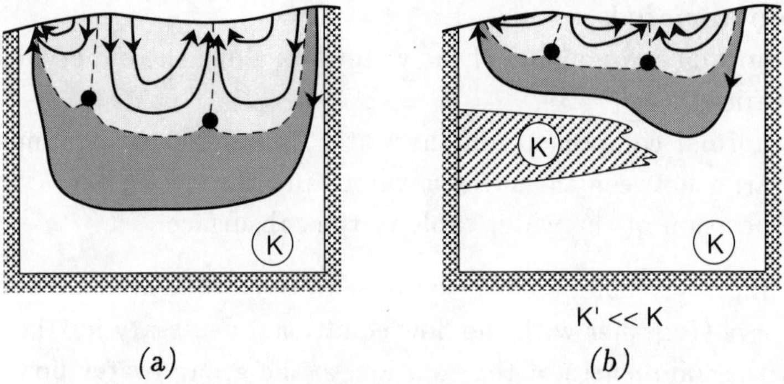

Figure 12. Subdivision of a hypothetical flow system into flow subsystems for two cases: (a) for a homogeneous subsurface, (b) for the same subsurface for which the existence of a less-pervious layer has been assumed; (•) – stagnation point, (– –) – boundary of a subsystem, (—) – flow path.

Consequently, one can add and calculate additional output variables for the flow subsystems like: traveling time distribution (isochrones), distribution of age etc. depending on the additional questions asked. Figure 12 illustrates a typical subdivision of a given flow system into flow subsystems. The above definition of a subsystem differs from the definition given in Section 5.1, where a subsystem was related to a spatial Fourier mode of the water table. As will be clear from the presentation in this chapter, the concept of 'system' is sufficiently flexible to assign more meaningful definitions to the notion 'subsystem.' Here the 'subsystem' is defined in relation to the groundwater velocity transporting dissolved mass, whereas in Section 5.1 the 'subsystem' is defined to assess the relation between horizontal spatial scale and vertical spatial scale (and time scale).

Example two

Goal: Knowing the flow system (velocity field) and the spatial distribution of pollution sources on the top boundary of the given groundwater system, find the evolution of the *pollution plume* within the subsoil.

Remark. Again a number of assumptions are hidden in the notion 'given groundwater system.' In particular it means that the flow system has been defined prior to solving the subsurface pollution problem. It may also mean that the three-dimensional geometry of the subsurface, the soil characteristics etc. are known or can be reasonably estimated. As before, the knowledge of the subsurface may originate from (measurements in) some real subsurface or is only a hypothetical knowledge generated for the sake of theoretical investigation. Whatever the source of information on the subsurface being considered, it does mean that all notions necessary for the systems analysis can be identified. In this case, however, additional information is needed: namely, an emission strength of pollution sources (as a function of time) and their distribution in space. The existence of pollution sources clearly indicates human interference in the groundwater system though it is assumed here that the pollution does not change the chemical nor hydraulic characteristics of the rock matrix within the subsurface.

System objects:

There are *two objects* in the system:

(i) a *flow domain*;
(ii) a *rock (solid) matrix*.

The objects are characterized by the following *attributes*:

Flow domain:
- velocity field;
- diffusion coefficient;
- dispersivity coefficient distribution;
- mass fraction of the pollutant in groundwater.

Rock matrix:
- mass fraction of the pollutant on the rock matrix;
- adsorption/desorption characteristics.

The *relations* between the attributes are:

(i) continuity of mass equation for the pollutant in the fluid phase;
(ii) continuity of mass equation for the pollutant on the solid phase;
(iii) Fick's law;
(iv) relationship between the velocity field and the dispersion coefficient;
(v) kinetics of the adsorption/desorption processes.

The relations are summarized in the convection-dispersion equation for the mass fraction of pollutant dissolved in the fluid phase and the mass balance equation for the mass fraction of pollutant on the solid phase. The special role is that of the velocity field within the subsurface; the groundwater velocity is incorporated in a natural way into the convective term of the convection-dispersion equation. For meaningful calculations of the pollutant mass fraction, the components of the velocity need to be known (calculated) *very accurately* prior to solving the convection-dispersion equation. Technically this is realized by solving the flow problem in the subsurface *first*, using for instance the velocity-oriented approach (see example one).

Environmental objects:

Also in this case there are two kinds of *objects* which are *not chosen* to be the objects of the system still interacting with the groundwater within the flow domain. The first are all the *impervious strata* that surround the given subsurface. The second are the *surface sources of pollution* (rivers, lakes, waste disposal sites, agricultural activities etc.) which can change the quality of the groundwater within the subsurface. For both types of the environment's objects the following *attributes* are specified:

(i) time intensities of the pollution sources;
(ii) space distributions of the pollution sources.

The *relations between the system and the environment* are represented by:

(i) known intensities of the pollutant sources along the boundary of the flow domain (assumed zero on the impervious strata).

A subdivision of the system's attributes (variables) into the four basic categories can be summarized as follows:

(i) ***state variables:***
- mass fraction of the pollutant in groundwater;
- mass fraction of the pollutant on the rock matrix.

(ii) ***input variables:***
- fluxes of pollutant from the sources along the flow domain boundaries.

The input variables come into the definitions of the boundary conditions that are, together with the initial conditions, the convection-dispersion equation for the groundwater and the mass balance equation for the solid matrix, necessary for the existence and the uniqueness of the solution to the pollution plume problem. There are no internal source terms for the pollution since only surface distributed sources have been assumed.

(iii) ***output variables:***
- time and space mass fraction of the pollutant in the groundwater;
- time and space mass fraction of the pollutant on the rock matrix.

The output variables are in this case identical with the state variables.

(iv) ***parameters:***
- velocity field;
- diffusion coefficient;
- dispersivity coefficient distribution;
- adsorption/desorption characteristics.

It should be noted here that the velocity field has become a parameter since there has been assumed to be no influence of pollution on groundwater flow.

Subsystems:

By solving the state equations and calculating the mass fraction distributions of the pollutant within the subsurface, one may ask additional questions:

(i) How is the pollutant transported by particular flow subsystems?
(ii) How do the flow subsystems interchange a mass of pollutant between each other?
(iii) How do the adsorption/desorption processes between the groundwater and the rock matrix delay the pollutant transport within the subsurface?

The questions actually define what is to be understood by the 'subsystems.' The subsystems can be either former flow subsystems or just water and rock matrix.

The first question can be answered by following the paths of each flow subsystem and assigning the calculated mass fraction to each point of the path at every moment of time. In particular, 'input-output' behavior of each flow subsystem can be calculated, thus evaluating how much and how fast the pollutant mass can be transported from the recharge to the discharge regions.

The second question can be answered by drawing imaginary boundaries between the flow subsystems (known from the flow systems analysis) and calculating the dispersive fluxes through these boundaries.

The last question can be answered by recalculating the fluxes from and into the rock matrix.

Concluding remark

It is clear from the two examples shown that, depending on the goal formulation, one may consider *two different systems for the same subsurface.*

Man-influenced systems

As for natural systems, also in the case of man-influenced systems, different goals can be formulated when considering groundwater as an object of practical activity. For instance, the objective could be to find out whether a given abstraction scheme may lead to polluting an exploited aquifer part of the subsurface from another part which is already polluted. On the other hand, it is sometimes important to know from which part of the flow domain the given abstraction well is supplied. This can again be related to the groundwater quality question. In another case, one needs to know whether it is at all possible to create an artificial local groundwater flow subsystem of the given shape. For example, one may consider the task of designing such a system in order to build a 'hydraulic isolation' of the waste disposal site. In all these examples some additional physical objects (like wells) must be added to the natural groundwater flow systems and the behavior of the resultant 'mixed' system needs to be evaluated. For this a mathematical description of the man-made objects and their physical interaction with the natural groundwater system are assumed to be known.

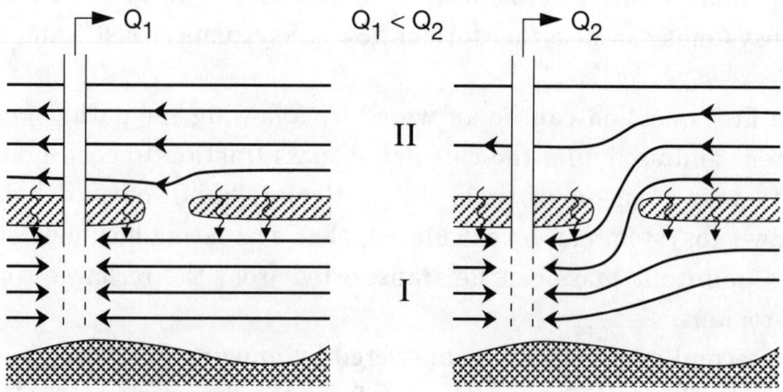

Figure 13. Two flow systems for two abstraction schemes.

Example three

Goal: To find out whether a given abstraction scheme may lead to polluting an exploited aquifer from another aquifer which is already polluted. The two aquifers may influence each other by means of seepage through an aquitard as well as by means of flow through the aquitard's discontinuity – see Figure 13.

Remark. As before, there are a number of assumptions hidden in the notions of 'exploited aquifer' and 'given abstraction scheme.' The former notion may mean that basic characteristics of the aquifer (hydraulic parameters, boundary conditions etc.) are known. It also means that all calculations that have been possible for the natural aquifer are, in principle, possible also in this case. In particular it can mean that the original flow field (i.e., before abstraction has started) can be calculated. The second notion indicates that the antropogenic stresses (pumping rates in this case) imposed on the aquifer are known completely. There is one hidden assumption rather typical for the problem solving situation in which different scenarios of human activities are being simulated. The assumption states that the boundary of the system does not change. This assumption needs to be checked every time the 'mixed' flow system is recalculated for a new abstraction scheme.

System objects:

There are *three objects* in the system:

(i) the aquifer being pumped (aquifer I);
(ii) the polluted aquifer (aquifer II);
(iii) the intermediate aquitard.

The three objects are characterized by the following *attributes*:

(i) flow system for aquifer I;
(ii) flow system for aquifer II;
(iii) flow system for the aquitard.

This means that all variables and the relations between the variables that have been defined for the natural system are also included in a definition of the 'mixed' system.

The *relations* between the attributes are as follows:

(i) the normal component of the groundwater flux between an aquifer and the aquitard is a continuous function;
(ii) the potential along the common boundary between an aquifer and the aquitard is a continuous function.

Environmental objects:

All the environmental objects that have been considered for the natural flow system are classified likewise for the 'mixed' system. Additionally, all the abstraction wells in aquifer I should belong to the set of external objects. It is assumed that they impose a stress on aquifer I and they themselves are not influenced by the groundwater flow phenomena. The pumping scheme (i.e., time series of pumping rates) is assumed to be known perfectly.

For the abstraction wells the following *attributes* are specified:

(i) abstraction rates as functions of time;
(ii) penetration of the wells into aquifer I;
(iii) locations of the wells.

The *relations between the system and the wells (which are part of the environment)* are represented by the mass convervation equation which states that:

> the amount of water that is abstracted from each well is equal to the amount of groundwater entering that well.

A subdivision of the system's attributes (variables) into the four basic categories can be summarized as follows:

(i) ***state variables:***
- potential field;
- three components of the velocity field.

Hence the state variables are exactly the same as for the natural system. This is true only if the assumption of the fixed boundary of the flow system can be retained. Otherwise additional state variables need to be introduced to describe a movement of the boundary.

(ii) ***input variables:***
- the same as for the natural system, plus
- the pumping rates of the abstraction wells.

(iii) ***output variable:***
- groundwater flux between the two aquifers.

The output variable enables one to judge whether aquifer I can possibly be polluted from aquifer II or not.

(iv) ***parameters:***
- all parameters defined for the natural flow system;
- penetration depths of the wells;
- locations of the wells.

It should be noted here that the wells in the example are simplified. They are considered to be described by the second type boundary condition for the classical groundwater flow equation; i.e., their entrance resistivity is neglected, and their nominal abstraction rates are always satisfied by the water entering them from the aquifer (see Section 8.4).

Subsystems:

By solving the state equations and calculating the water trajectories one can make a further subdivision of the 'mixed' flow domain into flow subsystems in the very same way as for the natural systems. In particular, when pollution is possible, groundwater paths that start from the polluted aquifer II and reach aquifer I can be calculated and shown, thus demonstrating the pollution pattern for a given pumping scheme (scenario). Naturally, one can calculate also additional output variables for the family of 'polluted' paths. For instance, the isochrones or the traveling time distribution for the paths that start from aquifer II and reach aquifer I (the isochrones) can be calculated and then used in decision-making.

Chapter 8

Numerical Modeling

8.1. Introduction to numerical approximation methods

Since exact analytical solutions to groundwater flow equations can only be found for idealized situations, the need for quantitative investigation of complex groundwater systems has led to the development and use of numerical methods. They offer approximations to the solutions of the classical or velocity-oriented flow equations that are consistent with the physical interpretations of the variables involved and sufficiently accurate for practical applications. The variables (fields) one is interested in are the fields $\boldsymbol{q}$ and ϕ, where $\boldsymbol{q}$ is the volumetric flow rate and ϕ the potential. The volumetric flow rate $\boldsymbol{q}$ is the quantity of particular interest but, since $\boldsymbol{q}$ is linearly related to the driving force density $\boldsymbol{e} = -\,\mathbf{grad}\,\phi$, we are equally satisfied when we are able to obtain a solution for $\boldsymbol{e}$. The vector field $\boldsymbol{e}$ can be obtained directly from the scalar field ϕ by taking the gradient, where ϕ is obtained from the relatively simple Laplace-type equation $\mathrm{div}(\underline{\boldsymbol{k}}\cdot\mathbf{grad}\,\phi) = 0$. The convenience of this scalar field is so great that we use it whenever there is any excuse not to do otherwise. However, numerical differentiation of the numerically determined scalar field ϕ is a source of many errors.

This section and the sections that follow are *not* meant to be a detailed or elementary introduction to numerical methods which could possibly be used for solving groundwater flow equations. It is, rather, a fast bridging between the theory presented in Chapters 1-6 and its practical applications. The Galerkin *finite element method* – the only numerical method which is actually used in this book – serves as a fast vehicle to reach the goals set by the real-life exploitation of groundwater systems. For an overview of other numerical methods that are commonly used for solving groundwater flow/transport problems the reader can be referred to many excellent textbooks (see, for example, Huyakorn and Pinder, 1983, or Kinzelbach, 1987).

The analysis presented below sharpens all the basic differences and difficulties that are encountered when either of the two equivalent theories of groundwater flow (the classical one and the velocity-oriented approach) are to be translated into numerical models.

Classical approach – general discussion

When a conventional first-order finite element method with piecewise linear continuous interpolations is applied to compute an approximation ϕ_Δ to the exact value ϕ, we find that the Sobolev seminorm $|\phi_\Delta - \phi|_0 = O(\Delta^2)$, where Δ is the maximum linear dimension of the elements. However, for the approximation $\boldsymbol{e}_\Delta$ to the exact value of $\boldsymbol{e} = -\,\mathbf{grad}\,\phi$ we find a piecewise constant approximation for which $|\boldsymbol{e}_\Delta - \boldsymbol{e}|_0 = O(\Delta)$ holds. This means that, when applying grid refinement, the approximation $\boldsymbol{e}_\Delta$ converges only linearly to the exact value $\boldsymbol{e}$. Consequently, when we are approaching the limit $\Delta \rightarrow 0$, we can expect relatively poor approximations $\boldsymbol{e}_\Delta$ to the exact values $\boldsymbol{e} = -\,\mathbf{grad}\,\phi$, even if the approximations ϕ_Δ to the exact values ϕ are sufficiently accurate. Similar considerations also hold for finite difference methods.

Another weakness of differentiation of numerically approximated potentials results from the finite number of significant digits in floating-point arithmetic. The following may serve as an example. Let us suppose that the computer uses floating-point words with seven significant digits. Due to the solution of the linear equations with, for example, a matrix having a condition number of 10^5, five significant digits are lost. The resulting numerically approximated potentials are then represented by only two significant digits. As a consequence the small potential differences in, say, the z-direction, contained in the third digit of the potentials, are lost in the numerical approximation of the potentials. On the other hand, the potential differences in, say, the x-direction, contained in the first two digits of the potential, are reflected with sufficient accuracy in the numerical approximation of the potentials. As a result, we find a meaningful approximation to the velocity component in the x-direction, and a totally wrong result for the velocity component in the z-direction (see also Section 3.5, Figure 4).

The best remedy to cure this error would be to use floating-point arithmetic with a sufficient number of significant digits (e.g., double precision). However, on some computers double precision calculations are cumbersome and relatively expensive in terms of memory requirement and computation time. Furthermore, when using iterative methods to solve the linear equations, we should not terminate iterations too early (say, after having calculated the first two significant digits), otherwise we would have the same trouble in finding the small velocity component in the z-direction. On the other hand, termination of the calculations after an adequate number of iterations will dramatically increase the computation time and is therefore more expensive.

From the above arguments it may be concluded that a numerical method that calculates velocity using the first significant digits for both the relatively large velocity component in horizontal directions and the relatively small velocity component in the vertical direction, may be considered as a more robust method than the conventional one.

Velocity-oriented approach – general discussion

In principle, there are a number of possible ways to overcome the difficulties encountered by numerical differentiation, and research in the field is gradually becoming popular in the geohydrological community. In this book we discuss the velocity-oriented approach in which the vectors $\boldsymbol{q}$ and $\boldsymbol{e}$ are expressed in terms of their three scalar components along the unit axes of the coordinate system appropriate for the boundary. The next step will then be to solve the resulting scalar component equations by the conventional finite element or finite difference techniques. But then we run into a complication. We will see that the three component equations for $\boldsymbol{q}$ and $\boldsymbol{e}$ do not, in general, separate into three equations for the three components alone. We are faced with a set of three coupled equations, each equation involving all three components. This means that the matrix obtained from discretization of the equations is three times as large as the matrix obtained from the classical approach. And, even worse, this matrix turns out to be not positive definite, which means that well-known iterative methods for solving systems of linear equations cannot be applied. Although it is not impossible to solve nonpositive definite systems, their solving involves relatively long computation times compared with solving matrix equations obtained from discretization of Laplace-type equations. Therefore, for practical applications, we should avoid nonpositive definite matrices.

The above complications can be removed by physical considerations related to perfect layering which approximately prevails in many sedimentary basins. Taking the Cartesian coordinate system with coordinates x, y and

z, and Cartesian boundary planes, we assume quasi-isotropy in the x-y planes, one principal axis of the conductivity tensor in the z-direction and layering of the porous medium (conductivity is only a function of z). An immediate implication of the last assumption for flow in porous medium is that e_x, e_y and q_z are continuously varying functions, even if the conductivity is a discontinuous function of z. This latter observation suggests that e_x, e_y and q_z are 'natural' components of the two vector fields $\boldsymbol{q}$ and $\boldsymbol{e}$. Indeed, if we derive equations for e_x, e_y and q_z, we obtain three uncoupled Laplace-type equations and boundary conditions resulting in three uncoupled systems of linear algebraic equations with three positive definite matrices. From a practical point of view this is a very important result since in most geohydrological situations the subsurface is layered, though with some deviations from perfect layering. In the latter case our task is reduced to finding a solution procedure to account for the deviations.

In conclusion, the approach presented here to avoid numerical differentiation is well-designed for approximately layered porous media which are, fortunately, frequently encountered in nature. In geohydrological applications, for which layering is considered a good approximation of reality, the method yields excellent results. However, the method is no longer attractive for problems where inhomogeneities in conductivity are equally distributed in all directions.

8.2. The Galerkin finite element method

Classical approach – finite element method

The Galerkin finite element approximation to the solution of the classical incompressible flow equation in terms of the potential:

$$\mathrm{div}(\underline{\boldsymbol{k}} \cdot \mathbf{grad}\, \phi) + S = 0 \tag{8.2.1}$$

where ϕ is the potential [Pa; $dbar \approx m$], $\underline{\boldsymbol{k}}$ is the conductivity tensor [$m^2 \cdot Pa^{-1} \cdot s^{-1}$; $m^2 \cdot dbar^{-1} \cdot d^{-1} \approx m \cdot d^{-1}$] and S is the intensity of the source [s^{-1}; d^{-1}], with boundary conditions of general form $\alpha_1 \phi + \alpha_2 q_n =$ $= 1$ (q_n being normal flux) can be obtained with the following procedure consisting of four steps:

(i) Flow domain Ω is subdivided into finite elements – in this case into tetrahedrons (see Section 9.1). The vertices of tetrahedrons form a finite element grid which discretizes the flow domain.

(ii) Family of basic functions $\{\Phi_i : i = 1, \ldots, N\}$ is defined in Ω – see Appendix E. Also the approximation to the solution of the flow equation (8.2.1) is assumed in the form of linear combinations of basic functions:

$$\hat{\phi}(x, y, z) = \sum_{j=1}^{N} \phi_j \cdot \Phi_j(x, y, z) \tag{8.2.2}$$

where ϕ_j is the unknown value of the potential in the j-th node of the finite element grid, $(j = 1, \ldots, N)$.

(iii) After substituting $\hat{\phi}$ for ϕ, equation (8.2.1) is multiplied by the i-th basic function and integrated over Ω which results in:

$$\int_\Omega \operatorname{div}(\underline{\boldsymbol{k}} \cdot \mathbf{grad}\, \hat{\phi})\ \Phi_i\ \mathrm{d}\Omega + \int_\Omega S\Phi_i\ \mathrm{d}\Omega = 0 \tag{8.2.3}$$

for $i = 1, \ldots, N$.

(iv) Approximation (8.2.2) is explicitly substituted in equations (8.2.3) and Gauss's theorem is applied to every equation. This results in N algebraic equations with N unknowns $\phi_1, \ldots, \phi_N$:

$$\begin{aligned}\int_{\partial\Omega} q_n \Phi_i\ \mathrm{d}A - \sum_{j=1}^{N} \phi_j \int_\Omega \Big[k_x\ (\partial\Phi_i/\partial x)(\partial\Phi_j/\partial x) + \\ + k_y\ (\partial\Phi_i/\partial y)(\partial\Phi_j/\partial y) + k_z\ (\partial\Phi_i/\partial z)(\partial\Phi_j/\partial z)\Big]\ \mathrm{d}\Omega + \\ + \int_\Omega S\Phi_i\ \mathrm{d}\Omega = 0, \qquad (i = 1, \ldots, N).\end{aligned} \tag{8.2.4}$$

Equations (8.2.4) can equivalently be written in a matrix form:

$$\underline{\boldsymbol{P}} \cdot \boldsymbol{\Phi} = \boldsymbol{b} \tag{8.2.5}$$

where $\boldsymbol{\Phi} = (\phi_1, \ldots, \phi_N)^T$ is the vector of unknowns, $\boldsymbol{b} = (b_1, \ldots, b_N)^T$ is the RHS-vector with its elements

$$b_i = \int_{\partial\Omega} q_n \Phi_i \, \mathrm{d}A + \int_{\Omega} S\Phi_i \, \mathrm{d}\Omega, \qquad (i = 1, \ldots, N) \tag{8.2.6}$$

and $\underline{\boldsymbol{P}} = \left\{ P_{ij} \right\}$ is the matrix of the geometric and hydraulic properties of the flow domain with its elements

$$\begin{aligned} P_{ij} &= \int_{\Omega} \Big[k_x \, (\partial\Phi_i/\partial x)(\partial\Phi_j/\partial x) + k_y \, (\partial\Phi_i/\partial y)(\partial\Phi_j/\partial y) + \\ &\quad + k_z \, (\partial\Phi_i/\partial z)(\partial\Phi_j/\partial z) \Big] \, \mathrm{d}\Omega = \\ &= \int_{\Omega} \Big[k_x \beta_i \beta_j + k_y \gamma_i \gamma_j + k_z \delta_i \delta_j \Big] \, \mathrm{d}\Omega, \ (i, j = 1, \ldots, N) \end{aligned} \tag{8.2.7}$$

where $\beta_i, \gamma_i, \delta_i$ are the coefficients associated with the i-th basic function Φ_i – see Appendix E, formulae (E5)-(E9).

If the conductivity $\underline{\boldsymbol{k}}$ can be considered elementwise constant, then analytical formulae can be used for calculating elements P_{ij} of the global matrix $\underline{\boldsymbol{P}}$:

$$P_{ij} = \sum_e{}' p^e_{lm}, \qquad (i, j = 1, \ldots, N) \tag{8.2.8}$$

where the prime ($'$) indicates the elementwise assembling operator which allows us to break the procedure of assembling the global matrix $\underline{\boldsymbol{P}}$ into smaller tasks of assembling local matrices $\underline{\boldsymbol{P}}^e$, and p^e_{lm}. Elements of the local matrix $\underline{\boldsymbol{P}}^e$ are given by analytical formula:

$$p^e_{lm} = V^e \left[k^e_x \beta^e_l \beta^e_m + k^e_y \gamma^e_l \gamma^e_m + k^e_z \delta^e_l \delta^e_m \right], \quad (l, m = 1, 2, 3, 4) \tag{8.2.9}$$

where V^e is the volume of the e-th finite element, $\underline{\boldsymbol{k}}^e = (k^e_x, k^e_y, k^e_z)$ is the conductivity of the e-th finite element, and β^e_l, γ^e_l, δ^e_l $(l = 1, 2, 3, 4)$ are the coefficients of the basic functions associated with the four nodes of the e-th finite element – see Appendix E.

Elements of the RHS-vector $\boldsymbol{b}$ can also be calculated elementwise, i.e:

$$\boldsymbol{b} = \sum_e{}' \boldsymbol{b}^e \tag{8.2.10}$$

where $\boldsymbol{b}^e = (b_1^e, b_2^e, b_3^e, b_4^e)$ are the local RHS-vectors and:

$$b_l^e = \int_{\partial\Omega^e} q_n \Phi_l \, \mathrm{d}A + \int_{\Omega^e} S\Phi_l \, \mathrm{d}\Omega \tag{8.2.11}$$

where $l = 1, 2, 3, 4$ are the local numbers of the basic functions associated with the e-th finite element, Ω^e is the part of the flow domain Ω occupied by the e-th element, and $\partial\Omega^e = \bar{\Omega}^e \cap \partial\Omega$ is the part of the flow domain boundary $\partial\Omega$ that coincides with the element e.

It is assumed in formula (8.2.11) that both functions q_n, representing normal volumetric flow rate along (some part of) the flow domain boundary, and S, representing wells or drains operating within Ω are available; i.e., they can be effectively estimated for a given hydrogeological situation. The Neumann-type boundary conditions (q_n-specified on some part of $\partial\Omega$) are incorporated in a natural way into the finite element formulation; see formula (8.2.6). In the case of boundary conditions of the Dirichlet-type along (some part of) the flow domain boundary, the term containing q_n is set to zero. More discussion on how the boundary conditions are taken into account and incorporated into equations (8.2.5) can be found in Sections 8.3 and 9.2.

Velocity-oriented approach (VOA) – finite element method

If the porous medium is isotropic, the VOA equations and the VOA boundary conditions can be written as follows:

$$\begin{cases} \operatorname{div}(k \; \mathbf{grad}\, e_x) & = k e_x \; \partial^2\alpha/\partial x^2 + k e_y \; \partial^2\alpha/\partial x\partial y + \\ & \quad + q_z \; \partial^2\alpha/\partial x\partial z \\ \operatorname{div}(k \; \mathbf{grad}\, e_y) & = k e_x \; \partial^2\alpha/\partial x\partial y + k e_y \; \partial^2\alpha/\partial y^2 + \\ & \quad + q_z \; \partial^2\alpha/\partial y\partial z \\ \operatorname{div}(k^{-1} \; \mathbf{grad}\, q_z) & = e_x \; \partial^2\alpha/\partial x\partial z + e_y \; \partial^2\alpha/\partial y\partial z + \\ & \quad - k^{-1} \; q_z \; \left(\partial^2\alpha/\partial x^2 + \partial^2\alpha/\partial y^2\right) \end{cases} \tag{8.2.12}$$

where e_x, e_y, q_z are the (transformed) components of the flow velocity and $\alpha = -\ln k = -\lambda$; see equations (3.4.2).

The Galerkin finite element approximation to equations (8.2.12) with hydrogeological boundary conditions of general form $\alpha_1\phi + \alpha_2 q_n = 1$ (q_n

being the normal flux) and additional boundary conditions required by the theory can be obtained with a procedure similar to that for the classical approach:

(i) The flow domain Ω is subdivided into finite elements – in this case tetrahedrons (see Chapter 9). The vertices of tetrahedrons form an N-element grid which discretizes the flow domain.

(ii) The family of basic functions $\{\Phi_i : i = 1, \ldots, N\}$ is defined in Ω – see Appendix E. Also the approximations to the solution of the flow equations (8.2.12) are assumed in the form of linear combinations of basic functions:

$$\begin{cases} \hat{e}_x(x,y,z) = \sum_{j=1}^{N} e_{xj}\ \Phi_j(x,y,z) \\ \hat{e}_y(x,y,z) = \sum_{j=1}^{N} e_{yj}\ \Phi_j(x,y,z) \\ \hat{q}_z(x,y,z) = \sum_{j=1}^{N} q_{zj}\ \Phi_j(x,y,z) \end{cases} \tag{8.2.13}$$

where e_{xj}, e_{yj}, q_{zj} are the unknown values of (transformed) velocity components in the j-th node of the finite element grid, $(j = 1, \ldots, N)$.

(iii) After substituting $\hat{e}_x, \hat{e}_y, \hat{q}_z$ for e_x, e_y, q_z, equations (8.2.12) are multiplied by the i-th basic function and integrated over Ω which results in:

$$\begin{aligned} &\int_\Omega \operatorname{div}(k\ \mathbf{grad}\,\hat{e}_x)\ \Phi_i\ \mathrm{d}\Omega = \tilde{b}_{xi} \\ &\int_\Omega \operatorname{div}(k\ \mathbf{grad}\,\hat{e}_y)\ \Phi_i\ \mathrm{d}\Omega = \tilde{b}_{yi} \\ &\int_\Omega \operatorname{div}(k^{-1}\ \mathbf{grad}\,q_z)\ \Phi_i\ \mathrm{d}\Omega = \tilde{b}_{zi}, \quad \text{for } i = 1, \ldots, N \end{aligned} \tag{8.2.14}$$

where:

$$\tilde{b}_{xi} = \int_{\Omega_i} \left[ke_x\ \partial^2\alpha/\partial x^2 + ke_y\ \partial^2\alpha/\partial x\partial y + q_z\ \partial^2\alpha/\partial x\partial z\right]\ \Phi_i\ \mathrm{d}\Omega$$

$$\tilde{b}_{yi} = \int_{\Omega_i} \left[ke_x\ \partial^2\alpha/\partial x\partial y + ke_y\ \partial^2\alpha/\partial y^2 + q_z\ \partial^2\alpha/\partial y\partial z\right]\ \Phi_i\ \mathrm{d}\Omega$$

$$\tilde{b}_{zi} = \int_{\Omega_i} \Big[e_x\ \partial^2\alpha/\partial x\partial z + e_y\ \partial^2\alpha/\partial y\partial z + \\ - k^{-1}\ q_z\ \left(\partial^2\alpha/\partial x^2 + \partial^2\alpha/\partial y^2\right)\Big]\ \Phi_i\ \mathrm{d}\Omega. \qquad (8.2.15)$$

(iv) Approximations (8.2.13) are explicitly substituted in equations (8.2.14) and Gauss's theorem is applied to every equation. This results in $3N$ algebraic equations with $3N$ unknowns $e_{x1},\ldots,e_{xN}$, $e_{y1},\ldots,e_{yN}$, $q_{z1},\ldots,q_{zN}$:

$$\int_{\partial\Omega_i} k\ \partial e_x/\partial n\ \Phi_i\ \mathrm{d}A + \\ - \sum_{j=1}^{N} e_{xj} \int_{\Omega_{ij}} k\ \Big[(\partial\Phi_i/\partial x)(\partial\Phi_j/\partial x) + (\partial\Phi_i/\partial y)(\partial\Phi_j/\partial y) + \\ + (\partial\Phi_i/\partial z)(\partial\Phi_j/\partial z)\Big]\ \mathrm{d}\Omega = \tilde{b}_{xi}, \quad (i = 1,\ldots,N) \qquad (8.2.16a)$$

$$\int_{\partial\Omega_i} k\ \partial e_y/\partial n\ \Phi_i\ \mathrm{d}A + \\ - \sum_{j=1}^{N} e_{yj} \int_{\Omega_{ij}} k\ \Big[(\partial\Phi_i/\partial x)(\partial\Phi_j/\partial x) + (\partial\Phi_i/\partial y)(\partial\Phi_j/\partial y) + \\ + (\partial\Phi_i/\partial z)(\partial\Phi_j/\partial z)\Big]\ \mathrm{d}\Omega = \tilde{b}_{yi}, \quad (i = 1,\ldots,N) \qquad (8.2.16b)$$

$$\int_{\partial\Omega_i} k^{-1}\ \partial q_z/\partial n\ \Phi_i\ \mathrm{d}A + \\ - \sum_{j=1}^{N} q_{zj} \int_{\Omega_{ij}} k^{-1}\ \Big[(\partial\Phi_i/\partial x)(\partial\Phi_j/\partial x) + (\partial\Phi_i/\partial y)(\partial\Phi_j/\partial y) + \\ + (\partial\Phi_i/\partial z)(\partial\Phi_j/\partial z)\Big]\ \mathrm{d}\Omega = \tilde{b}_{zi}, \quad (i = 1,\ldots,N) \qquad (8.2.16c)$$

where $\Omega_{ij} = \Omega_i \cap \Omega_j$ (intersection of supports for Φ_i and Φ_j) and $\partial\Omega_i = \partial\Omega \cap \Omega_i$.

Equations (8.2.16) can equivalently be written in a matrix form:

$$\underline{\boldsymbol{E}}_x \boldsymbol{\cdot} \boldsymbol{e}_x = \boldsymbol{b}_x \tag{8.2.17a}$$

$$\underline{\boldsymbol{E}}_y \boldsymbol{\cdot} \boldsymbol{e}_y = \boldsymbol{b}_y \tag{8.2.17b}$$

$$\underline{\boldsymbol{Q}}_z \boldsymbol{\cdot} \boldsymbol{q}_z = \boldsymbol{b}_z \tag{8.2.17c}$$

where

$$\left.\begin{array}{l} \boldsymbol{e}_x = (e_{x1}, \ldots, e_{xN}) \\ \boldsymbol{e}_y = (e_{y1}, \ldots, e_{yN}) \\ \boldsymbol{q}_z = (q_{z1}, \ldots, q_{zN}) \end{array}\right\} : \quad \text{vectors of unknowns}$$

$$\left.\begin{array}{l} \boldsymbol{b}_x = (b_{x1}, \ldots, b_{xN}) \\ \boldsymbol{b}_y = (b_{y1}, \ldots, b_{yN}) \\ \boldsymbol{b}_z = (b_{z1}, \ldots, b_{zN}) \end{array}\right\} : \quad \text{RHS-vectors with}$$

$$\begin{aligned} b_{xi} &= -\tilde{b}_{xi} + \textstyle\int_{\partial\Omega_i} k\ \partial e_x/\partial n\ \Phi_i\ dA \\ b_{yi} &= -\tilde{b}_{yi} + \textstyle\int_{\partial\Omega_i} k\ \partial e_y/\partial n\ \Phi_i\ dA \\ b_{zi} &= -\tilde{b}_{zi} + \textstyle\int_{\partial\Omega_i} k^{-1}\ \partial q_z/\partial n\ \Phi_i\ dA \end{aligned} \tag{8.2.18a}$$

and $\tilde{b}_{xi}$, $\tilde{b}_{yi}$, $\tilde{b}_{zi}$ given by formulae (8.2.15).

$$\left.\begin{array}{l} \underline{\boldsymbol{E}}_x = \left\{E_{xij}\right\} \\ \underline{\boldsymbol{E}}_y = \left\{E_{yij}\right\} \\ \underline{\boldsymbol{Q}}_z = \left\{Q_{zij}\right\} \end{array}\right\} : \quad \begin{array}{l}\text{matrices of geometric and hydraulic} \\ \text{properties of the flow domain with}\end{array}$$

$$
\begin{aligned}
E_{xij} &= E_{yij} = \\
&= \int_{\Omega} k \Big[(\partial\Phi_i/\partial x)(\partial\Phi_j/\partial x) + (\partial\Phi_i/\partial y)(\partial\Phi_j/\partial y) + \\
&\qquad + (\partial\Phi_i/\partial z)(\partial\Phi_j/\partial z) \Big] \, d\Omega = \\
&= \int_{\Omega} k \Big[\beta_i\beta_j + \gamma_i\gamma_j + \delta_i\delta_j \Big] \, d\Omega, \; (i,j = 1,\ldots,N). \quad (8.2.18b)
\end{aligned}
$$

Formulae for Q_{zij} can readily be obtained by exchanging k^{-1} for k in (8.2.18b).

Similarly, as for the classical approach, all the integrations necessary for assembling the matrices $\underline{\boldsymbol{E}}_x$, $\underline{\boldsymbol{E}}_y$, $\underline{\boldsymbol{Q}}_z$ and the RHS-vectors $\boldsymbol{b}_x$, $\boldsymbol{b}_y$, $\boldsymbol{b}_z$ can be done elementwise. In addition, considerable simplifications in the assembling procedure are possible when the conductivity k can be considered constant in each finite element. Then the formulae similar to these (8.2.8)-(8.2.11) apply for calculating the corresponding local matrices.

Boundary conditions for the VOA theory of groundwater flow have especially simple form when (segments of) the boundary can be considered horizontal or vertical. As for the classical approach, the VOA boundary conditions are formulated in terms of specified values of e_x, e_y, q_z or their normal derivatives or linear combinations of the two. Once specified they can be readily used within the finite element approximation of the VOA equations.

The VOA boundary conditions are then formulated as follows:

Top/bottom boundary (assumed horizontal)
Position of the boundary is characterized by the normal vector $\underline{\boldsymbol{n}} = (0, 0, n_z)$ having only one nonzero component n_z – see Figure 14.

If ϕ is specified on the boundary then:

$$
\begin{cases}
e_x = -\partial\phi/\partial x \\
e_y = -\partial\phi/\partial y \\
\partial q_z/\partial n = n_z \left(\partial/\partial x \;\; (k \; \partial\phi/\partial x) + \partial/\partial y \, (k \; \partial\phi/\partial y) \right)
\end{cases}
\quad (8.2.19)
$$

i.e., the first type boundary conditions need to be specified for the horizontal components of the flow and the second type boundary conditions must be prescribed for the vertical component.

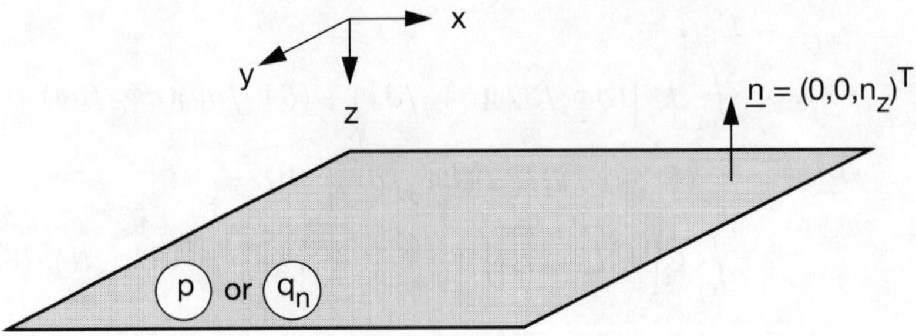

Figure 14. Boundary conditions on the horizontal boundary.

If q_n is specified on the boundary then:

$$\begin{cases} \partial e_x/\partial n = \partial/\partial x\left(k^{-1}\ q_n\right) \\ \partial e_y/\partial n = \partial/\partial y\left(k^{-1}\ q_n\right) \\ q_z = n_z q_n \end{cases} \tag{8.2.20}$$

i.e., the second type boundary conditions for horizontal components of the flow and the first type boundary conditions for the vertical component must be specified.

Side boundary (assumed vertical)
Position of this boundary is characterized by the normal vector $\underline{n}$ = $= (n_x, n_y, 0)$ having one or two nonzero components. In addition to this a local system of coordinates (ξ_1, ξ_2, ξ_3) can be introduced with ξ_2 assumed vertical and ξ_3 normal to the boundary – see Figure 15.

If ϕ is specified on the boundary then:

$$\begin{cases} q_z = -k\ \partial\phi/\partial\xi_2 \\ e_y = -n_x\ \partial\phi/\partial\xi_1 \\ \partial e_x/\partial\xi_3 + k^{-1}\ (\partial k/\partial\xi_3)\ e_x = \\ \quad = n_x\ k^{-1}\ \{\partial/\partial\xi_1\,(k\ \partial\phi/\partial\xi_1) + \partial/\partial\xi_2\,(k\ \partial\phi/\partial\xi_2)\} \end{cases} \tag{8.2.21}$$

i.e., the first type boundary conditions must be specified for q_z and e_y components of flow whereas for e_x the third type boundary conditions need to be prescribed.

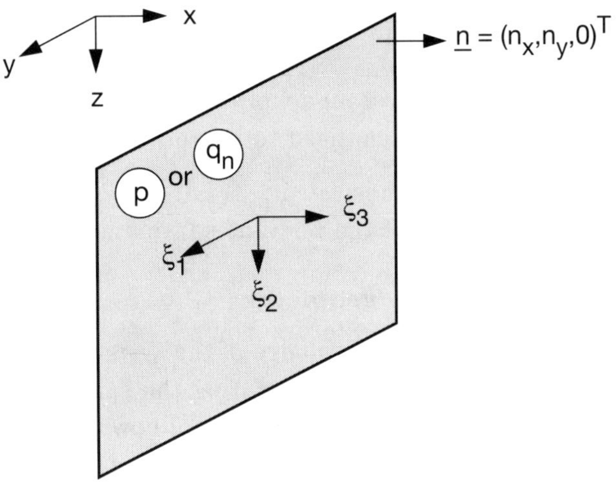

Figure 15. Boundary conditions on the side (vertical) boundary.

If q_n is specified on the boundary then:

$$\begin{cases} e_x = n_x \; k^{-1} \; q_n \\ \partial e_y / \partial n = n_x \; \partial(k^{-1} \; q_n)/\partial \xi_1 \\ \partial q_z / \partial n + k \; (\partial k^{-1} / \partial \xi_3) \; q_z = k \; \partial(k^{-1} \; q_n)/\partial \xi_2 \end{cases} \tag{8.2.22}$$

i.e., the first, second and third type boundary conditions need to be specified for e_x, e_y and q_z, respectively.

Within the finite element procedure all three types of boundary conditions can be routinely incorporated into the formulation of the resultant algebraic equations (8.2.17). The *first type boundary conditions* are usually taken into account by forcing the solver not to modify the unknowns having their values specified. *The second type boundary conditions* are already included in the finite element formulation – see formulae (8.2.18a). Incorporating the *third type boundary conditions* is generally more difficult for the VOA flow equations. For the special case of horizontally layered subsurface they can, however, be treated relatively simply – see Section 10.3.

Algebraic equations (8.2.17), after being solved by any solver, result in the sought approximations of the (transformed) groundwater flow components e_x, e_y, q_z or, after recalculation, in the original components q_x, q_y and q_z.

8.3. Numerical characterization of the flow domain

When constructing a numerical model of the groundwater system, two major characteristics of the system need to be known (approximated):

(i) overall geometry of the system;
(ii) spatial distribution of the relevant parameters (e.g., conductivity).

Model of the flow domain geometry

In most practical situations a knowledge of the systems geology and hydrogeology provides the basis for discriminating the spatial extent of what is later considered the flow domain. A problem of how to validate experimentally the assumed geometrical model of the subsurface system lies outside the scope of this book.

At present it is common practice to specify the top and bottom boundaries of the system as being horizontal, whereas the side boundaries are usually assumed to be vertical. As the planar view of such a system can be of arbitrary shape, the overall geometry of the flow domain can be called a *generalized cylinder*. For this form of the system a specific finite element grid generator can be applied – see Section 9.1.

Model of conductivity

When conductivity k changes in space arbitrarily analytical formulae that result from the Galerkin finite element formulation (Section 8.2) cannot be applied and, in general, numerical integrations must be performed in order to calculate elements of the global matrices $\underline{\boldsymbol{P}}$, $\underline{\boldsymbol{E}}_x$, $\underline{\boldsymbol{E}}_y$ or $\underline{\boldsymbol{Q}}_z$, as well as the corresponding RHS vectors. Still, if a linear model of conductivity can be assumed, some other analytical formulae can be found and the numerical integration avoided. Indeed, if $k = k(x, y, z)$ is elementwise linear then

$$k^e(x, y, z) = \sum_{l=1}^{4} k_l^e \, \Phi_l(x, y, z), \qquad (e = 1, \ldots, N_e) \tag{8.3.1}$$

where k_l^e $(l = 1, 2, 3, 4)$ are the values of conductivity at four nodes of the e-th element, Φ_l $(l = 1, 2, 3, 4)$ is the l-th basic function, and N_e is the number of elements.

With the above definition one can calculate integrals for the classical and VOA equations (8.2.7) and (8.2.18).

Classical approach – assembling flow equations

By assuming isotropy and substituting (8.3.1) to (8.2.7)-(8.2.8) one obtains the following expression for elements p_{ij}^e of the *local matrix* $\underline{P}^e$:

$$p_{ij}^e = \int_{\Omega^e} \sum_{l=1}^{4} k_l^e \left[\beta_i\beta_j + \gamma_i\gamma_j + \delta_i\delta_j\right]^e \Phi_l \, d\Omega =$$

$$= \left[\beta_i\beta_j + \gamma_i\gamma_j + \delta_i\delta_j\right]^e \sum_{l=1}^{4} k_l^e \int_{\Omega^e} \Phi_l \, d\Omega.$$

Hence,

$$p_{ij}^e = \left[\beta_i\beta_j + \gamma_i\gamma_j + \delta_i\delta_j\right]^e \sum_{l=1}^{4} k_l^e / 4$$

or

$$p_{ij}^e = \left[\beta_i^e\beta_j^e + \gamma_i^e\gamma_j^e + \delta_i^e\delta_j^e\right] \hat{k}^e, \quad (i,j = 1,2,3,4) \qquad (8.3.2)$$

where $\beta_i^e, \gamma_j^e, \delta_i^e$ are the coefficients related to the i-th basic function restricted to the e-th element and $\hat{k}^e = 1/4 \sum_{l=1}^4 k_l^e$ is the (arithmetic) mean conductivity within the e-th element (tetrahedron).

The elements of the global matrix $\underline{P}$ can be calculated elementwise out of the element p_{ij}^e given by formula (8.3.2). Model (8.3.1) does not change boundary conditions for the classical formulation of groundwater flow but it does for the following VOA formulation.

Velocity-oriented approach – assembling flow equations

Assembling formulae for the global matrices $\underline{E}_x$ and $\underline{E}_y$ are identical to those for the matrix $\underline{P}$; see (8.2.7) and (8.2.18). Therefore, only the assembling of the matrix $\underline{Q}_z$ is discussed here.

Since k^{-1} is the parameter for the third equation of (8.2.12) and linear model (8.3.1) is assumed for the conductivity k, the global matrix $\underline{Q}_z$ can be assembled out of the following local elements in the following way:

$$Q_{ij} = \sum_{e}{}' \, q_{ij}^e \qquad (8.3.3)$$

where

$$q_{ij}^e = \left[\beta_i^e \beta_j^e + \gamma_i^e \gamma_j^e + \delta_i^e \delta_j^e\right] J_e \tag{8.3.4}$$

$$J_e = \int_{\Omega^e} \mathrm{d}\Omega \Big/ \sum_{l=1}^{4} k_l^e \, \Phi_l(x,y,z). \tag{8.3.5}$$

A closed formula has been found for calculating the integral (8.3.5) analytically. It reads (see Nawalany, 1989):

$$J_e = 6V^e \; (1/2) \sum_{l=1}^{4} k_l^e \ln k_l^e / \Pi_{m \neq l}(k_l - k_m). \tag{8.3.6}$$

It can be shown that formula (8.3.6) has the following good features:

(i) It is positive.

(ii) Integral J_e is independent of cyclic permutation of the indices of nodes in the element.

(iii) It changes reciprocally if conductivity k is scaled by a factor m, i.e.,

$$J_e(mk_1, mk_2, mk_3, mk_4) = m^{-1} \; J_e(k_1, k_2, k_3, k_4). \tag{8.3.7}$$

(iv) In the limiting process it gives the proper formulae for J_e when for instance $k_1 \to k_2$, or $k_1 \to k_2 \neq k_3 \to k_4$, etc. In the particular case, when $k_1 = k_2 = k_3 = k_4$, the formula has its limit $k_1 \to k_2 = k_3 = k_4 = k_0$ equal to

$$J_e(k_0, k_0, k_0, k_0) = V_e / k_0. \tag{8.3.8}$$

Analytical formulae for all possible cases have been derived by Nawalany (1989b). In the computer realization of the VOA equations, assembling of matrix $\boldsymbol{Q}_z$ can be proceeded by checking which of the cases takes place and hence which formula must be used for a given element. For practical reasons a simpler solution has been taken: unless $k_1 = k_2 = k_3 = k_4$ (formula 8.3.8) or all k_1, k_2, k_3, k_4 are different *per se* (formula 8.3.6), the nodal values of k which are equal to each other are changed (disturbed) by a small percentage of their values so that the equality does not hold. Then formula (8.3.6) applies without the danger of real overflow. Computer experiments have shown that, by changing values of k by 1% (which is acceptable for practitioners), one does not change the value of J_e by more than 1%.

8.4. Numerical characterization of man's activity

When considering numerical models for regional flow, the major human activity that needs to be taken into consideration is the abstraction of groundwater from wells. In the classical approach the wells can be adequately approximated by the Dirac-delta functions in the right-hand side of the continuity equation. Such approximations, although seemingly far from the way real wells do operate, are successfully used in many numerical models. If S represents an intensity of the source/sink (as in equation 8.2.1) it can be related to the yield of the well through the following formulae:

$$S(x,y,z) = Q\ \delta(x - x_p,\ y - y_p,\ z - z_p) \tag{8.4.1}$$

where Q $[m^3 \cdot s^{-1}]$ is the production of the well, (x_p, y_p, z_p) is a location of the well's screen and $\delta(\cdot,\cdot,\cdot)$ is the three-dimensional Dirac-delta function – see Figure 16-(a).

If, on the other hand, a screen of a vertically drilled well with lateral position (x_w, y_w) must be described more realistically (i.e., by taking into account its finite vertical length) then:

$$S(x,y,z) = Q\ \delta(x - x_w,\ y - y_w)\ \eta(z; z_t, z_b)/L_w \tag{8.4.2}$$

where L_w is the length of the screen $[m]$, $\delta(\cdot,\cdot)$ is the two-dimensional Dirac-delta function $[m^{-2}]$ and $\eta(z; z_t, z_b)$ is the step function equal to zero everywhere except in the interval $[z_t, z_b]$ in which $\eta = 1$ $[-]$ (see Figure 16-(b)).

The above approach is capable of keeping the global water balance of the groundwater system (with operating wells) satisfied. Moreover, the 'far distance' approximation of the potential can still be sufficiently good. Naturally, when approaching wells closer the approximation cannot reproduce a behavior of the depression cone realistically.

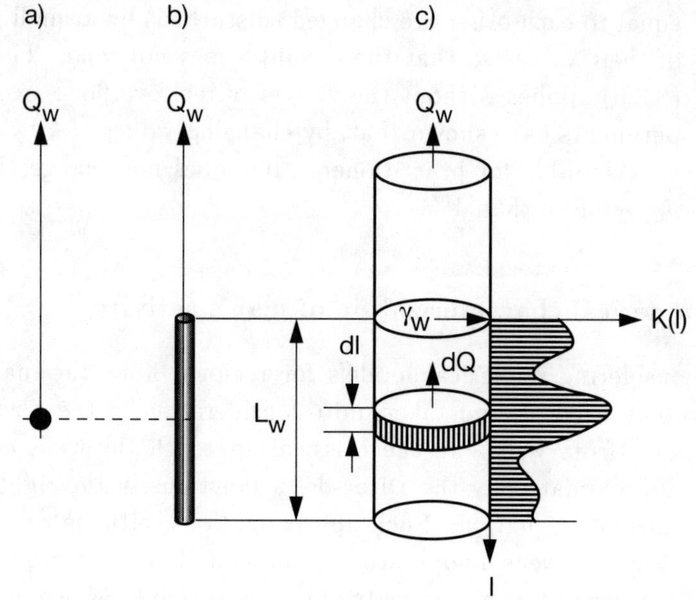

Figure 16. Three models of the well.

The problem of representing the operating abstraction wells in the VOA approach is more difficult than in the classical one. In principle, wells are lying outside the flow domain (outside the porous medium); therefore they should be treated like external boundaries. Consequently, when trying to model wells in a framework of the finite element technique, one must remove several elements (tetrahedrons) from a discretization grid since some of them are lying outside the porous medium. This creates complications, for some of the connections between vertices of the tetrahedrons must be invalidated. Still, it can be done relatively easily. It is more difficult to judge what kinds of boundary conditions need to be specified along the sides (walls) of the validated tetrahedron. A detailed discussion of the *general* boundary conditions for wells will not be presented here.

In the heuristic approach shown below (see Figure 16-(c)) it was assumed that:

- (x_w, y_w) coordinates of an operating well coincide with (x, y) coordinates of some finite element nodes;
- yield Q of the well is known;
- vertical positions of the top and bottom of the well's screen are

known (z_t, z_b); so is the length of the screen $L_w = |z_t - z_b|$ and its diameter r_w;

- total yield of the well is related to the conductivity distribution around the well and to a gradient of potential:

$$Q = \int_0^{L_w} 2\pi r_w \left(-k(l)\ \partial\phi/\partial r\right)_{r=r_w}\ \mathrm{d}l \tag{8.4.3}$$

- *assumption* is made that $\partial\phi/\partial r = -e_r$ is constant along a perimeter of the well. This is the Dupuit-approximation, which is accurate for fully penetrating wells in aquifers; see Section 4.2.

Then:

$$(e_r)_{r=r_w} = -\left(\partial\phi/\partial r\right)_{r=r_w} = (Q/2\pi r_w) \Big/ \int_0^{L_w} k(l)\ \mathrm{d}l. \tag{8.4.4}$$

Consequently, an infinitesimal contribution of a $\mathrm{d}l$-length of the screen to the total yield is equal to:

$$\mathrm{d}Q = Q\ k(l)\ \mathrm{d}l/(L_w \bar{k}) \tag{8.4.5}$$

where $\bar{k}$ is the average conductivity along the screen and

$$\bar{k} = L_w^{-1} \int_0^{L_w} k(l)\ \mathrm{d}l. \tag{8.4.6}$$

Actually, a finite version of formula (8.4.5) is normally used; the screen is subdivided into N_s sections, each having a length equal to $\Delta l = L_w/N_s$. Outside the screen the corresponding values of the conductivity are equal to k_i, $i = 1, \ldots, N_s$. Each section of the screen is then considered the point sink which can be handled easily when solving potential equations (classical formulation) for the *local system*:

$$(\Delta Q_i) = Q/L_w\ k_i(\Delta l)_i/\bar{k}, \qquad (i = 1, \ldots, N). \tag{8.4.7}$$

- After the potentials are calculated in a vicinity of the well, the values of e_x, e_y and q_z in the neighboring nodes can be correspondingly evaluated using Darcy's law.
- When solving the VOA equations a *simplistic assumption* is made that the precalculated values of e_x, e_y and q_z in the vicinity of the well can be used as the Dirichlet-type boundary conditions.

Chapter 9

Computer Aspects

9.1. Grid generator

An immediate consequence of the generalized cylinder-like geometry of the hydrogeological system is that it is relatively easy to cover a whole space of the flow domain with three-dimensional finite elements. The simplest possible three-dimensional elements – tetrahedrons – can be chosen to serve the purpose. First, the whole system is scaled-down to the cube $[0,1] \times \times [0,1] \times [0,1]$ and then the tetrahedral grid can be generated within the cube in three stages:

Stage 1: A *one-dimensional grid* is being generated along the *vertical* axis. Nodes represent discretization points in which abrupt changes in conductivity or in boundary conditions take place. A one-dimensional grid can be so refined that a ratio between two consecutive intervals between the nodes does not exceed some specified value, say, 1.5. Positions of the nodes define positions of the horizontal planes that are to be used for the ultimate three-dimensional grid.

Stage 2: A *planar* (triangular) finite element grid is generated. This grid is refined over the *horizontal* plane in order to fulfill the accuracy requirements asked by the user. The refinement is done automatically by the generator in the vicinity of edges or wells. A quad-tree approach is applied to the problem of generating the finite element planar grid; details of the technique can be found in Yerry and Shephard, 1983; and Nawalany, 1986c. Figure 17 shows an example of the refined planar grid with some densification around the well.

Stage 3: A *three-dimensional* (tetrahedral) finite element *grid* is generated out of the vertical and planar ones. The triangular elements of the top plane are simply repeated on every horizontal plane down

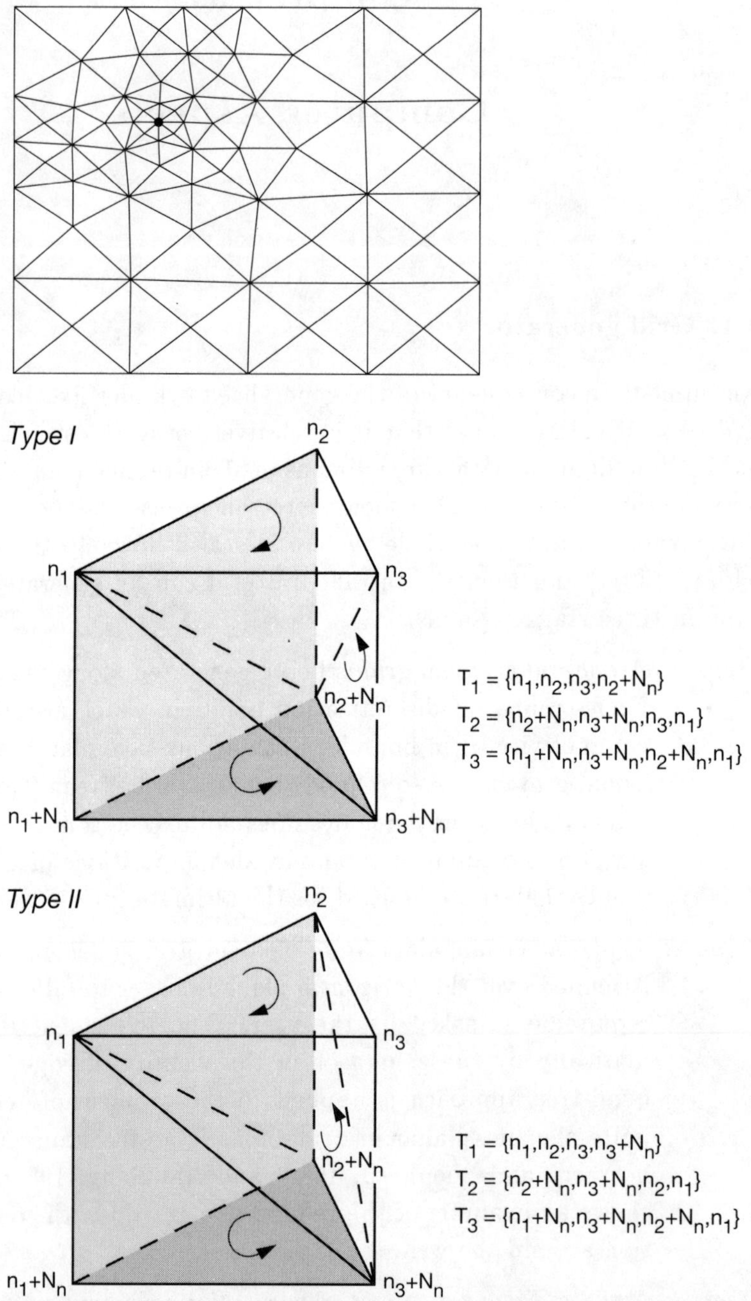

Figure 17. Example of the planar triangular grid with some refinement around a well (top) and two types of pentahedrons that can be generated under each triangle.

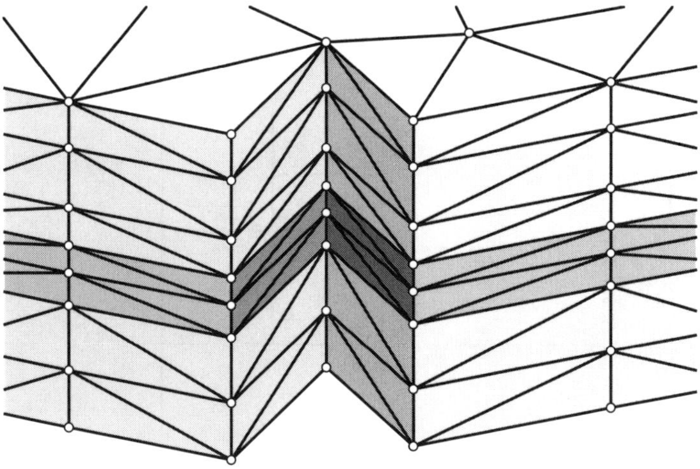

Figure 18. Three-dimensional grid.

to the bottom plane. Below each triangle on the horizontal plane a pentahedron is generated. Figure 17 shows the two types of pentahedral elements together with their subdivisions into three tetrahedrons. Both types are used in the generator to make the resultant tetrahedral finite element grid consistent. If N_n is the number of nodes on each horizontal plane and N_z is the number of horizontal planes, then the total number of nodes in the grid is $N = N_n N_z$. Nodes are numerated plane-by-plane. Figure 18 shows a part of the resultant three-dimensional finite element grid.

Quad-tree technique

The graph-theoretical approach – so-called quad-tree technique – is used for generating a planar finite element grid. Before giving details, three observations must be made. Firstly, it is sufficient to generate a rectangular grid. The triangular grid can be obtained by dividing each rectangle into, say, four triangles. Secondly, with this technique it is also sufficient to generate a grid densified around selected points of the flow domain. The grid for the entire domain can be obtained by the concatenation of these partial grids. And thirdly, it can be assumed that a horizontal projection of the groundwater system lies within some rectangle. The latter can be scaled down to the square $[0, 1] \times [0, 1]$ and the square can be easily covered by finite elements using the quad-tree technique. Finite elements generated outside the flow domain boundary can be discarded where necessary.

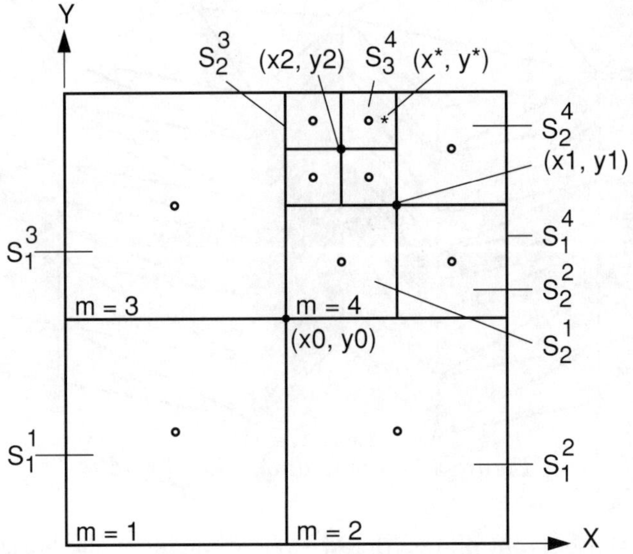

Figure 19. Subdivision of S_0.

Let (x^*, y^*) represent a point within the square $S_0 = [0,1] \times [0,1]$. The point is said to be 'covered' by some subsquare $S_k \subset S_0$ with a given accuracy ϵ $(0 < \epsilon < 1)$ if $(x^*, y^*) \in S_k$, and the lengths of the sides of S_k do not exceed ϵ. In order to cover a given point (x^*, y^*) with some square $S_k \subset S_0$ with the accuracy ϵ, S_0 is first subdivided into four squares and then each square is checked to determine to which one the point (x^*, y^*) belongs. Subsequent subdivisions of S_0 result in a descending series of squares $S_0 \supset S_1^{m_1} \subset S_2^{m_2} \subset \cdots \subset S_n^{m_n}$, $m_n \in \{1, 2, 3, 4\}$, satisfying the following criteria:

$$\begin{aligned} &\text{(i)}\ (x^*, y^*) \in S_n^{m_n} \\ &\text{(ii)}\ \ldots 2^{-n} < \epsilon \leq 2^{-(n-1)}. \end{aligned} \qquad (9.1.1)$$

The subdivisions are illustrated in Figure 19. The procedure of whether or not to subdivide the squares into four subsquares can be uniquely related to the procedure of constructing the *quad-tree* (see Shephard and Law, 1984; and Nawalany, 1986b). The most appealing feature of the quad-tree technique is its simplicity, which lies in replacing a problem of subsequent square subdivisions by a much simpler operation of concatenating binary series with logical operations.

Before presenting the techniques of generating the series $(S_k)_{k=0,\dots,n}$ and the corresponding quad-tree, some notation needs to be introduced:

s^L	– length of the sides of S_L $(L = 0, 1, \dots)$;
(x_L, y_L)	– coordinates of the center of S_L;
m	– number of the subsquare ($m = 1, 2, 3$ or 4);
$\mathbf{e}_k$	– unit vector having zeros everywhere except for '1' on the k-th position;
$\mathbf{t}^L$	– binary representation of the quad-tree that corresponds to S_L for which L can be interpreted as the generation number.

An algorithm for the recursive quad-tree generator can be presented as follows:

(i) $L = 0$

$s^0 = 1, \; x_0 = 0.5, \; y_0 = 0.5, \; S_0 = [0, 1] \times [0, 1]$

$n_f^0 = 1$

$\mathbf{t}^0 = (1 \mid \dots 0 \dots)$

(ii) $L := L + 1$

if $2^{-L} < \epsilon$ **then stop**

otherwise

$m_x = \text{sign}\,(x^* - x_{L-1}), \quad m_x \in \{-1, +1\}$

$m_y = \text{sign}\,(y^* - y_{L-1}), \quad m_y \in \{-1, +1\}$

$m = 2 + m_y + [(1 + m_x)/2], \quad m \in \{1, 2, 3, 4\}$

$s^L = 0.5 * s^{L-1}$

$x_L = x_{L-1} + m_x \cdot s^L$

$y_L = y_{L-1} + m_y \cdot s^L$

$S_L = \left[x_L - s^L/2, \; x_L + s^L/2\right] \times \left[y_L - s^L/2, \; y_L + s^L/2\right]$

$n_f^L = 4n_f^{L-1} - 3 + m$

$\mathbf{t}^L = \mathbf{t}^{L-1} + \mathbf{e}(n_f^L)$

goto (ii).

For practical applications *the reduced representation of the tree* $\boldsymbol{t}_r =$ $= (n_f^0, n_f^1, \ldots, n_f^n)$ is generated instead of $\boldsymbol{t}^n$. In the latter, only *positions of ones* in $\boldsymbol{t}^n$ are memorized; this saves computer memory tremendously.

After generating the quad-trees t_j for some points (x_j^*, y_j^*), $(j =$ $= 1, 2, \ldots)$ within the square $[0,1] \times [0,1]$, the quad-trees can be concatenated to get the global quad-tree for the set of all points. The example in Figure 20 shows the concatenation of two rectangular grids G_1 and G_2 and the concatenation of the corresponding trees in both the graphical and analytical forms. It must be remarked here that the primitive rectangular grid (as illustrated by Figure 19) is not suitable for transfer to the triangular grid since some nodal points would be located on the sides of triangles. In order to construct a compatible (or *balanced*) triangular tree it is sufficient to generate the rectangular tree in such a way that every two 'sons without sons,' which are neighbors to each other, differ at most by one generation. This can be done by a balancing procedure. Generation of the triangular grid from the rectangular grid is, in principle, very simple except for the interfaces between the neighboring rectangles which differ in their generations by one. For such cases additional interconnections between the nodes need to be generated.

The quad-tree technique presented above has proved to be very efficient computationally.

Representing boundaries and wells

Once the tetrahedral finite element grid is generated for the whole three-dimensional flow domain, a need for removing some of the elements may arise. There are two cases when a finite element must be removed: either it is outside the groundwater system boundary or it is within the well (i.e., outside the porous medium). Naturally, after discarding some of the finite elements, the remaining grid must remain consistent.

Concatenation of grids

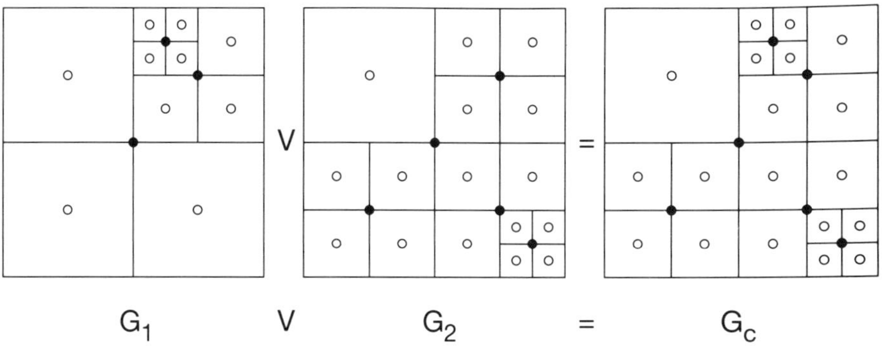

Concatenation of quad-trees

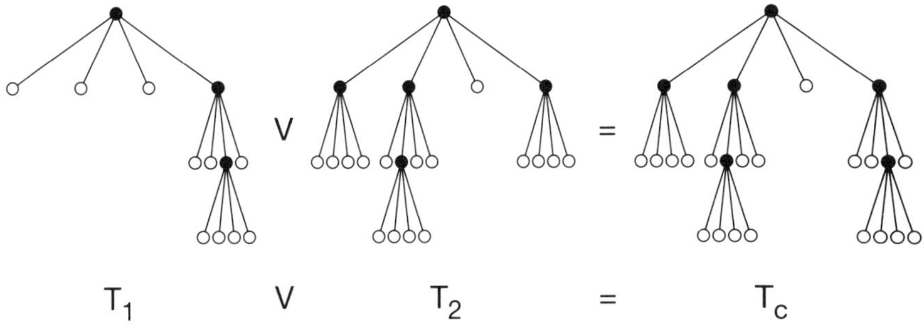

Concatenation of the trees' binary representations

$$\underline{t}_1 = (1 \mid 0001 \mid 0000\ 0000\ 0000\ 0010 \mid \dots 0 \dots)$$

$$\underline{t}_1 = (1 \mid 1101 \mid 0000\ 0100\ 0000\ 0000 \mid \dots 0 \dots)$$

$$\underline{t}_1 \vee \underline{t}_2 = (1 \mid 1101 \mid 0000\ 0100\ 0000\ 0010 \mid \dots 0 \dots) = \underline{t}_c$$

Figure 20. Concatenation of the two rectangular grids.

9.2. Solvers

There are specific numerical problems related to a solution of algebraic equations that arise when applying the Galerkin finite element method to either classical or VOA groundwater flow equations. The way the equations are solved can be clearly explained when discussing the classical equation solver first and then generalizing the results for the VOA solver.

It can be deduced from the finite element assembling procedure for the classical flow equations ($\underline{\boldsymbol{P}} \cdot \boldsymbol{\Phi} = \boldsymbol{b}$) – see paragraph 8.2 – that the resultant algebraic equations (having a general form $A\boldsymbol{x} = \boldsymbol{b}$) have the matrix A which is:

(i) large;
(ii) sparse;
(iii) symmetric;
(iv) positive definite.

The consequences of characteristics (i)-(iv) on the choice of the solver are considerable. Firstly, because the matrix is large it is impractical to use the Gauss elimination-type solvers since they operate on the whole matrix and their computational costs are rather high. Consequently, the iterative methods are considered more economic, especially when PC-computers are used for solving the groundwater flow problems. Secondly, because the matrix is sparse, the entries of matrix A may be kept in a computer memory in the compact form, i.e., by memorizing only the nonzero entries. Naturally, an efficient way of retrieving the elements of matrix A needs to be designed. The compact vector representation of a sparse matrix is discussed briefly below. Thirdly, the symmetry and positive-definiteness of matrix A makes the conjugate gradient iterative method a good candidate to be chosen as the solver for the groundwater flow equations. Actually, this is the ICCG method – the special, very fast, version of the CG method – which is used nowadays throughout the groundwater flow models. The major steps and important features of the method are recalled below for completeness. The ICCG algorithm consists of three steps:

(i) ***Incomplete Cholesky decomposition.*** For a given matrix A, a lower triangular matrix L is computed with the Cholesky algorithm with the additional condition of setting equal to zero those elements L_{ij} which correspond to zero elements in the lower triangular part of matrix A. This is done formally by introducing a set P of all the indices (i, j) for which the corresponding elements of A are zero, that is:

$$P = \left\{ (i, j) \mid A_{ij} = 0; \; i, j = 1, \ldots, N \right\}.$$

The modified (incomplete) Cholesky decomposition algorithm can be written as follows (see Meijerink and Van der Vorst, 1977, 1983; and Kershaw, 1978):

$$\begin{aligned}
&\textbf{for } j = 1, 2, \ldots, N \\
&L_{jj} = \left(A_{jj} - \sum_{k=1}^{j-1} L_{jk}^2 \right)^{1/2} \\
&\textbf{for } i = j + 1, \ldots, N \\
&L_{ij} = 0 \\
&\textbf{if } (i, j) \notin P \textbf{ then } L_{ij} = \left(A_{ij} - \sum_{k=1}^{j-1} L_{ik} L_{jk} \right) \Big/ L_{jj}.
\end{aligned} \tag{9.2.1}$$

The obvious advantage of using L is that the matrix needs very little storage in the computer memory. Moreover, when the compact vector representation of matrix A is used (see the description that follows), the structural information is already available since matrices A and L have exactly the same pattern of nonzero entries by the definition of the incomplete Cholesky decomposition. Another advantage of using matrix L is that there is an improvement in the convergence behavior of the following preconditioned CG method.

(ii) ***preconditioned conjugate gradient method.*** The ICCG method is an iterative method which starts from an arbitrary estimate $\boldsymbol{x}_0$ of the solution $\boldsymbol{\alpha}$ for the equation $A\boldsymbol{x} = \boldsymbol{b}$, and proceeds from one estimate $\boldsymbol{x}_i$ to the next one ($\boldsymbol{x}_{i+1}$) as follows:

$$\left|\begin{array}{l} \boldsymbol{x}_0 : \textbf{arbitrary} \\ \boldsymbol{r}_0 = \boldsymbol{b} - A\boldsymbol{x}_0 \\ \boldsymbol{p}_0 = (LL^T)^{-1}\, \boldsymbol{r}_0 \\ \textbf{for } i = 0, 1, \ldots \\ \alpha_i = \left(\boldsymbol{r}_i, (LL^T)^{-1}\, \boldsymbol{r}_i\right) \Big/ (\boldsymbol{p}_i, A\boldsymbol{p}_i) \\ \boldsymbol{x}_{i+1} = \boldsymbol{x}_i + \alpha_i \boldsymbol{p}_i \\ \boldsymbol{r}_{i+1} = \boldsymbol{r}_i - \alpha_i A\boldsymbol{p}_i \\ \beta_i = \left(\boldsymbol{r}_{i+1}, (LL^T)^{-1}\, \boldsymbol{r}_{i+1}\right) \Big/ \left(\boldsymbol{r}_i, (LL^T)^{-1}\, \boldsymbol{r}_i\right) \\ \boldsymbol{p}_{i+1} = (LL^T)^{-1}\, \boldsymbol{r}_{i+1} + \beta_i \boldsymbol{p}_i. \end{array}\right. \qquad (9.2.2)$$

A product $\boldsymbol{v} = (LL^T)^{-1}\, \boldsymbol{r}$ in formulae (9.2.2), with L being a triangular matrix, and $\boldsymbol{r}$ the known vector, can be easily calculated in two steps:

$$\text{(a)}\quad L\boldsymbol{y} = \boldsymbol{r} \quad \text{(forward substitution)} \qquad (9.2.3a)$$

$$\text{(b)}\quad L^T\boldsymbol{v} = \boldsymbol{y} \quad \text{(backward substitution)} \qquad (9.2.3b)$$

(iii) ***Stop criterion.*** Iterations (9.2.2) are continued until the distance between two consecutive iteratives becomes smaller than some prescribed small number ϵ. However, some better stopping criteria are possible (e.g., the criterion based on estimating the smallest eigenvalue of A, see Kaasschieter, 1987).

Calculations performed for many examples (see, for instance, Nawalany, 1986a,b,c, 1988) show that the ICCG method is fast and robust for large, positive-definite and symmetric matrices, encountered in groundwater flow models. This can be explained partly by a reduction of the matrix condition number and partly by clustering the eigenvalues of the system's matrix. It was shown by Van der Vorst (1982) that the rate of convergence of the CG method increases when cond (A) becomes smaller. From the formula for the i-th residual of the CG method:

$$\| \boldsymbol{x}_i - \boldsymbol{\alpha} \|_A^2 < 4\left(\sqrt{c} - 1/\sqrt{c} + 1\right)^{2i} \; \| \boldsymbol{x}_0 - \boldsymbol{\alpha} \|_A^2 \qquad (9.2.4)$$

where $c = \lambda_{max}/\lambda_{min}$; λ_{max} and λ_{min} are the largest and smallest eigenvalues of A, respectively, and $\boldsymbol{\alpha}$ is the true solution (i.e., $A\boldsymbol{\alpha} = \boldsymbol{b}$) it is seen that by 'squeezing' the spectrum of A (i.e., by making $\lambda_{max}/\lambda_{min}$ closer to 1) the upper-bound of the residual in (9.2.4) becomes smaller. However, this upper-bound is still considered by Van der Vorst (1982) to be pessimistic. He indicates that a favorable distribution of the smallest eigenvalues of A is an important factor since this may cause a rapid convergence of the ICCG method. It has been shown by many authors (see for instance, A. Peters *et al.*, 1988) that a selection of an efficient preconditioning matrix for the CG solver for random sparse systems of algebraic equations is a processor dependent problem. While for scalar processors ICCG is decisively superior to any other technique on vector processors, algorithms with lower convergence rates but higher vectorization potential may work better.

Vector representation of the sparse matrices

When representing any sparse matrix in the computer memory one needs to store only the nonzero elements of the matrix and additionally some structural information which uniquely determines the elements' position within the matrix. Obviously, for symmetric matrices it is sufficient to store only the lower (or the upper) triangular submatrix. It must also be observed here that keeping the whole bands (or halfbands) of the large sparse matrix elements is restricted to computers having vast virtual memories, and can be justified only by considerable gain in the computation time. In the following two structures for representing sparse matrices are considered: the forward structure and the backward structure. The latter has been introduced in order to solve the backward substitution part of the equation $(L \cdot L^T)\ \boldsymbol{v} = \boldsymbol{r}$ (see formulae (9.2.3)) efficiently. Computer experiments have shown that the use of the backward structure in representing a sparse upper triangular matrix L has speeded-up the ICCG solver by a factor of 5.

Similar structures have also been developed for handling sparse matrices on vector computers (see Zijl, Nawalany and Pasveer, 1986).

The forward structure

Let A be a symmetric sparse matrix having its nonzero elements randomly distributed over the matrix (i.e., no specific structure is assumed), and let the diagonal elements of A be nonzero elements. matrix A can be represented by its rowwise condensed form $\boldsymbol{a}$ containing all (and only) nonzero entries existing in subsequent rows of the lower triangular submatrix of A. The dimension L_a of the vector $\boldsymbol{a}$ can be expressed in the following way:

$$L_a = \sum_{i=1}^{N} k_i \tag{9.2.5}$$

where N is the number of rows (columns) in the matrix A, and k_i the number of nonzero elements in the i-th row of the lower triangular part of A.

To make the representation $\boldsymbol{a}$ of A unique, two pointer vectors, $\boldsymbol{p_c}$ and $\boldsymbol{p_d}$, must additionally be defined. The first vector contains the numbers of the columns of subsequent nonzero elements of matrix A represented in the vector $\boldsymbol{a}$. Consequently, the dimension of $\boldsymbol{p_c}$ is equal to L_a. The second vector, $\boldsymbol{p_d}$, contains the positions of the diagonal elements of A in the vector $\boldsymbol{a}$. In other words, it contains the following integer numbers:

$$\boldsymbol{p_d} = \left\{ \nu_j \mid \nu_j = \sum_{i \leq j} k_i;\ j = 1, \ldots, N \right\}. \tag{9.2.6}$$

Instead of introducing the formal notations for vectors $\boldsymbol{a}$ and $\boldsymbol{p_c}$ the following example is proposed to illustrate the principle of the forward structure for the sparse matrix representation.

Example 1

$$A = \begin{pmatrix} 1 & 0 & 3 & 0 \\ 0 & 1 & 0 & 4 \\ 3 & 0 & 2 & 8 \\ 0 & 4 & 8 & 1 \end{pmatrix}. \tag{i}$$

Here $N = 4$, $k_1 = 1$, $k_2 = 1$, $k_3 = 2$ and $k_4 = 3$. Consequently, the dimension of the vector $\boldsymbol{a}$ (and $\boldsymbol{p_c}$) is equal to:

$$L_a = k_1 + k_2 + k_3 + k_4 = 7.$$

Vector $\boldsymbol{a}$ contains all the nonzero elements of the lower triangular submatrix of A arranged rowwise, hence:

$$\boldsymbol{a} = (1, 1, 3, 2, 4, 8, 1) \tag{ii}$$

and $\boldsymbol{p}_c$ represents the numbers of the columns corresponding to the elements of $\boldsymbol{a}$:

$$\boldsymbol{p}_c = (1, 2, 1, 3, 2, 3, 4). \tag{iii}$$

Vector $\boldsymbol{p}_d$, representing the cumulative positions of the diagonal elements of A in the vector $\boldsymbol{a}$, has the following elements:

$$\boldsymbol{p}_d = (1, 2, 4, 7). \tag{iv}$$

From the above definitions it can easily be seen that a recovering of the diagonal elements of A is simple. Indeed, $A_{ii} = \boldsymbol{a}(\nu_i) = \boldsymbol{a}(\boldsymbol{p}_d(i))$.

In order to recover a nondiagonal nonzero element A_{ij} (where $j < i$) from the ($\boldsymbol{a}$, $\boldsymbol{p}_c$, $\boldsymbol{p}_d$) representation of matrix A, one must first find the range (ν_{i-1}, ν_i). This can be read from $\boldsymbol{p}_d$ since $\nu_i = \boldsymbol{p}_d(i)$. If $\nu_i = 1$ then, of course, $j = 1$ and $A_{ij} = \boldsymbol{a}(1)$. If $\nu_i > 1$ then the range of elements $\boldsymbol{p}_c$ must be searched to find which element of $\boldsymbol{p}_c$ is equal to the given j. This results in the number ν which satisfies the following conditions $\nu \in (\nu_{i-1}, \nu_i)$ and $\boldsymbol{p}_c(\nu) = j$. Hence $A_{ij} = \boldsymbol{a}(\nu)$.

For example, a recovery of the value of the A_{43} element of matrix A from its vector representation starts from defining the range (ν_3, ν_4). Here $\nu_3 = \boldsymbol{p}_d(3) = 4$ and $\nu_4 = \boldsymbol{p}_d(4) = 7$. The elements $\boldsymbol{p}_c(5)$ and $\boldsymbol{p}_c(6)$ are then being compared with $j = 3$. Since $\boldsymbol{p}_c(6) = 3$, the corresponding ν is set equal to 6. Therefore $A_{43} = \boldsymbol{a}(6) = 8$, which is in agreement with definition (i).

Finding the element A_{ij} that corresponds to a given element $\boldsymbol{a}(\nu)$ is also possible. The number of the column is readily obtainable from $\boldsymbol{p}_c$, that is $j = \boldsymbol{p}_c(\nu)$. The number '$i$' of the row to which the element belongs satisfies the following condition:

$$i = \min_{j \le n \le N} \{n : \boldsymbol{p}_d(n-1) < \nu \le \boldsymbol{p}_d(n)\}. \tag{9.2.7}$$

To find (i, j) which corresponds to $\nu = 5$ in the above example, one must recover j first. Here $j = \boldsymbol{p}_c(5) = 2$. By inspecting the elements of $\boldsymbol{p}_d$ one can see that the first 'n' for which the condition in (9.2.7) is satisfied is $n = 4$. This n must be substituted for i, that is $i = 4$. Therefore $\boldsymbol{a}(5) = A_{42} = 4$, which is again in agreement with definition (i).

The above example indicates that the direct and reverse procedures of recovering the elements of the matrix A require some searching. Direct searching is always limited to examining very few nonzero elements of the vector $\boldsymbol{a}$. If N_e denotes a maximum number of neighbors for any node in a given finite element grid then the number of matrix elements within the range searched never exceeds N_e, i.e.,

$$|\nu_i - \nu_{i-1} - 1| \leq N_e; \quad i = 2, \ldots, N. \tag{9.2.8}$$

The reverse search, although algorithmically simple, may be slightly more time-consuming.

The backward structure

In order to solve efficiently the backward substitution part of the algebraic equations $(LL^T)\,\boldsymbol{v} = \boldsymbol{r}$ (see formula (9.2.3b)) an additional structure can be introduced. The structure $(\boldsymbol{a}', \boldsymbol{p}'_c, \boldsymbol{p}'_d)$ is simply the same as the one used for representing the lower triangular part of the sparse matrix A except that:

- It applies to the upper triangular part of matrix A.
- The cumulative positions k_i of the diagonal elements of A are calculated backwards in terms of rows and columns, i.e.,

$$\boldsymbol{p}'_d = \left(\nu_j \mid \nu_j = \sum_{i \leq j} k'_i;\ j = N, \ldots, 1 \right) \tag{9.2.9}$$

  where k'_i is the number of nonzero elements in the i-th row of matrix A right to and including diagonal element A_{ii}; $i = 1, \ldots, N$.
- Instead of using a real compact vector $\boldsymbol{a}'$ to memorize the upper part of A, the integer vector $\boldsymbol{l}'$ is used which contains the positions of subsequent elements of $\boldsymbol{a}'$ in the vector $\boldsymbol{a}$. Such an approach saves computer memory.

The following example illustrates a concept of the backward structure.

Example 1'

In the example the very same matrix A is used as before, for which the backward cumulative positions of diagonal elements are as follows: $k'_4 = 1,\ k'_3 = 3,\ k'_2 = 5,\ k'_1 = 7$ and hence

$$\boldsymbol{p}'_d = (1,3,5,7). \tag{iv'}$$

Because of the symmetry of A, also $\sum_{i=1}^{N} k'_i = L_a$ here. If the backward compact vector $\boldsymbol{a}'$ were to be used to memorize the upper part of A, it would look like this:

$$\boldsymbol{a}' = (1,8,2,4,1,3,1). \tag{ii'}$$

Instead, the integer vector $\boldsymbol{l}'$ can be found:

$$\boldsymbol{l}' = (7,6,4,5,2,3,1) \tag{ii''}$$

for which the following relation holds:

$$\boldsymbol{a}'(\nu') = \boldsymbol{a}(\boldsymbol{l}'(\nu')), \quad \text{for } \nu = 1,\ldots,L_a.$$

Since the columns' numbers that correspond to the elements of $\boldsymbol{a}'$ are the rows' numbers for elements of $\boldsymbol{a}$ (which are not memorized), there is a necessity to define a separate vector $\boldsymbol{p}'_c$ for them. In this example:

$$\boldsymbol{p}'_c = (4,4,3,4,2,3,1). \tag{iii'}$$

This completes a description of the forward and backward compact structures for the lower and/or upper part of matrix A.

Now the question can be posed of how the efficiency of the ICCG method can be extended for solving the VOA equations. The latter consists of three Laplace-like coupled equations for e_x, e_y and q_z (or q_x, q_y, q_z) which, if uncoupled, could be solved numerically in the same way as the classical equation for the potential. Therefore, the problem is reduced to uncoupling the VOA equations. This can be done either by

- decoupling the VOA equations physically, or by
- decoupling the VOA equations artificially.

Physical decoupling can take place only if there are physical conditions which make the right-hand-side for each variable's equation independent of other variables. For instance, when the hydraulic parameters are perfectly layered, the equations for e_x, e_y and q_z are decoupled (see also Section 10.3).

If the VOA equations are decoupled, the (ICCG) solver can be used three times in solving the equations for the three components of the flow field one after another. In general, however, the VOA equations are physically coupled and decoupling can be done only artificially; the right-hand sides of the VOA equations can be calculated by solving the classical equation approximating values of q_x, q_y and q_z and using them to evaluate the right-hand sides of the VOA equations. It is also possible to solve the coupled system of equations in which matrix A includes the coupling terms. Then, however, matrix A is not positive definite, so that the equivalent system $A^T A\boldsymbol{x} = A^T \boldsymbol{b}$ can be solved instead ($A^T A$ is symmetric and positive definite; see Section 4.6). This method has been applied by Kuppen (1988). It turns out that the method is more accurate, but also more demanding with respect to computer resources, than artificial decoupling. Of course, application of the VOA equations makes sense only if the subsurface is not strongly heterogeneous; see Section 4.5.

9.3. Calculation of flow paths

In order to calculate the trajectory of a water particle in the flow domain Ω one must:

(i) choose the starting point for the path;
(ii) solve the equation of motion numerically;
(iii) decide upon the stopping criterion for the particle.

One can also calculate the length of the trajectory and the traveling (or residence) time of the water particle in the flow domain. By calculating traveling or residence times for a sufficiently large number of trajectories one can calculate the traveling (residence) time distribution of the whole groundwater system (subsystem).

Starting point

Obviously, any point in the recharge region $\partial\Omega^r$ can be chosen as the starting point (x_0, y_0, z_0) for the path – see, for instance, Figure 21 (Chapter 10). Generally, any point within the flow domain can be chosen as well.

Moving the particle

The *equation of motion* of the water particle consists of three ordinary differential equations:

$$\begin{cases} \dot{x} = v_x(x, y, z) \\ \dot{y} = v_y(x, y, z) \\ \dot{z} = v_z(x, y, z) \end{cases} \tag{9.3.1}$$

where v_x, v_y, v_z are the velocity components, $(x, y, z) = (x(t), y(t), z(t))$ is the position of the particle at the moment of time t and the dot '•' indicates the time derivative.

The components v_x, v_y and v_z can readily be computed from the estimates of e_x, e_y and q_z obtained when solving the groundwater flow equations numerically:

$$\begin{aligned} v_x &= k(x, y, z)\ \hat{e}_x(x, y, z)/\theta(x, y, z) \\ v_y &= k(x, y, z)\ \hat{e}_y(x, y, z)/\theta(x, y, z) \\ v_z &= \hat{q}_z(x, y, z)/\theta(x, y, z). \end{aligned} \tag{9.3.2}$$

where θ is the total porosity.

Using the linear model for the conductivity (8.3.1) and the linear approximations (8.2.13) for e_x, e_y and q_z, one can equivalently write formulae (9.3.2) for any finite element Ω^e:

$$\begin{aligned} v_x^e &= \left(\sum_{l=1}^{4} k_l^e\ \Phi_l(x, y, z)\right) \left(\sum_{l=1}^{4} e_{xl}^e\ \Phi_l(x, y, z)\right) \Big/\ \theta(x, y, z) \\ v_y^e &= \left(\sum_{l=1}^{4} k_l^e\ \Phi_l(x, y, z)\right) \left(\sum_{l=1}^{4} e_{yl}^e\ \Phi_l(x, y, z)\right) \Big/\ \theta(x, y, z) \\ v_z^e &= \sum_{l=1}^{4} q_{zl}^e\ \Phi_l(x, y, z)\ \Big/\ \theta(x, y, z), \quad (e = 1, \ldots, N_e) \end{aligned} \tag{9.3.3}$$

where $k_1^e, \ldots, k_4^e$ are the values of the conductivity at the four nodes of the e-th tetrahedron, and $e_{xl}^e, e_{yl}^e, q_{zl}^e$ are the (transformed) components of a volumetric flux at the l-th node of the e-th tetrahedron, $(l = 1, 2, 3, 4)$.

After introducing the following notation:

$$\begin{cases} \boldsymbol{y} = (x, y, z)^T \\ \boldsymbol{F} = (v_x, v_y, v_z)^T \end{cases} \tag{9.3.4}$$

the equation of motion together with the initial condition (i.e., starting points) can be written shortly as:

$$\left| \begin{array}{l} \dot{\boldsymbol{y}}(t) = \boldsymbol{F}(\boldsymbol{y}(t)) \\ \boldsymbol{y}(0) = \boldsymbol{y}_0 \end{array} \right. \tag{9.3.5}$$

where t is the independent variable (e.g., time), $\boldsymbol{y}$ is the water particle position vector $\boldsymbol{y} \in R^3$, $\boldsymbol{y}_0$ is the initial value of $\boldsymbol{y}$, and $\boldsymbol{F}$ is the known function (known velocity field).

The England method (see England, 1969) offers the effective algorithm of proceeding from one position of the water particle – say from $\boldsymbol{y}(t)$ to the next position $\boldsymbol{y}(t + \Delta t)$ – by solving sets of nonlinear ordinary differential equations. The method combines the accuracy of the fourth-order Runge-Kutta method with the adaptive mechanism for changing the increment of the independent variable t. The variable t is interpreted here as time. The criterion of changing Δt is based on the estimate of the approximation error E. At each instant of time the estimate of E is calculated and Δt is diminished or enlarged depending on whether E is within or outside the given range of accuracy (E_1, E_2). This mechanism gives the desired flexibility of control of the approximation error without keeping the increment of time Δt constant. Consequently, in most instances, a compromise between a reasonable computational time and an accepted range of the solution error can be found.

The solution to the problem (9.3.5) is estimated for some discrete values of the independent variable $t_1, t_2, \ldots$. The method can be summarized as follows:

- *Parameters of the method:* Δt^0 is the starting value of the increment of the independent variable; E_1, E_2 are the lower and upper boundaries of the approximation error's acceptance interval

($E_1 < E_2$ and usually $E_2/E_1 \sim 10$).
- *Starting value:* $\boldsymbol{y}(0) = \boldsymbol{y}_0$ (known or estimated).
- For $i = 0, 1, 2, \ldots$

(a) $\boldsymbol{y}(t_{i+1}) = \boldsymbol{y}(t_i) + 1/6\ (\boldsymbol{k}_1 + 4\boldsymbol{k}_3 + \boldsymbol{k}_4)$ (9.3.6)

where

$$\begin{aligned}
\boldsymbol{k}_1 &= \Delta t\ \boldsymbol{F}(\boldsymbol{y}(t_i)) \\
\boldsymbol{k}_2 &= \Delta t\ \boldsymbol{F}(\boldsymbol{y}(t_i) + 1/2\ \boldsymbol{k}_1) \\
\boldsymbol{k}_3 &= \Delta t\ \boldsymbol{F}(\boldsymbol{y}(t_i) + 1/4\ (\boldsymbol{k}_1 + \boldsymbol{k}_2)) \\
\boldsymbol{k}_4 &= \Delta t\ \boldsymbol{F}(\boldsymbol{y}(t_i) - \boldsymbol{k}_2 + 2\boldsymbol{k}_3).
\end{aligned}$$

This is the basic fourth-order Runge-Kutta formula that allows one to proceed from the previous time instant t_i to the next time instant $t_{i+1} = t_i + \Delta t$.

(b) $\boldsymbol{E} = 1/90\ (4\boldsymbol{k}_3 - \boldsymbol{k}_1 + 17\boldsymbol{k}_4 - 23\boldsymbol{k}_5 + 4\boldsymbol{k}_7 + \boldsymbol{k}_8)$ (9.3.7)

where

$$\begin{aligned}
\boldsymbol{k}_5 &= \Delta t\ \boldsymbol{F}(\boldsymbol{y}(t_{i+1})) \\
\boldsymbol{k}_6 &= \Delta t\ \boldsymbol{F}(\boldsymbol{y}(t_{i+1}) + 1/2\ \boldsymbol{k}_5) \\
\boldsymbol{k}_7 &= \Delta t\ \boldsymbol{F}(\boldsymbol{y}(t_{i+1}) + 1/4\ (\boldsymbol{k}_5 + \boldsymbol{k}_6)) \\
\boldsymbol{k}_8 &= \Delta t\ \boldsymbol{F}(\boldsymbol{y}(t_{i+1}) + \\
&\quad + 1/6\ (\boldsymbol{k}_1 - 96\boldsymbol{k}_2 + 92\boldsymbol{k}_3 - 121\boldsymbol{k}_4 + 144\boldsymbol{k}_5 + 6\boldsymbol{k}_6 - 12\boldsymbol{k}_7)).
\end{aligned}$$

The error formula (9.3.7) is used for checking which of the three possibilities holds, whether:

(α) $|\boldsymbol{E}| < E_1$ or
(β) $E_2 < |\boldsymbol{E}|$ or
(γ) $E_1 < |\boldsymbol{E}| < E_2$.

If (α) is satisfied, i.e., the computational error is small, Δt is doubled ($\Delta t\ :=\ 2\ \Delta t$) and computations are continued from (a) for the new instant of time, i.e., $t = t_{i+1}$. If (β) is satisfied, i.e., the computational error is too big, then Δt is halved ($\Delta t := \Delta t/2$) and computations are repeated for the same instant of time, i.e., $t = t_i$ starting from point (a). If (γ) holds, i.e., the computational error is within the acceptance interval, then Δt remains the same and computations proceed to (c):

$$\text{(c)}\ \boldsymbol{y}(t_{i+2}) = \boldsymbol{y}(t_{i+1}) + 1/6\ (\boldsymbol{k}_5 + 4\boldsymbol{k}_7 + \Delta t\boldsymbol{\cdot}\boldsymbol{F}(\boldsymbol{y}(t_{i+1}) - \boldsymbol{k}_6 + 2\boldsymbol{k}_7)). \tag{9.3.8}$$

It should be noted that the procedure 'costs' as much as the Runge-Kutta method (i.e., four evaluations of $\boldsymbol{F}$ per one time step) as long as the error $|\boldsymbol{E}|$ belongs to the interval (E_1, E_2). Whenever $|\boldsymbol{E}| < E_1$ the time increment Δt is enlarged and the algorithm is superior to all other methods that keep Δt constant.

Stopping criterion

As the *stopping criterion* the following relation can be used:

$$(x, y, z) \in \partial\Omega^d = \partial\Omega \setminus \partial\Omega^r \setminus \partial\Omega^0$$

which means that, whenever the particle 'hits' the boundary's discharge region $\partial\Omega^d$, the traveling process is stopped ($\partial\Omega^0$: part of the flow domain boundary being impervious; $\partial\Omega^r$: recharge region). The moment of time for which the stopping criterion is satisfied is simply a residence time T_r for a particle, i.e., $T_r = t_s$. If necessary, the length of the trajectory L_t can be calculated according to the formula:

$$L_t = \int_0^{T_r} \sqrt{\dot{x}^2(t) + \dot{y}^2(t) + \dot{z}^2(t)}\ dt. \tag{9.3.9}$$

All examples presented in Chapter 10 are illustrated using sets of groundwater paths to show the physical correctness or limitations of the theories applied.

Chapter 10

Examples and Applications

Testing new ideas, like the velocity-oriented approach to groundwater flow, always needs some courage. For years the classical approach has been used successfully by hydrogeologists and civil engineers, especially when applied to hydrogeological problems which involve the quantity rather than the quality of the groundwater. For instance, abstraction schemes, seepage analysis and protection, water balance calculations etc. routinely use classical flow equations, and the everyday experience shows that the approach is sufficiently accurate for these purposes. The routine is also something of an obstacle when new problems (e.g., prediction of pollution transport) call for a qualitative change in the way the groundwater flow needs to be described. The VOA equations, being more sophisticated mathematically, set higher requirements for all the stages of computer calculations. When preprocessing the input (field) data one needs not only to handle all the three-dimensional information but above all to translate physical boundary conditions formulated in terms of potentials into the specific boundary conditions required by the VOA theory. This must be done by the VOA code automatically since there is no simple intuition that would facilitate their formulation just from field experience. In all the VOA flow models made until now the translation of boundary conditions is, hence, transparent for the codes' users. Solving the VOA equations has two distinct features, one of which can be considered advantageous whereas the second sometimes creates difficulties. The first feature – the Laplace-like form of the VOA equations – allows one to apply all the existing standard finite element or finite difference codes (practically) without changing anything. The second feature of the VOA equations is their possible coupling. It means that the right-hand sides of the equations may contain terms depending on the unknowns q_x, q_y, q_z. As discussed in Section 9.2, in the case of physical coupling, the VOA equations need to be artificially decoupled, thus making some changes in existing codes inevitable. As opposed to these additional computing problems, the postprocessing is easier when the velocity field is estimated with the VOA theory. The estimates of q_x, q_y, q_z are calculated for all discretization nodes and consequently only an interpolation between the nodes is needed (*no differentiation* of numerically estimated variables as for the classical approach!). The examples presented in the following sections have been calculated with the finite element package FLOSA (Flow Systems Analysis). This package has been developed by Nawalany (1989a; 1989b) to analyze the differences between the two approaches to groundwater flow, and to demonstrate the feasibility and the practical usefulness of the VOA theory at the same time. The examples chosen also show how different in their nature could be the practical problems in which the VOA theory can be applied.

10.1. Comparison between the classical and the VOA numerical approximations of velocity fields

Two quasi-three-dimensional solutions have been derived to provide a reference for numerical solutions to the classical and the VOA equations. They are quasi-three-dimensional because they offer changes of state variables only in the x- and z-directions. The third direction y is supposed to be a direction in which there is no movement of water. The Tóth solution (see Tóth, 1963) is repeated here for the sake of completeness, while the rectangular cavity solution was derived (see Nawalany, 1989) to offer another example for the comparison.

Tóth's problem

Assumptions

(i) The flow domain is of rectangular shape: $\Omega = [0, L] \times [0, D]$.
(ii) It is assumed that all the boundaries are impervious except for the upper one on which the potential $f(x)$ is specified; see Figure 21.
(iii) The conductivity of the porous medium is homogeneous, i.e., k = constant.
(iv) There are the steady-state conditions of flow, hence $\phi = \phi(x, z)$.

Flow equation

$$\partial^2\phi/\partial x^2 + \partial^2\phi/\partial z^2 = 0, \qquad \text{for } (x, z) \in \Omega. \tag{10.1.1}$$

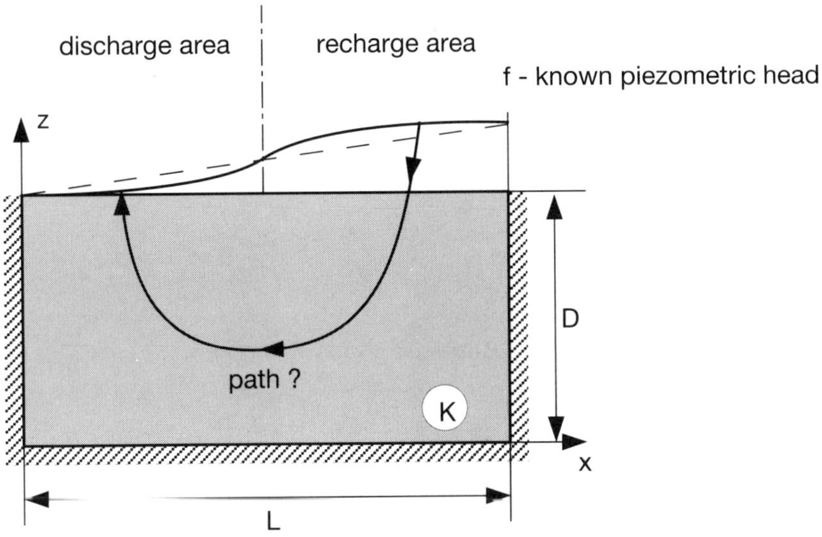

Figure 21. Flow domain and coordinate system.

Boundary conditions

(1) $\phi(x, z) = f(x)$, for the upper boundary (10.1.2)

(2) $\partial\phi/\partial n = 0$, for the rest of the boundary. (10.1.3)

The specified function $f(x)$ is assumed to be equal to some sinusoidal wave superimposed on a linear trend:

$$f(x) = D + cx - cL/2\pi \sin(2\pi x/L) \tag{10.1.4}$$

where c is the slope of the linear trend.

The form of (10.1.4) guarantees that f is consistent with both left and right boundary conditions, i.e., $\partial f/\partial x = 0$ for $x = 0$ and for $x = L$.

Solution in Ω

The sought solution for $\phi(x,z)$ is given by the following formula:

$$\begin{aligned}\phi(x,z) &= D + cL/2 + \\ &+ 8cL/\pi^2 \sum_{m=1,3,\ldots}^{\infty} (1-\cos\ m\pi)/(m^2(m^2-4)) \cdot \\ &\cdot \cos\ (\lambda_m x)\ \cosh\ (\lambda_m z)/\cosh\ (\lambda_m D)\end{aligned} \tag{10.1.5}$$

where $\lambda_m = m\pi/L,\ (m = 1, 3, \ldots)$.

Velocity field

By applying Darcy's law we can calculate the volumetric flow rate in Ω for the Tóth problem:

$$\begin{aligned}q_x(x,z) &= k\ 8cL/\pi^2 \sum_{m=1,3,\ldots}^{\infty} (1-\cos m\pi)/(m^2(m^2-4)) \cdot \\ &\cdot \lambda_m\ \sin\ (\lambda_m x)\ \cosh\ (\lambda_m z)/\cosh\ (\lambda_m D) \\ q_z(x,z) &= -k\ 8cL/\pi^2 \sum_{m=1,3,\ldots}^{\infty} (1-\cos m\pi)/(m^2(m^2-4)) \cdot \\ &\cdot \lambda_m\ \cos\ (\lambda_m x)\ \sinh\ (\lambda_m z)/\cosh\ (\lambda_m D).\end{aligned} \tag{10.1.6}$$

Solutions to the classical and the VOA equations for the Tóth problem have also been approximated numerically using the three-dimensional finite element method and then compared with the analytical solution (10.1.6) – see Figures 22–23. The comparisons, given in terms of water trajectories, show a superiority of the VOA solutions over the classical solutions especially in regions of low velocity, for instance, in the vicinity of stagnation points. (The classical approach does not result in one stagnation point, but in a stagnation element.)

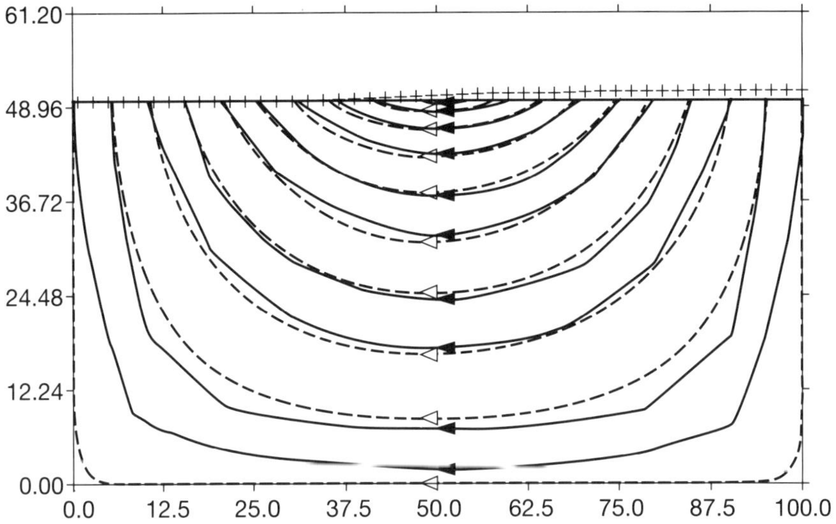

Figure 22a. *Trajectories of water particles calculated from the analytical solution (– –) and with the classical approach (—) for N_n = = 121 and M = 5 (N_n – number of nodes on the horizontal plane, M – number of horizontal planes).*

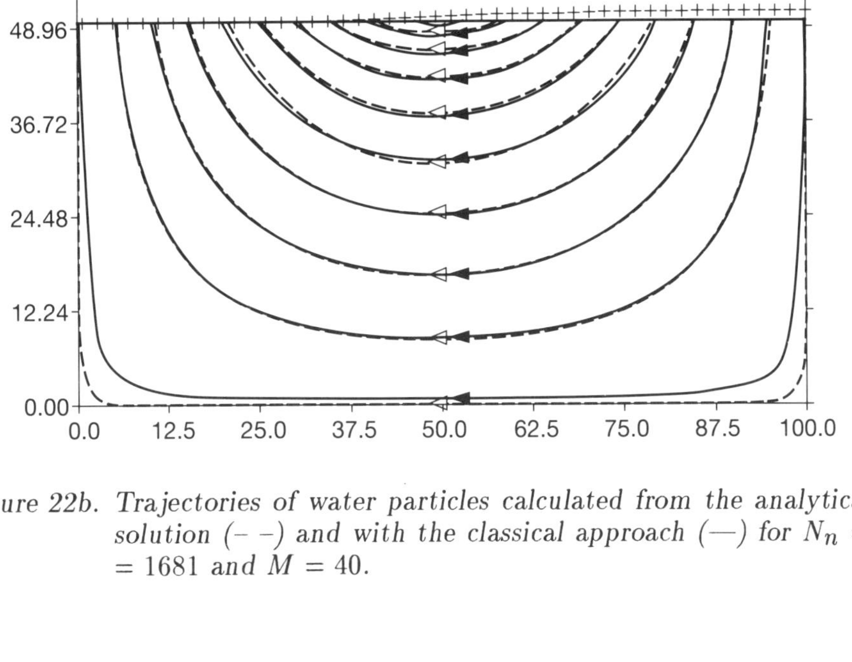

Figure 22b. *Trajectories of water particles calculated from the analytical solution (– –) and with the classical approach (—) for N_n = = 1681 and M = 40.*

Rectangular cavity problem

Assumptions

(i) The flow domain is of rectangular shape: $\Omega = [-L, L] \times [-D_2, D_1]$.
(ii) It is assumed that all the boundaries are impervious except for the upper halves of the eastern and the western boundaries on which a volumetric flow rate h is specified – see Figure 24.
(iii) The conductivity of the porous medium is homogeneous, i.e., k = constant.
(iv) There are steady-state conditions of flow, hence the potential $\phi = \phi(x, z)$.

Flow equation

$$\partial^2\phi/\partial x^2 + \partial^2\phi/\partial z^2 = 0, \quad \text{for } (x, z) \in \Omega. \tag{10.1.7}$$

Boundary conditions

$$\begin{aligned} &(1) \quad -k\ \partial\phi/\partial n = -h, \quad \text{for } (x, z) \in \{-L\} \times [0, D_1] \\ &(2) \quad -k\ \partial\phi/\partial n = +h, \quad \text{for } (x, z) \in \{L\} \times [0, D_1] \\ &(3) \quad \partial\phi/\partial n = 0, \quad \text{for the rest of the boundary.} \end{aligned} \tag{10.1.8}$$

Solution of the flow equation

Because of the symmetry of the flow domain we consider two subregions Ω_1 and Ω_2 in which the appropriate formulae

$$\phi_1(x, z), \quad \text{for } \Omega_1 = [0, L] \times [0, D_1]$$

and

$$\phi_2(x, z), \quad \text{for } \Omega_2 = [0, L] \times [-D_2, 0]$$

are to be found for $\phi(x, z)$.

Remark: Solutions for regions Ω_1' and Ω_2' can be calculated by substituting $x' := -x$ and $h' := -h$.

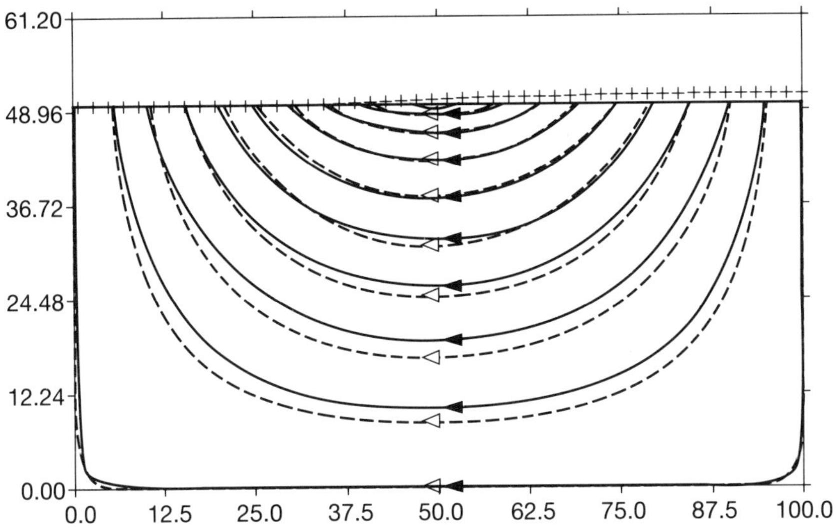

Figure 23a. *Trajectories of water particles calculated from the analytical solution (– –) and with the velocity-oriented approach (—) for* $N_n = 41$ *and* $M = 5$.

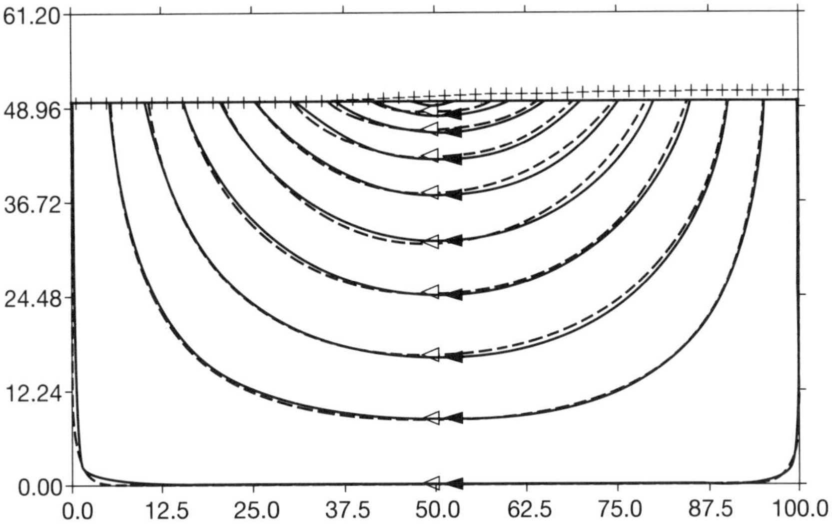

Figure 23b. *Trajectories of water particles calculated from the analytical solution (– –) and with the velocity-oriented approach (—) for* $N_n = 121$ *and* $M = 5$.

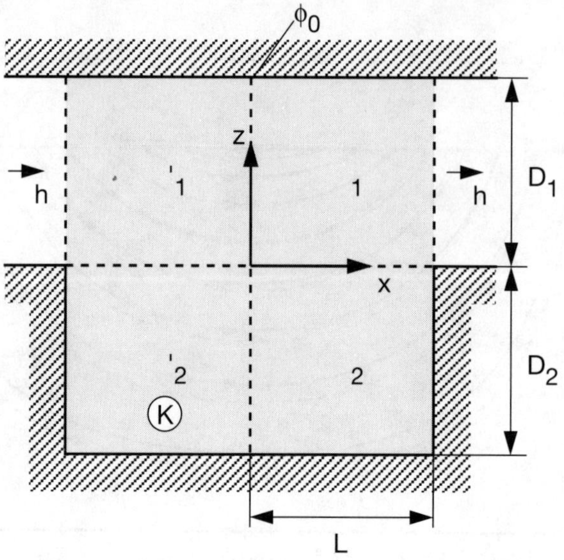

Figure 24. Flow domain and coordinate system.

Solutions in Ω_1 and Ω_2

$$\phi_1(x,z) = \phi_0 - hx/k + \sum_{k=1}^{\infty} A_k^1 \sin(\lambda_k x)\ \cosh\left[\lambda_k(D_1 - z)\right]. \quad (10.1.9a)$$

$$\phi_2(x,z) = \phi_0 - \sum_{k=1}^{\infty} A_k^2 \sin(\lambda_k x)\ \cosh\left[\lambda_k(D_2 + z)\right]. \quad (10.1.9b)$$

where

$$\lambda_k = (2k-1)\ \pi/2L, \quad (k = 1, 2, \ldots). \quad (10.1.10)$$

$$A_k^1 = 8L(-1)^{k+1}\ h/(\pi^2\ (2k-1)^2\ k) \cdot$$
$$\cdot\ \{\cosh(\lambda_k D_1) + \cosh(\lambda_k D_2)\sinh(\lambda_k D_1)/\sinh(\lambda_k D_2)\}^{-1} \quad (10.1.11)$$

$$A_k^2 = A_k^1 \sinh(\lambda_k D_1)/\sinh(\lambda_k D_2). \quad (10.1.12)$$

Figure 25 shows how the potential field changes within the rectangular cavity.

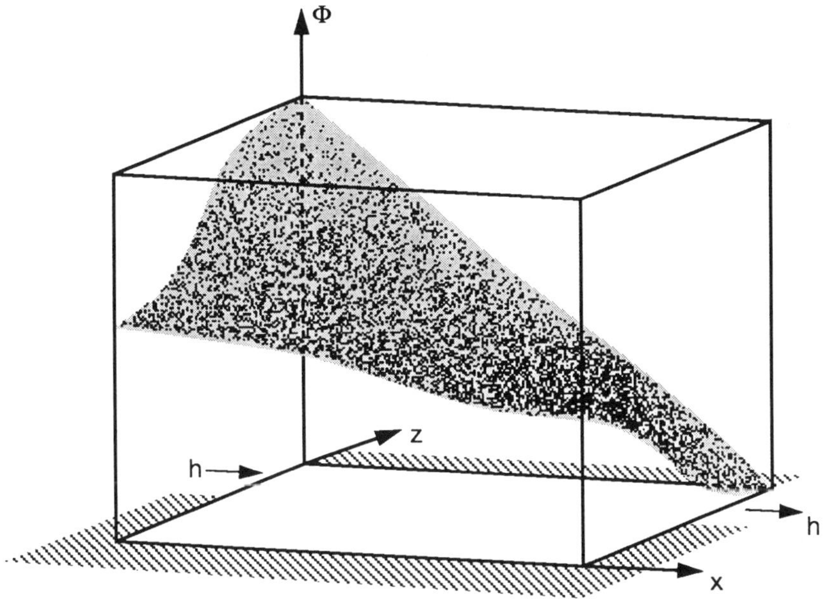

Figure 25. Analytical solution $\phi(x, z)$ over rectangular cavity; $L = 1\ m$, $D_1 = D_2 = 1\ m$, $k = 1\ m^2 \cdot dbar^{-1} \cdot s^{-1}$ and $h = 1\ m \cdot s^{-1}$.

Velocity field

By applying Darcy's law we can calculate the volumetric flow rate in regions Ω_1 and Ω_2:

$$\begin{cases} q_x^1(x,z) = h - k \sum_{k=1}^{\infty} A_k^1 \lambda_k \cos(\lambda_k x) \cosh\left[\lambda_k (D_1 - z)\right] \\ q_x^2(x,z) = k \sum_{k=1}^{\infty} A_k^1 \lambda_k \cos(\lambda_k x) \cosh\left[\lambda_k (D_2 + z)\right] \cdot \\ \quad \cdot \sinh(\lambda_k D_1)/\sinh(\lambda_k D_2) \\ q_z^1(x,z) = k \sum_{k=1}^{\infty} A_k^1 \lambda_k \sin(\lambda_k x) \sinh\left[\lambda_k (D_1 - z)\right] \\ q_z^2(x,z) = k \sum_{k=1}^{\infty} A_k^1 \lambda_k \sin(\lambda_k x) \sinh\left[\lambda_k (D_2 + z)\right] \cdot \\ \quad \cdot \sinh(\lambda_k D_1)/\sinh(\lambda_k D_2). \end{cases} \qquad (10.1.13)$$

Three-dimensional trajectories of the water particles, calculated with the England method (see Section 9.3), are shown in Figure 26.

Solutions to the classical and the VOA equations for the rectangular cavity problem have also been approximated numerically using the three-dimensional finite element method and then compared with the analytical solution (10.1.13) – see Figure 27. Again, the comparisons, given in terms of water particle trajectories, show that the VOA solutions are better than the classical ones especially in the region of low velocities.

In addition to the Tóth problem and the rectangular cavity problem for which the VOA approximation can be compared with the analytical solutions, a series of examples is given to illustrate differences in numerically approximated velocity fields obtained for the same physical situations using the classical and the VOA equations. Examples show – see Figures 28 and 29 – that the velocity-oriented model reaches stable solutions (i.e., solutions independent of the grid size) even for quite rare discretization grids, while the traditional simulator converges to the exact solution considerably slower. Also, for almost stagnant flows, trajectories obtained by the velocity-oriented approach exhibit much better physical behavior. As can be observed, the streamlines slide along impervious boundaries and smoothly change directions of movement at corners. Since boundaries between subsystems often pass stagnation points, this makes the approach more suitable for discriminating flow systems. In addition to the higher-order velocity approximation obtained by the velocity-oriented simulator, calculated trajectories may also be more reliable for flows that sharply change directions, e.g., close to pumping wells or along boundaries between strata having a high contrast in their conductivities – see Figure 30.

The above series of examples illustrate the accuracy and efficiency of the velocity-oriented approach. They also indicate perspectives of the approach in a vast area of environmental problems where decisions are based on discrimination of the groundwater flow systems followed by accurate simulations of pollution transport processes. For the latter, a velocity field plays a key role.

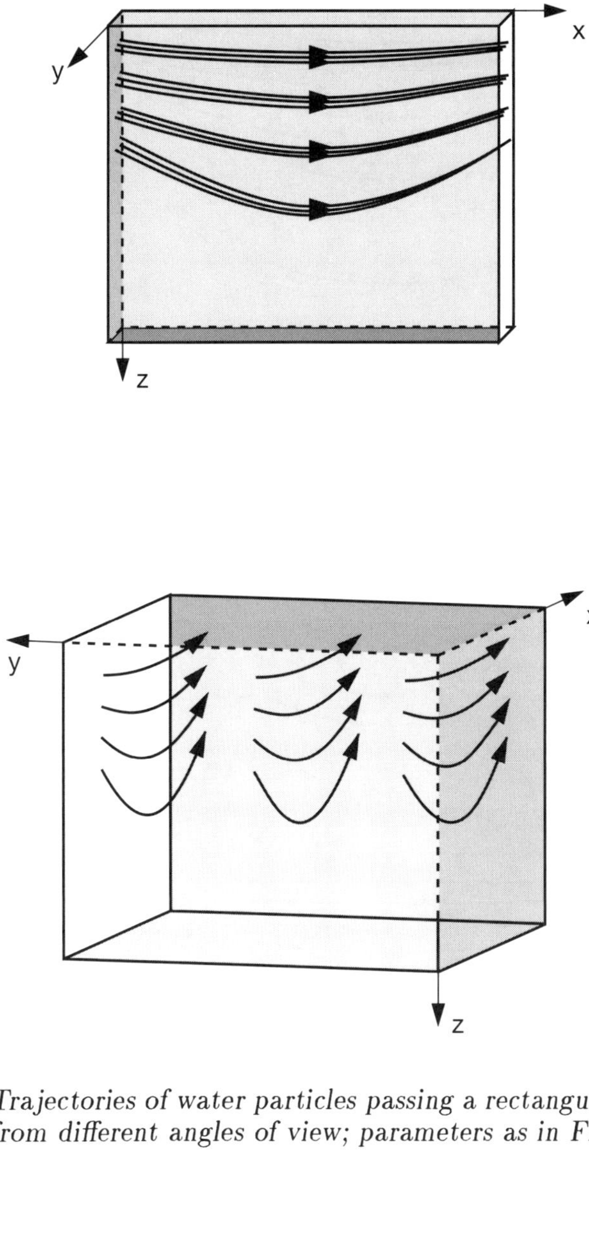

Figure 26. Trajectories of water particles passing a rectangular cavity seen from different angles of view; parameters as in Figure 25.

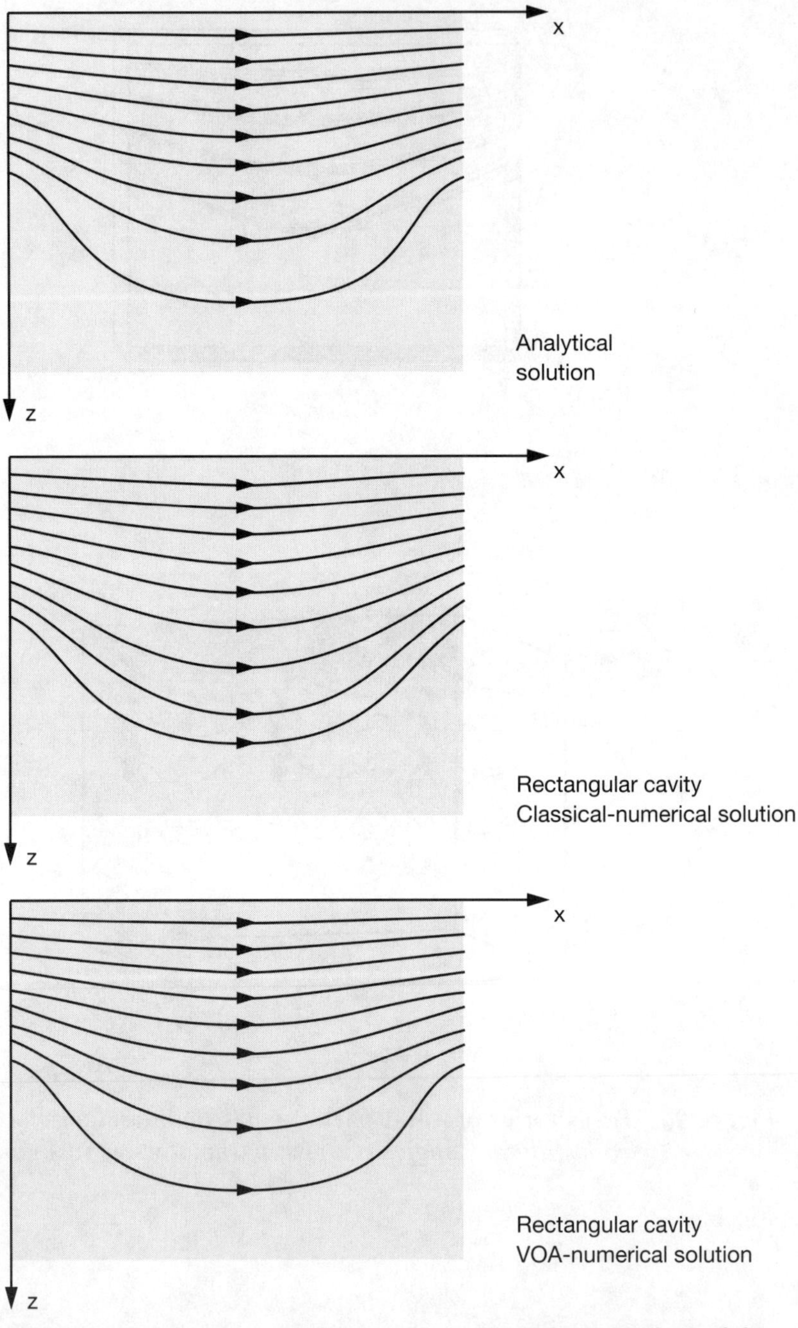

Figure 27. Trajectories of water particles calculated with an analytical formula (10.1.13) (above), with the classical method (by numerical differentiation) (in the middle) and with the VOA equations solved numerically by the finite element (bottom).

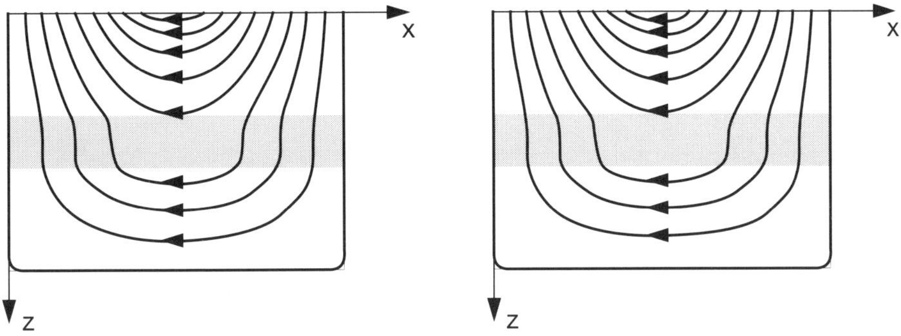

Figure 28a. *VOA trajectories for a groundwater flow induced by the water table conditions on the top boundary. The semipervious horizontal layer located in the middle partly isolates two aquifers. Calculations performed for* $N_1 = 300$ *(left) and* $N_2 = 1200$ *(right) finite element discretization nodes.*

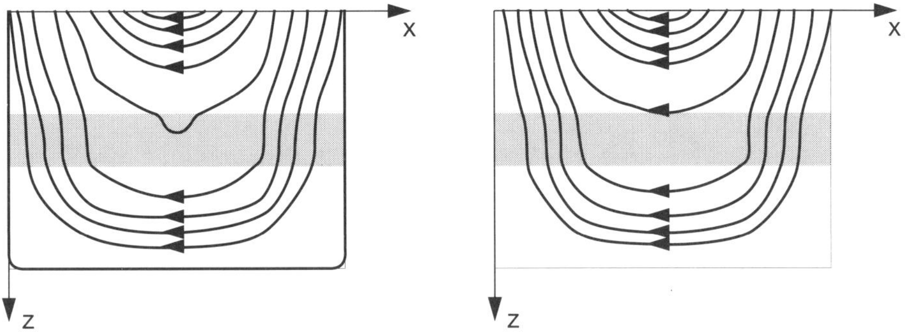

Figure 28b. *Classical trajectories for a groundwater flow induced by the water table conditions on the top boundary. The semipervious horizontal layer located in the middle partly isolates two aquifers. Calculations performed for* $N_1 = 300$ *(left) and* $N_2 = 1200$ *(right) finite element discretization nodes.*

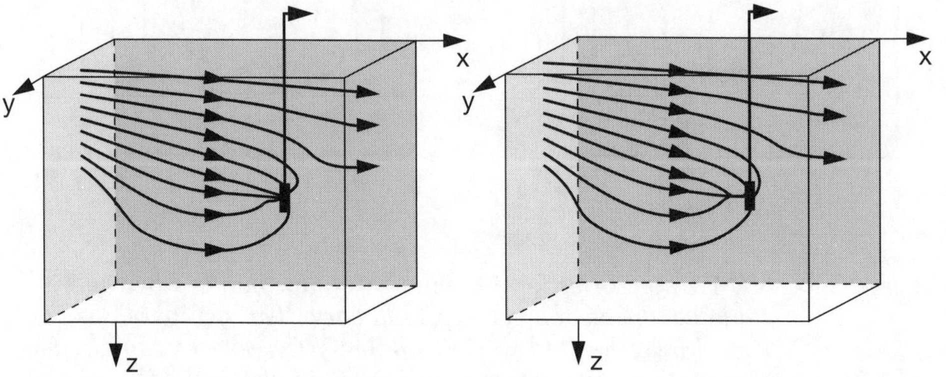

Figure 29a. *VOA trajectories for a groundwater flow induced by flow boundary conditions imposed on part of the vertical boundary and pumping rate of a well. Calculations performed for* $N_1 =$ $= 300$ *(left) and* $N_2 = 1200$ *(right) finite element discretization nodes.*

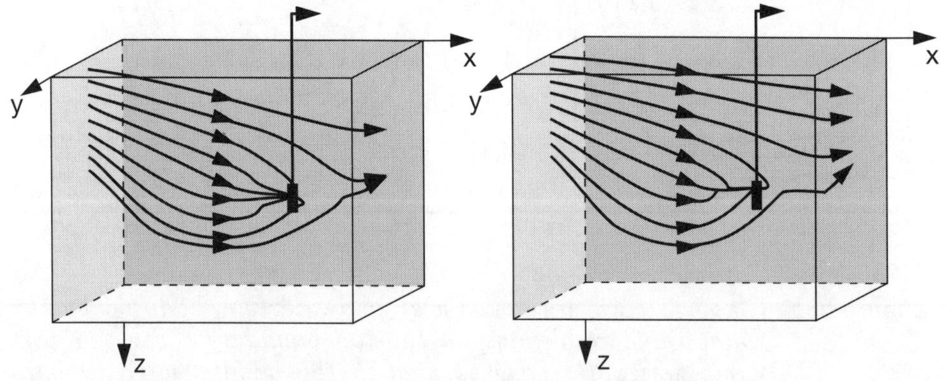

Figure 29b. *Classical trajectories for a groundwater flow induced by flow boundary conditions imposed on part of the vertical boundary and pumping rate of a well. Calculations performed for* $N_1 =$ $= 300$ *(left) and* $N_2 = 1200$ *(right) finite element discretization nodes.*

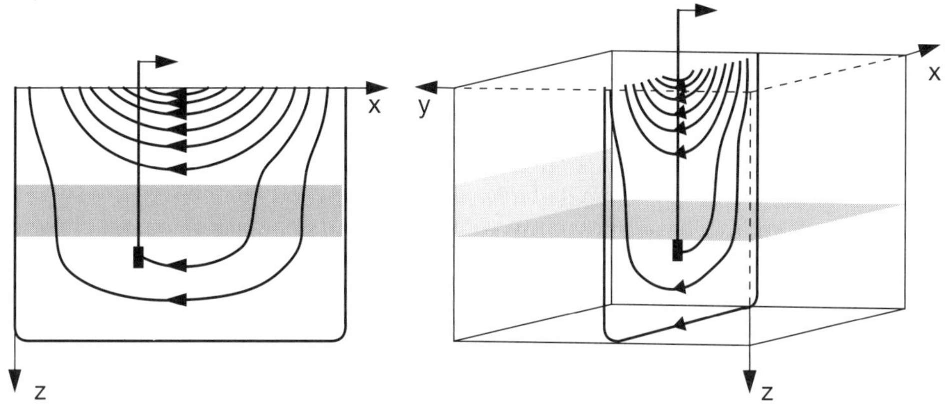

Figure 30. *Well located below a semipervious layer (anisotropy factor* $-1 : 100$*) can intercept some trajectories.*

10.2. Regional vs. local computations of groundwater flow

A dilemma of how to calculate groundwater flow on a regional scale, and accurately estimate local flow at the same time, was and still is manifesting itself in most of the papers on computer methods applied to hydrogeology and environmental protection. Grid refinement and high-order polynomial approximations used within the framework of the finite element method, though straightforward, require in many instances prohibitively large computer memories and/or long computing times.

There is, however, an alternative based on a decomposition of the groundwater system being simulated into its regional and local components. In such a hierarchical approach one can first compute the regional flow in terms of potential or the velocity components, and then descend into any of the existing local systems (e.g., into the vicinity of wells). Boundary conditions necessary for calculating flow in the local system can be specified from the numerical solution (potential) calculated for the regional system. A numerical method used in the local system can obviously be changed to satisfy the requirements of accuracy.

The approach has been applied to a realistic case in which a three-dimensional groundwater flow has been calculated for some hydrogeological regional system driven by the natural recharge and by the abstraction from a number of wells. This was done using the *classical three-dimensional finite element solver* with the potential as a primary variable. In the vicinity of wells, several local systems have been distinguished. For their boundaries, the calculated regional potentials have become boundary conditions. Finally, the *velocity-oriented approach* was used to calculate the velocity field accurately in the vicinity of wells.

By combining the regional and the local solutions one could answer questions on the residence time distributions, calculate accurate inverse trajectories, etc. The case calculated was supported by a three-dimensional graphical postprocessor which allowed visualization of water particles' trajectories.

For the two kinds of systems two different sets of state variables have been considered and analyzed:

(i) Potential was assumed to be a primary state variable in the regional flow system, while the components of the volumetric flow rate (obtainable through Darcy's law) were merely the secondary state variables.

(ii) Contrary to the regional approach, there were components of the volumetric flow rate used as the primary state variables in the local flow system while the potential was the secondary variable.

To illustrate how the decomposition works, the classical flow equation has been solved for a small 'regional system,' whereas the VOA equations have been used for the 'local system' shown in Figure 31.

The following set of parameters has been used (see Nawalany, 1990):

'Regional system'

origin of the system: $(x_{r0}, y_{r0}, z_{r0}) = (0,0,0)$

size of the system: $(L_x, L_y, L_z) = (100\ m,\ 100\ m,\ 10\ m)$

conductivity: $k = 10\ m^2 \cdot dbar^{-1} \cdot d^{-1}$

porosity: $\theta = 0.2$

number of wells: $N_w = 3$

positions of the wells: $(x_{w1}, y_{w1}, z^t_{w1}, z^b_{w1}) = (20\ m,\ 20\ m,\ -4\ m,\ -8\ m)$

$(x_{w2}, y_{w2}, z^t_{w2}, z^b_{w2}) = (40\ m,\ 60\ m,\ -1\ m,\ -4\ m)$

$(x_{w3}, y_{w3}, z^t_{w3}, z^b_{w3}) = (70\ m,\ 80\ m,\ -1\ m,\ -6\ m)$

production rates: $(Q_1, Q_2, Q_3) = (1000\ m^3 \cdot d^{-1}, 500\ m^3 \cdot d^{-1},\ 500\ m^3 \cdot d^{-1})$

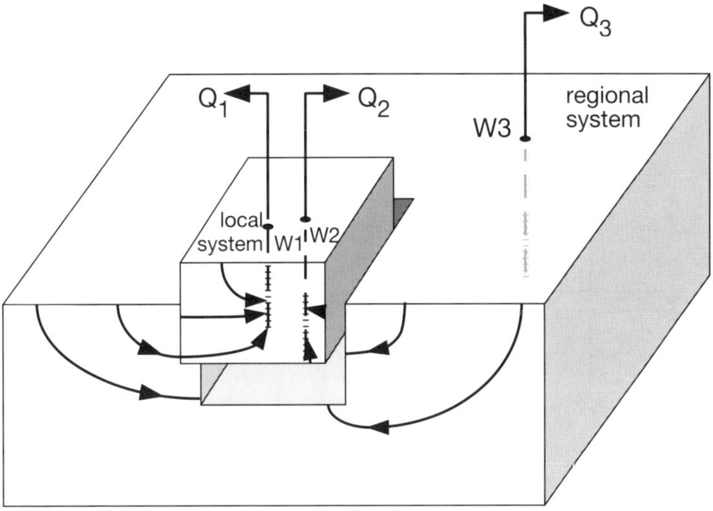

Figure 31. Regional and local groundwater flow systems. The local system is extracted from the regional one.

boundary conditions: for W, N, E, S, B : $q_n = 0\ m{\bullet}d^{-1}$
for T: specified potential in the form of a quadratic spline function compatible with other boundary conditions on the top boundary.

'Local system'

origin of the system:	$(x_{l0}, y_{l0}, z_{l0}) = (10\ m,\ 10\ m,\ 10\ m)$
size of the system:	$(L_{xl}, L_{yl}, L_{zl}) = (55\ m,\ 55\ m,\ 8\ m)$
conductivity:	$k = 10\ m^2{\bullet}dbar^{-1}{\bullet}d^{-1}$
porosity:	$\theta = 0.2$
number of wells:	$N_w = 2$
positions of the wells:	as before
production rates:	as before
boundary conditions:	for W, N, E, S, B and T: potential specified from the regional solution.

Figures 32a and 32b show the distributions of the potential along the horizontal planes at the 5 m depths in the *regional* and *local systems* respectively. Figures 33a and 33b present a set of pathlines which start on the top boundary and are either attracted by the wells in the local system or leave the system.

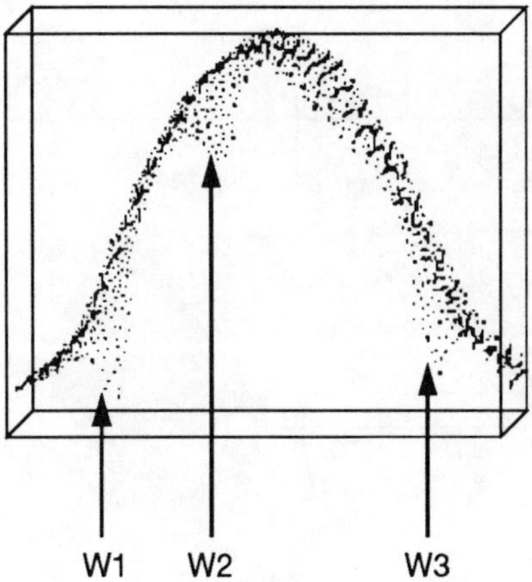

Figure 32a. Potential distribution in the regional system for $z = = -5\ m$. Depression cones caused by the wells w_1, w_2 and w_3 are clearly visible.

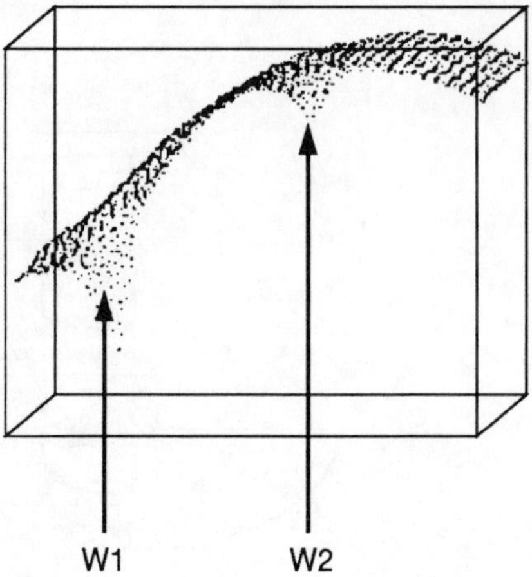

Figure 32b. Potential distribution in the local system for $z = = -5\ m$. The picture is distorted due to normalizing the local system.

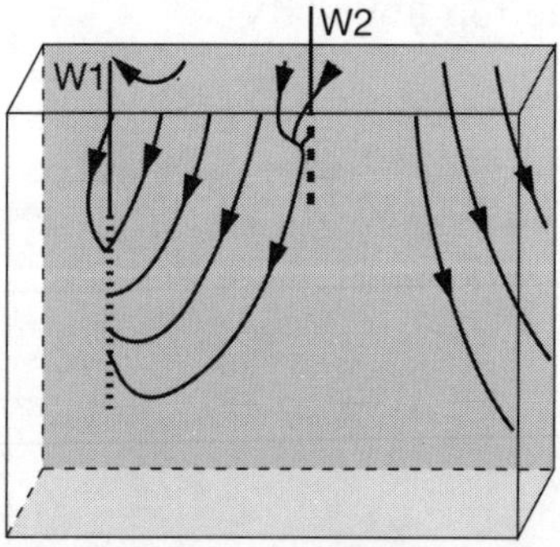

Figure 33a. Water trajectories in the local system. Some trajectories are captured by wells.

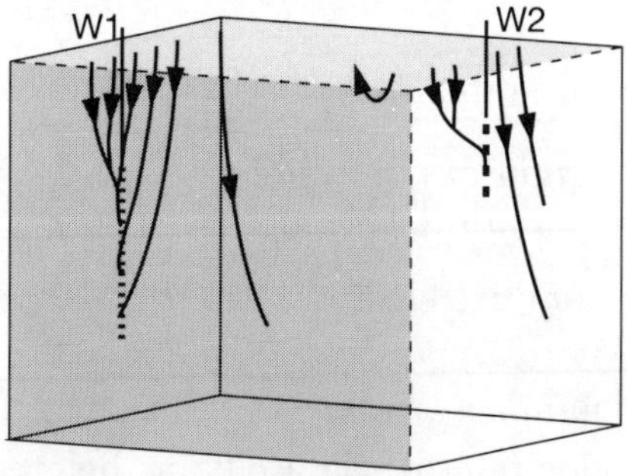

Figure 33b. The same trajectories seen from another angle of view.

Remarks

(1) By introducing global and local systems it becomes possible to overcome the unfavorable geometric aspect ratio between the dimensions of the flow domain and the distances between wells. Also calculating accurate pathlines within a local system is becoming feasible.

(2) By distributing wells, i.e., by representing their screens as a series of point sinks/sources, one can obtain a more realistic picture of the potentials and pathlines in the three-dimensional flow domain (see Section 8.4).

(3) Introducing the linear model for the conductivity makes it possible to deal with heterogeneity and have an analytical (and quick) method of assembling the VOA equations (see Section 8.3).

Inverse pathlines

The inverse pathlines calculations are related to the specific decision-making problem. The problem can be characterized shortly by the following question: 'If the water abstracted by a given well is polluted, where does the pollution come from?' The problem can therefore be named 'the polluter detection problem.' To find the origin of the water abstracted by the well one needs to follow the water trajectories backwards. By choosing starting points of sufficiently many water paths in the vicinity of the well's screen, and following the England algorithm inversely in time, one can discriminate all the recharge areas (sources) from which the well is supplied. For this, an accurate approximation of the velocity field is needed, otherwise numerical instability of the inverse trajectory may occur. Therefore the VOA may become indispensible. Figure 34 illustrates the inverse path computations for a well being polluted from some waste disposal site. Naturally, all the time characteristics, like the traveling time distribution or age distribution of the water being abstracted, can readily be calculated.

10.3. Hydraulic isolation of waste disposal sites

While solving the groundwater flow equations with the velocity-oriented approach, one must solve three coupled PDEs for three velocity components v_x, v_y and v_z as the unknowns. In general, a coupling results both from the complicated subsurface geometry and from the boundary conditions. However, when a geometry is simple (the subsurface is perfectly

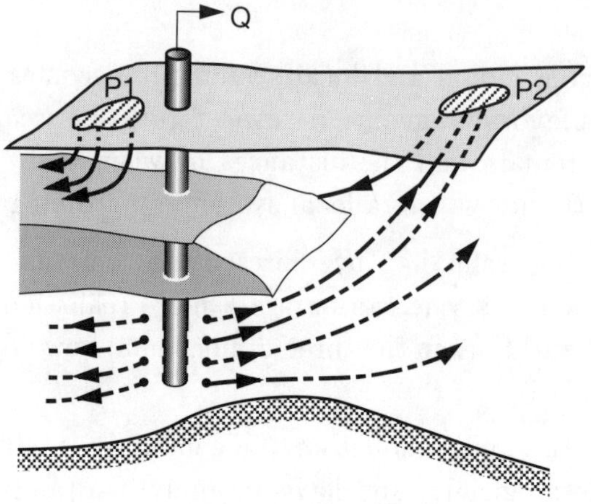

Figure 34. Inverse trajectories (dashed lines) used to detect the polluter.

layered and its boundaries can be considered either horizontal or vertical), only the third type boundary condition causes coupling. Since any decoupling is beneficial from a numerical and computational point of view, the attempt can be made to keep the VOA equations for v_x, v_y and v_z separated and solved in sequence. Decoupling is possible after introducing an additional equation for the potential ϕ and solving it numerically as the first equation. The example shows the hydraulic isolation of the waste disposal site (WDS) in which a local groundwater flow system was artificially created below the bottom of the WDS. A system of simple rectangular ditches with an assumed difference in their water levels was used to create the local flow system. When investigating the impact of different geometries and the bottom resistivities of the ditches on the effectiveness of the WDS hydraulic isolation, very accurate calculations of the velocity field have become necessary. The assumption that water in ditches interacts with groundwater through some semipervious layer of bottom sediments resulted in introducing the third type boundary condition into the problem formulation, thus causing coupling of the VOA equations. The linear finite element method was used to solve the equations with a discretization grid changing its spatial density by two orders of magnitude. Adding the equation for ϕ and subsequent decoupling of the VOA equations was what made all the computations feasible. An efficient postprocessor based on the England method (see Section 9.3) was used to calculate and then visualize the trajectories of water particles below the WDS (see Nawalany *et al.*,

1991).

The basic assumption of the *perfectly layered subsurface* can be expressed in terms of the horizontal and the vertical conductivities:

$$k_h = k_h(z) \text{ and } k_z = k_z(z). \tag{10.3.1}$$

For subsurfaces that fulfill assumption (10.3.1) the VOA equations of the groundwater flow become considerably simpler, i.e., the equations for the horizontal and the vertical components of the groundwater flow can then be decoupled. Still the simple geometry needs to be retained for decoupling. In addition to assumption (10.3.1) it is also assumed that the potential ϕ in the subsurface is a continuous function of the x- and z-coordinates up to the second derivative inclusive. In particular, it satisfies the well-known relationship:

$$\partial^2\phi/\partial x\partial z = \partial^2\phi/\partial z\partial x. \tag{10.3.2}$$

The system of coordinates (x, z) is oriented as follows: x-axis is pointing eastwards and the z-axis is pointing upwards. The boundary of the flow domain Ω is assumed to be piecewise vertical or horizontal, though Ω is not necessarily convex. The vertical boundaries are abbreviated here as V and the horizontal boundaries as H.

The classical groundwater flow equation is expressed in terms of the potential ϕ; that is:

$$\begin{aligned}&\partial/\partial x\ (k_h\ \partial\phi/\partial x) + \partial/\partial z\ (k_z\ \partial\phi/\partial z) = 0,\\ &\text{for all } (x, z) \in \Omega.\end{aligned} \tag{10.3.3}$$

The unique solution for $\phi = \phi(x, z)$ exists in Ω provided one of the following boundary conditions is specified on each segment of the flow domain Ω:

I-type b.c. (potential specified)

$$\phi = \begin{cases} f(x), & \text{on } H\text{-segments of } \partial\Omega \\ f(z), & \text{on } V\text{-segments of } \partial\Omega \end{cases} \tag{10.3.4}$$

where $f(\cdot)$ is the known (specified) function.

***II*-type b.c. (volumetric flow rate normal to $\partial\Omega$ specified)**

$$q_n = \begin{cases} h(x), & \text{on } H\text{-segments of } \partial\Omega \\ h(z), & \text{on } V\text{-segments of } \partial\Omega \end{cases} \tag{10.3.5}$$

where $h(\cdot)$ is the known (specified) function.

***III*-type b.c. (volumetric flow rate normal to $\partial\Omega$ specified in terms of ϕ)**

$$q_n = (\phi - \phi_0)/c \tag{10.3.6}$$

where ϕ_0 is the specified value of the potential 'outside' Ω; it can be a function of x on H-segments of $\partial\Omega$ and of z on V-segments of $\partial\Omega$; c is the 'resistivity' parameter known or estimated.

After introducing a notation:

$$e_x = -\partial\phi/\partial x = q_x/k_h \tag{10.3.7}$$

the VOA flow equations can be written as follows:

$$\partial/\partial x\ (k_h\ \partial e_x/\partial x) + \partial/\partial z\ (k_z\ \partial e_x/\partial z) = 0, \quad \text{for all } (x,z) \in \Omega \tag{10.3.8}$$

$$\partial/\partial x\ \left(k_z^{-1} \partial q_z/\partial x\right) + \partial/\partial z\ \left(k_h^{-1}\ \partial q_z/\partial z\right) = 0, \quad \text{for all } (x,z) \in \Omega. \tag{10.3.9}$$

Both equations (10.3.8) and (10.3.9) are of elliptic type and hence their unique solutions exist provided the appropriate boundary conditions are additionally defined. The boundary conditions for the VOA theory are discussed below for H-horizontal and for V-vertical segments of $\partial\Omega$, respectively. Components of the unit vector normal to $\partial\Omega$ are denoted as $\boldsymbol{n} = (n_x, n_z)$.

VOA BOUNDARY CONDITIONS FOR HORIZONTAL BOUNDARIES

I-type b.c. (potential specified)

In this case:

$$\phi = f(x), \text{ where } f \text{ is the known function.}$$

Boundary conditions for equations (10.3.8) and (10.3.9) are therefore as follows:

$$\begin{cases} e_x = -\partial f/\partial x, & \text{(i.e., } I\text{-type b.c. for } e_x) \\ \partial q_z/\partial n = n_z k_h\ \partial^2 f/\partial x^2, & \text{(i.e., } II\text{-type b.c. for } q_z). \end{cases} \tag{10.3.10}$$

II-type b.c. (volumetric flow rate normal to $\partial\Omega$ specified)

Similarly for:

$$q_n = h(x), \text{ where } h \text{ is the known function,}$$

one obtains:

$$\begin{cases} q_z = n_z h(x), & \text{(i.e., } I\text{-type b.c. for } q_z) \\ \partial e_x/\partial n = k_z^{-1}\ \partial h/\partial x, & \text{(i.e., } II\text{-type b.c. for } e_x). \end{cases} \tag{10.3.11}$$

III-type b.c. (volumetric flow rate normal to $\partial\Omega$ specified in terms of ϕ)

In this case:

$$q_n = (\phi - \phi_0)/c.$$

Consequently:

$$\begin{cases} \partial e_x/\partial n + c^{-1} k_z^{-1}\ e_x = 0, & \text{(i.e., } III\text{-type b.c. for } e_x) \quad (10.3.12a) \\ \partial q_z/\partial x + n_z\ c^{-1}\ e_x = 0, & \text{(i.e., } I\text{-type b.c. for } q_z). \quad (10.3.12b) \end{cases}$$

where it has been assumed that ϕ_0 and c are constants.

Remark

In principle, boundary conditions for q_z can be calculated after equation (10.3.8) for e_x is solved with the boundary condition (10.3.12a). Then

$$q_z(x) = -c^{-1}\, n_z \int e_x(x)\, \mathrm{d}x + const.$$

The constant *const* is, however, unknown unless some additional information is given, for instance, when the total flow through a given (horizontal) segment of the boundary is specified. In a case where ϕ is known, the boundary condition for q_z can be readily calculated from a simple formula:

$$q_z = n_z(\phi - \phi_0)/c. \tag{10.3.12c}$$

VOA BOUNDARY CONDITIONS FOR VERTICAL BOUNDARIES

Boundary conditions for a vertical boundary are listed below in a similar way as for the horizontal boundary.

I-type b.c. (potential specified)

For:

$$\phi = f, \text{ where } f \text{ is the known function,}$$

one obtains the sought boundary conditions for equations (10.3.8) and (10.3.9):

$$\begin{cases} q_z = -k_z\, \partial f/\partial z, & \text{(i.e., } I\text{-type b.c. for } q_z) \\ \partial e_x/\partial n = n_x k_h^{-1}\ \partial/\partial z\ [k_z\, \partial f/\partial z], & \\ & \text{(i.e., } II\text{-type b.c. for } e_x). \end{cases} \tag{10.3.13}$$

II-type b.c. (volumetric flow rate normal to $\partial\Omega$ specified)

If q_n is known, i.e.,

$$q_n = h(z), \text{ where } h \text{ is the known function,}$$

then:

$$\begin{cases} e_x = n_x k_h^{-1}\, h, & \text{(i.e., } I\text{-type b.c. for } e_x) \\ \partial q_z/\partial n = k_z\, \partial(k_h^{-1}\, h)/\partial z & \\ & \text{(i.e., } II\text{-type b.c. for } q_z). \end{cases} \tag{10.3.14}$$

III-type b.c. (volumetric flow rate normal to $\partial\Omega$ specified in terms of ϕ)

In this case, i.e., when:

$$q_n = (\phi - \phi_0)/c$$

one obtains:

$$\begin{cases} \partial q_z/\partial n + c^{-1}k_h^{-1}\, q_z = -n_x k_z k_h^{-1}\, (\partial k_h/\partial z)\, e_x, & \\ \qquad \text{(i.e., } III\text{-type b.c. for } q_z) & (10.3.15\text{a}) \\ \partial(k_h\, e_x)/\partial z = -n_x c^{-1} k_z^{-1}\, q_z & \\ \qquad \text{(i.e., } I\text{-type b.c. for } e_x). & (10.3.15\text{b}) \end{cases}$$

where it has been assumed that ϕ_0 and c are constants.

The last formulae clearly indicate coupling between equation (10.3.8) and (10.3.9) due to the existence of the b.c. of the III-rd type on the vertical boundary of the flow domain Ω.

Remark

If ϕ is known (calculated), equations (10.3.8) and (10.3.9) can be decoupled:

$$e_x = n_x k_h^{-1}\, (\phi - \phi_0)/c. \tag{10.3.16}$$

Decoupling the VOA equations of groundwater flow has been successfully used for calculating the effectiveness of hydraulic isolation of waste disposal sites. The latter is one of the methods of protecting groundwater against pollution from uncontrolled (nonisolated) WDS. Technically wells, ditches or drains are used to create a local groundwater flow system that forms a hydraulic envelope enclosing the WDS from beneath. A system of two parallel ditches on both sides of the WDS has been tested as a possible means of hydraulic isolation. A simple geometry of the WDS and of the ditches has been assumed as shown in Figures 35 and 36. The WDS comprises a part of the groundwater flow domain boundary on which constant normal flux is prescribed. The ditches make a part of the boundary with the third type boundary conditions specified.

The effectiveness of the hydraulic isolation tested depends on: differences in water levels in the ditches, their bottom resistivities c and the geometric aspect ratios between the distances involved. Figure 35 shows the situation when the hydraulic isolation is not effective, i.e., when the hydraulic envelope is not closed due to a nonsufficient (too small) difference between the water levels in the ditches. In that case, flowlines originating at the bottom of the WDS join the regional groundwater flow system. When a difference in the water levels becomes considerably larger, the envelope is created and consequently every flowline starting from the WDS is ultimately discharged into one of the ditches (see Figure 36).

In all computer simulations of the hydraulic isolation described above, the classical flow equation (10.3.3) has been added to the VOA equations, thus making the decoupling of the equations for v_x and v_z possible. Linear finite elements have been applied to discretize the flow equations (10.3.3), (10.3.8) and (10.3.9) within the domain Ω. The fast ICCG solver for the resultant algebraic equations and the England method for the trajectories' calculations have been used for calculating the velocity field and for post-processing.

10.4. Contaminant transport

A detrimental impact of uncontrolled waste disposals on groundwater does manifest itself through the actual or foreseen contamination patterns. The main factors which determine the origin and fate of pollutants washed out from waste disposals can be summarized as follows:

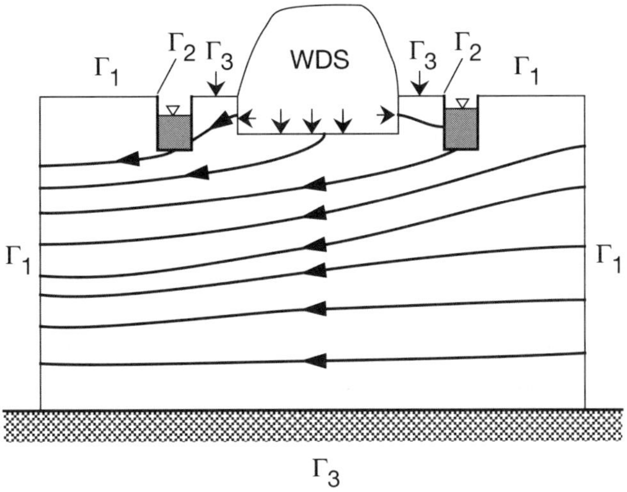

Figure 35. Flowlines beneath the WDS for the noneffective hydraulic isolation (Γ_1 : I-type b.c., Γ_2 : II-type b.c., Γ_3 : III-type b.c.).

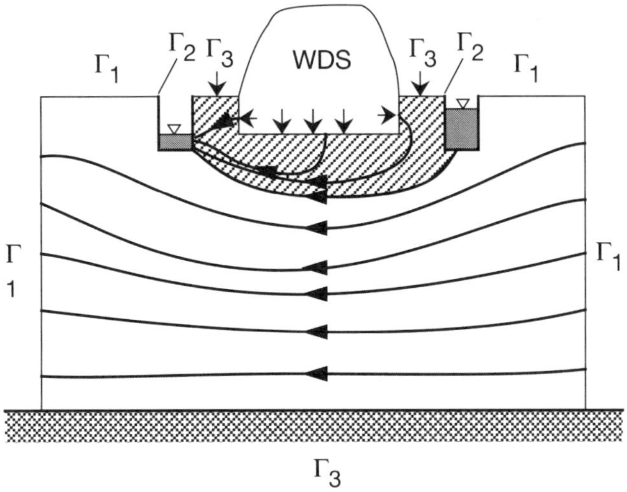

Figure 36. The hydraulic isolation is effective – the envelope is closed (Γ_1 : I-type b.c., Γ_2 : II-type b.c., Γ_3 : III-type b.c.).

- type of waste;
- wash-out mechanism;
- interaction of the pollutants with solids;
- chemical reactions in the water body;
- convection;
- hydrodynamic dispersion.

In the example described below the last two phenomena, responsible for a mass transport, are investigated.

The convection-dispersion equation (CDE) is assumed to describe sufficiently well a state of the aquifer polluted by some soluble substance washed-out from a waste disposal. The aim is to build a three-dimensional numerical model that solves the convection-dispersion equation with the boundary conditions characteristic for waste disposals.

The state variable that describes a space-time evolution of solute transport in porous medium is the mass fraction ω of the pollutant expressed in kilograms of the solute mass per unit mass of water-solute mixture (see Section 6.7). It is assumed that a velocity field $\boldsymbol{v} = \boldsymbol{q}/\theta$ is calculated by the three-dimensional velocity-oriented groundwater flow simulator prior to mass fraction calculations. Here θ is the total porosity, *not* the effective porosity; see Sections 6.2 and 6.6. Time evolution of a solute mass fraction ω in the flow domain Ω is described by the following CDE (see Chapter 6):

$$\partial(\theta\rho\omega)/\partial t + \operatorname{div}(\theta\rho\boldsymbol{v}) = \operatorname{div}(\theta\rho\underline{\boldsymbol{D}}\boldsymbol{\cdot}\,\mathbf{grad}\,\omega) \tag{10.4.1a}$$

Under the condition that $\operatorname{div}\boldsymbol{q} = 0$ is a good approximation to the continuity equation, equation (10.4.1a) simplifies to:

$$\begin{aligned}&\theta\;\partial\omega/\partial t + \operatorname{div}(\omega\boldsymbol{q}) = \operatorname{div}(\theta\underline{\boldsymbol{D}}\boldsymbol{\cdot}\,\mathbf{grad}\,\omega)\\ &\text{for every } \boldsymbol{x} = (x,y,z) \in \Omega \text{ and } t \geq 0\end{aligned} \tag{10.4.1b}$$

with *initial conditions*: $\omega(\boldsymbol{x},0) = \omega_0(\boldsymbol{x})$, for $\boldsymbol{x} \in \Omega$, and
with *boundary conditions*, which can be one of the three types:

$$\begin{aligned}&\text{(i) } \textit{Dirichlet type: } \omega(\boldsymbol{x},t) = \mathring{\omega}(\boldsymbol{x},t), && \text{for } \boldsymbol{x} \in \Gamma_1;\\ &\text{(ii) } \textit{Neumann type: } -\underline{\boldsymbol{n}}\boldsymbol{\cdot}(\underline{\boldsymbol{D}}\boldsymbol{\cdot}\,\mathbf{grad}\,\omega) = \mathring{q}_n(\boldsymbol{x},t), && \text{for } \boldsymbol{x} \in \Gamma_2;\\ &\text{(iii) } \textit{Cauchy type: } v_n\omega - \underline{\boldsymbol{n}}\boldsymbol{\cdot}(\underline{\boldsymbol{D}}\boldsymbol{\cdot}\,\mathbf{grad}\,\omega) = \mathring{q}_n(\boldsymbol{x},t), && \text{for } \boldsymbol{x} \in \Gamma 3,\end{aligned} \tag{10.4.2}$$

where $\overset{\circ}{\omega}(\boldsymbol{x},t)$ and $\overset{\circ}{q}_n(\boldsymbol{x},t)$ are known functions specified on the boundary $\partial\Omega$, $\Gamma_i \subset \partial\Omega$ $(i = 1,2,3)$, $\Gamma_i \cap \Gamma_j = \emptyset$, for $i \neq j$ and $\Gamma_1 \cup \Gamma_2 \cup \Gamma_3 = \partial\Omega$.

Contaminant transport in Ω is influenced by the velocity of water in at least two ways:

(i) through the *convective flux* $\boldsymbol{v}\omega$, and
(ii) through the *dispersive flux* $-\underline{\boldsymbol{D}}\cdot\, \mathbf{grad}\, \omega$

where $\underline{\boldsymbol{D}}$ is a function of $\boldsymbol{v}$.

When the mass fraction of the contaminant is approximated with some numerical method, the accuracy of the velocity approximation used has an obvious impact on the results.

Let us consider the convection terms first. If there would be no dispersion nor diffusion, solute particles and water particles should follow the same trajectories. Although the trajectories' spatial distributions would be precisely the same, the time characteristics (such as age or traveling time distributions) could differ because of physical or chemical interactions of solute with the soil matrix or simply because of the isotope effect. Any change of a velocity field causes the time characteristics to change accordingly. The picture becomes even more complicated when the second velocity dependent process – dispersion – is to be taken into account. The elements of the dispersion tensor depend on the velocity components in the following way (see Section 6.3):

$$D_{ij} = (D_m + \alpha_T \mid \boldsymbol{v} \mid)\, \delta_{ij} + (\alpha_L - \alpha_T)\, v_i v_j / \mid \boldsymbol{v} \mid, \quad (i,j = 1,2,3) \qquad (10.4.3)$$

where α_T and α_L are the transversal and longitudinal dispersion lengths, respectively, and D_m is the molecular diffusion coefficient.

Since a dispersion process by its very nature allows the solute particles to depart from convective trajectories even for 'exact' velocity fields, the relationship sought between numerically calculated velocities and time characteristics of contaminant transport is strongly influenced by the nonlinear character of formula (10.4.3).

Numerical approximation to the solution of the CDE

A particular scenario has been considered in which a waste disposal site (WDS) was located on the top of a flow domain. The WDS was assumed to be easily penetrable for the precipitation, so it did not disturb the original recharge nor the groundwater velocity field within the aquifer. It was

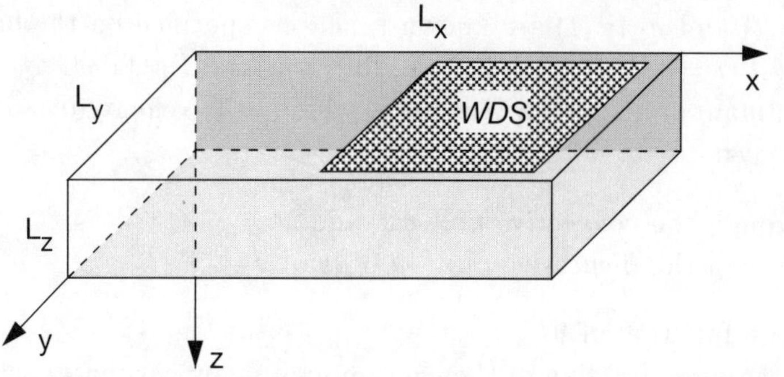

Figure 37. Flow domain and a system of coordinates.

assumed, however, that water percolating to the flow domain was characterized by a constant mass fraction of some soluble pollutant. The VOA simulator has been used to 'generate' an accurate velocity field.

To solve equation (10.4.1b) numerically the centered scheme of the alternating direction method (ADM) proposed by Frind (1984) has been adopted. The algorithm proposed below follows the major steps of Frind's publication (see also Nawalany, 1992).

(i) The flow domain is of rectangular shape $\Omega = [0, L_x] \times [0, L_y] \times [0, L_z]$.

(ii) *The flow domain* is *discretized* with a set of nodes, which are equidistant in each direction, i.e.,

$$\Delta x = 2a = L_x/(N_x - 1)$$
$$\Delta y = 2b = L_y/(N_y - 1)$$
$$\Delta z = 2c = L_z/(N_z - 1)$$

where N_x, N_y and N_z are numbers of nodes in the x-, y-, and z-directions, respectively. Consequently, the total number of nodes is $N_n = N_x N_y N_z$. Any node $n = (i, j, k)$ has its coordinates given by the following formulae:

$$\begin{cases} x_n = \Delta x(i-1), \ (i = 1, \ldots, N_x) \\ y_n = \Delta y(j-1), \ (j = 1, \ldots, N_y) \\ z_n = \Delta z(k-1), \ (k = 1, \ldots, N_z). \end{cases} \tag{10.4.4}$$

(iii) Since the Galerkin finite element technique is used to approximate a solution to the convection-dispersion equation, a family of *basic functions* is defined. For every node $n = (i, j, k)$ there are three basic functions $w_i^n(x, y, z)$, $w_j^n(x, y, z)$ and $w_k^n(x, y, z)$. Their supports are as follows:

$$\begin{aligned}
\Omega_{ijk}^x &= \Omega \cap [x_n - 2a,\ x_n + 2a] \times \\
&\quad \times [y_n - b,\ y_n + b] \times [z_n - c,\ z_n + c] \\
\Omega_{ijk}^y &= \Omega \cap [x_n - a,\ x_n + a] \times \\
&\quad \times [y_n - 2b,\ y_n + 2b] \times [z_n - c,\ z_n + c] \\
\Omega_{ijk}^z &= \Omega \cap [x_n - a,\ x_n + a] \times \\
&\quad \times [y_n - b,\ y_n + b] \times [z_n - 2c,\ z_n + 2c].
\end{aligned} \tag{10.4.5}$$

The basic functions themselves are defined in a natural way:

$$w_i^n(x, y, z) \hat{=} w_i(x) = \begin{cases} \Phi_a(x - x_n), & \text{for } (x, y, z) \in \Omega_{ijk}^x, \\ 0, & \text{otherwise,} \\ & (i = 1, \ldots, N_x) \end{cases}$$

$$w_j^n(x, y, z) \hat{=} w_j(y) = \begin{cases} \Phi_b(y - y_n), & \text{for } (x, y, z) \in \Omega_{ijk}^y, \\ 0, & \text{otherwise,} \\ & (j = 1, \ldots, N_y) \end{cases}$$

$$w_k^n(x, y, z) \hat{=} w_k(z) = \begin{cases} \Phi_c(z - z_n), & \text{for } (x, y, z) \in \Omega_{ijk}^z, \\ 0, & \text{otherwise,} \\ & (k = 1, \ldots, N_z) \end{cases}$$

(10.4.6)

where

$$\Phi_d(t) = 1 - |t|/2d, \text{ for } t \in [-2d, 2d]. \tag{10.4.7}$$

(iv) *Alternating direction method (centered scheme).* The solution to the problem (10.4.1b)-(10.4.2) can be approximated numerically with the 'splitting operator' technique. First an approximation is introduced; it is assumed that the x-, y- and z-directions coincide with the principal axes of the dispersion tensor $\underline{\boldsymbol{D}}$. In that case only the diagonal terms of the dispersion tensor are nonzero. After introducing the following notations:

$$\begin{aligned} \mathbb{L}_x &= \partial/\partial x \ (D_{xx}\ \partial/\partial x) \\ \mathbb{L}_y &= \partial/\partial y \ (D_{yy}\ \partial/\partial y) \\ \mathbb{L}_z &= \partial/\partial z \ (D_{zz}\ \partial/\partial z) \end{aligned} \tag{10.4.8}$$

the sought approximation ω^{m+1} of the mass fraction ω for the $(m+1)$-th instant of time can be found by breaking the centered scheme into three smaller steps in time (see Frind, 1984):

Step 1:

$$\begin{aligned} &3/2\ \mathbb{L}_x\omega^{m+(1/3)} - 1/2\ \mathbb{L}_x\omega^m + \mathbb{L}_y\omega^m + \mathbb{L}_z\omega^m = \\ &\quad = (\omega^{m+(1/3)} - \omega^m)/(\Delta t/3) \end{aligned} \tag{10.4.9a}$$

Step 2:

$$\begin{aligned} &\mathbb{L}_x\omega^{m+(1/3)} + 3/2\ \mathbb{L}_y\omega^{m+(2/3)} + \\ &\quad -1/2\ \mathbb{L}_y\omega^{m+(1/3)} + \mathbb{L}_z\omega^{m+(1/3)} = \\ &\quad = (\omega^{m+(2/3)} - \omega^{m+(1/3)})/(\Delta t/3) \end{aligned} \tag{10.4.9b}$$

Step 3:

$$\begin{aligned} &\mathbb{L}_x\omega^{m+(2/3)} + \mathbb{L}_y\omega^{m+(2/3)} + \\ &\quad +3/2\ \mathbb{L}_z\omega^{m+1} - 1/2\ \mathbb{L}_z\omega^{m+(2/3)} = \\ &\quad = (\omega^{m+1} - \omega^{m+(2/3)})/(\Delta t/3) \end{aligned} \tag{10.3.9c}$$

where:

$$\begin{aligned} &\omega^{m+\beta} = \omega(\boldsymbol{x},\ t^{m+\beta}) = \omega(\boldsymbol{x},\ (m+\beta)\ \Delta t), \\ &\beta = 0,\ 1/3,\ 2/3,\ 1, \\ &m = 0, 1, 2, \ldots\ . \end{aligned} \tag{10.4.10}$$

(v) *Galerkin finite element method.* Finite element equations that correspond to scheme (10.4.9) are derived only for step 1. Corresponding formulae for step 2 and step 3 can be obtained by permuting the spatial variables x, y, z and shifting the time indices by 1/3. After fixing the indices j and k and multiplying the scheme by basic function $w_i(x)$ for step 1 (by $w_j(y)$ for step 2, by $w_k(z)$ for step 3) followed by integration over the flow domain Ω, we obtain a set of equations:

$$\iiint\limits_{\Omega} dxdydz \; \Big\{3/2 \; \mathbb{L}_x \omega^{m+(1/3)} - 1/3 \; \mathbb{L}_x \omega^m + \\ + \mathbb{L}_y \omega^m + \mathbb{L}_z \omega^m - (\omega^{m+(1/3)} - \omega^m)/(\Delta t/3)\Big\} \cdot \\ \cdot w_i(x) = 0, \quad (i = 1, \ldots, N_x). \tag{10.4.11}$$

Introducing the following notations:

$$\begin{aligned} I_{xx}^m &= \iiint\limits_{\Omega} dxdydz \; \partial/\partial x \; (D_x \; \partial\omega^m/\partial x) \; w_i(x) = \\ &= \iiint\limits_{\Omega_{ijk}^x} dxdydz \; \partial/\partial x \; (D_x \; \partial\omega^m/\partial x) \; w_i(x) = \\ &= \int_{-2a}^{2a} dx \int_{-b}^{b} dy \int_{-c}^{c} dz \; \partial/\partial x \; (D_x \; \partial\omega^m/\partial x) \; w_i(x) \\ K_{yx}^m &= \iiint\limits_{\Omega} dxdydz \; \partial/\partial y \; (D_y \; \partial\omega^m/\partial y) \; w_i(x) \qquad (10.4.12) \\ K_{zx}^m &= \iiint\limits_{\Omega} dxdydz \; \partial/\partial z \; (D_z \; \partial\omega^m/\partial z) \; w_i(x) \\ L_x^m &= \iiint\limits_{\Omega} dxdydz \; \omega^m \; w_i(x) \\ M_{xx}^m &= \iiint\limits_{\Omega} dxdydz \; \partial/\partial x \; (v_x \omega^m) \; w_i(x) \end{aligned}$$

$$N_{yx}^{m} = \iiint_{\Omega} dxdydz\ \partial/\partial y\ (v_y \omega^m)\ w_i(x)$$

$$N_{zx}^{m} = \iiint_{\Omega} dxdydz\ \partial/\partial z\ (v_z \omega^m)\ w_i(x)$$

we can write equation (10.4.11) in the following form:

$$\begin{aligned} &3/2\ \left(I_{xx}^{m+(1/3)} - M_{xx}^{m+(1/3)}\right) - 1/2\ (I_{xx}^{m} - M_{xx}^{m}) + \\ &+ \left(K_{yx}^{m} - N_{yx}^{m}\right) + (K_{zx}^{m} - N_{zx}^{m}) + \\ &- (3/\Delta t)\ \left(L_{x}^{m+(1/3)} - L_{x}^{m}\right) = 0. \end{aligned} \tag{10.4.13}$$

The integrals I, K, L, M and N can readily be obtained by integrating first in the explicit directions (y and z) and then integrating by parts in the x-direction.

The resultant algebraic equations (10.4.13), and similar formulae for steps 2 and 3, are tridiagonal because of the basic functions introduced. Consequently, a quick and robust tridiagonal solver can be used to find the approximation for ω in flow domain Ω.

Results

'Real time' animation of the pollution cloud was performed using the ADM solver for the CDE. The computed mass fraction was represented by glistering dots having their 'screen density' proportional to the values of ω calculated by the solver. Figure 38 shows an extent of the cloud for four consecutive moments of time. With the simulator it was also possible to estimate how long it takes for the pollutant to be washed out from the aquifer or diluted below some assumed safety level after the source (a waste disposal site) has been stopped.

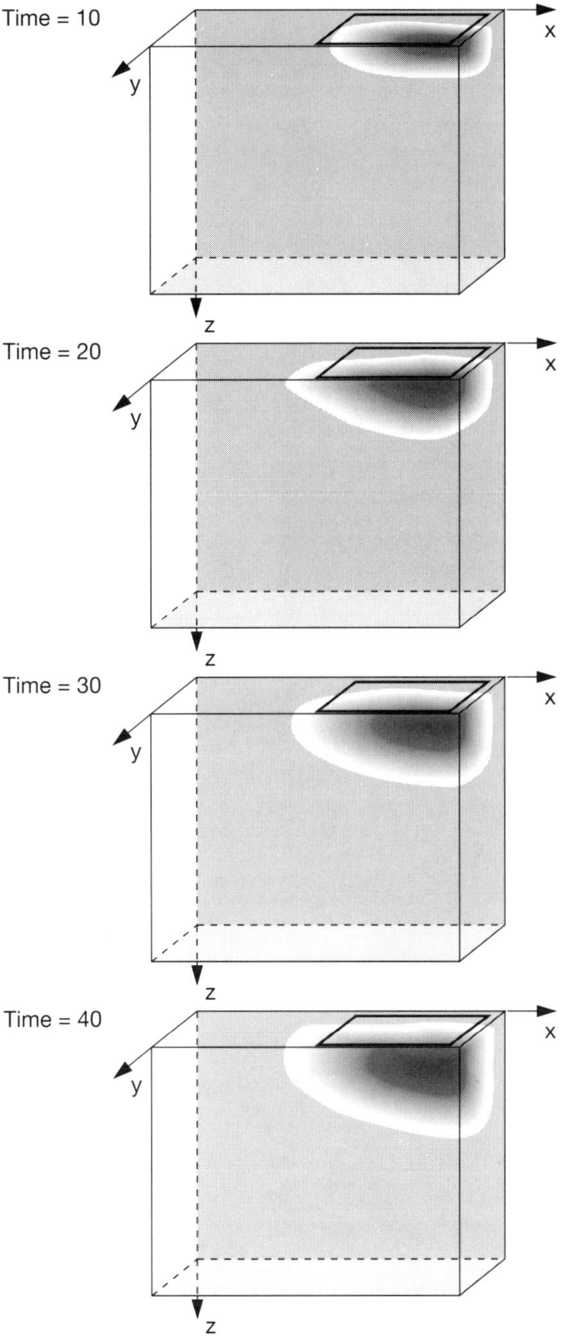

Figure 38. Time-evolution of a solute plume.

Bibliography

Ababou, R., Three Dimensional Flow in Random Porous Media, Ph.D. Thesis, Massachusetts Institute of Technology, Vol. I, pp. 211-220 (January, 1988).

Abramowitz, M. and Stegun, I.A., *Handbook of Mathematical Functions*, Dover Publications, Inc., New York (1972).

Appelo, C.A.J., *Geochemistry, groundwater and pollution*, (in press), Balkema (1992).

Atkin, R.J. and Craine, R.E., Continuum theories of mixtures: basic theory and historical development, *Quart. J. Mech. Appl. Math.*, Vol. 29, Part 2, pp. 209-244 (1976).

Bear, J., *Dynamics of Fluid in Porous Media*, Elsevier, New York (1972).

Bear, J., *Hydraulics of Groundwater*, McGraw-Hill International Book Company, New York (1979).

Bear, J. and Verruijt, A., *Modeling Groundwater Flow and Pollution: With Computer Programs for Sample Cases*, D. Reidel Publishing Company, Dordrecht (1987).

Bensoussan, A., Lions, J.L., and Papanicolaou, G., *Asymptotic Analysis for Periodic Structures*, North-Holland Publishing Company, Amsterdam (1978).

Bervoets, A.F., Flow Pattern Analysis for a Potential Defined by a Fourier Series, MSc. Thesis, TNO Institute of Applied Geoscience, Report No. OS-91-30A (1991).

Bervoets, A.F., Zijl, W., and Van Veldhuizen, M., Spatial scale analysis of groundwater flow patterns for transport in sedimentary basins, European Simulation Symposium on Modelling and Control of Water Resources Systems and Global Changes, Ghent (November 6-8, 1991).

Birch, C. and Cobb, J.B., Jr., *The Liberation of Life: From the Cell to the Community*, Cambridge University Press, Cambridge (1981).

Butkov, E., *Mathematical Physics*, World Student Series Edition, Addison-Wesley Publishing Company, Reading, PA (1973).

Carlslaw, H.S. and Jaeger, J.C., *Conduction of Heat in Solids*, Oxford

University Press, Oxford (1959).

Dagan, G., *Flow and Transport in Porous Formations*, Springer-Verlag, Berlin (1989).

Daly, H.E. and Cobb, J.B., Jr., *For the Common Good: Redirecting the Economy towards Community, the Environment, and a Sustainable Future*, Green Print, The Merlin Press, London (1990).

Davis, J.C., *Statistics and Data Analysis in Geology*, John Wiley & Sons, Inc., New York (1986).

Dunford, N. and Schwartz, J.T., *Linear Operators II: Spectral Theory*, Interscience Publishers, New York (1963).

Engelen, G.B. and Jones, G.P., (eds.), Developments in the Analysis of Groundwater Flow Systems, Wallingford, IAHS Publication No. 163 (1986).

England, R., Error estimates for Runge-Kutta type solutions to systems of ordinary differential equations, *Comp. J.*, Vol. 12.5 (1969).

Frind, E.O., The principal direction technique for advective-dispersive transport simulation in three dimensions, Proc. 5th Int. Conf. on Finite Elements in Water Resources, Burlington, Vermont, June (1984).

Gelhar, L.W., Stochastic subsurface hydrology from theory to application, *Water Resour. Res.*, Vol. 22, No. 9, pp. 135-145 (1986).

Ghosh, D.P., The Application of Linear Filter Theory to the Direct Interpretation of Geoelectrical Resistivity Measurements, Ph.D Thesis, Delft University of Technology, Delft (1970).

Green, D.H. and Wang, H.F., Specific storage as a poroelastic coefficient, *Water Resour. Res.*, Vol. 26, No. 7, pp. 1631-1637 (1990).

Griffin, D.R., Process theology as empirical, rational, and speculative: Some reflections on method, *Process Studies*, Vol. 19, No. 2 (1990).

Haldorsen, H.H. and Lake, L.W., A new approach to shale management in field scale models, *SPEJ*, pp. 427-457 (August, 1984).

Harbaugh, J.W. and Merriam, D.F., *Computer Applications in Stratigraphic Analysis*, John Wiley & Sons, Inc., New York (1968).

Hassanizadeh, S.M., Modeling species transport by concentrated brine in aggregated porous media, *Transp. Porous Media*, Vol. 3, pp. 299-318 (1988).

Huyakorn, P.S. and Pinder, G.F., *Computational Methods in Subsurface Flow*, Academic Press, New York (1983).

James, W.R., FORTRAN IV Program using double Fourier series for surface fitting of irregularly spaced data, *Kansas Geological Survey Com-*

puter Contributions, Vol. 5 (1966).

JPT, *Journ. of Pet. Technol.*, pp. 2019-2056 (September, 1982).

Kaasschieter, E.F., A practical termination criterion for the conjugate gradient method, Reports of the Faculty of Math. and Inf., No. 87-62, Delft University of Technology, Delft (1987).

Kershaw, D.S., The incomplete Cholesky-conjugate gradient method for the iterative solution of systems of linear equations, *J. Comp. Phys.*, Vol. 26, pp. 43-65 (1978).

Kinzelbach, W., *Numerische Methoden zur Modellierung des Transports von Schadstoffen im Grundwasser*, Oldebourg-Verlag, Munich (1987).

Klir, G.J., *An Approach to General Systems Theory*, Van Nostrand Reinhold Company, New York (1969).

Koefoed, O., *Geosounding Principles, 1: Resistivity Sounding Measurements*, Elsevier Scientific Publishing Company, Amsterdam (1979).

Kuppen, W.J.J.M., De gepreconditioneerde methode der geconjugeerde gradiënten toegepast op de lineaire vergelijkingen verkregen uit de vectorveld-methode [The method of preconditioned conjugate gradients applied to the linear equations obtained from the vector field method], MSc. Thesis, TNO Institute of Applied Geoscience, Report No. OS-88-24 (1988).

Lage, J.L. and Bejan, A., Numerical study of forced convection near a surface covered with hair, *Int. J. Heat and Fluid Flow*, Vol. 11, No. 3, pp. 242-248 (1990).

Laszlo, E., *Introduction to Systems Philosophy*, Gordon & Breach, New York (1972).

Laszlo, E., *Systems Science and World Order*, Pergamon Press, Oxford (1983).

Lovelock, J.E., *Gaia, A New Look at Life on Earth*, Oxford University Press, Oxford (1979).

Lovelock, J.E., *The Ages of Gaia, A Biography of Our Living Earth*, Oxford University Press, Oxford (1988).

Maas, C., The use of matrix differential calculus in problems of multiple-aquifer flow, *J. Hydrol.*, Vol. 88, pp. 43-67 (1986).

Maas, C., Groundwater flow to a well in a layered porous medium 1: Steady flow, *Water Resour. Res.*, Vol. 23, pp. 1675-1681 (1987a).

Maas, C., Groundwater flow to a well in a layered porous medium 2: Nonsteady multiple-aquifer flow, *Water Resour. Res.*, Vol. 23, pp. 1683-1688 (1987b).

Meijerink, J.A. and Van der Vorst, H.A., An iterative solution method for linear systems of which the coefficient matrix is a symmetric M-matrix, *Math. of Comput.*, Vol. 31, No. 137, January (1977).

Meijerink, J.A. and Van der Vorst, H.A., Guidelines for the usage of incomplete decomposition in solving sets of linear equations as occur in practical problems, Report TR-9, Academic Computer Centre, Utrecht (1983).

Millington, R.J. and Quirk, J.P., Permeability of porous solids, *Trans. Faraday Soc.*, Vol. 57, pp. 1200-1207 (1961).

Mitchell, J.K., *Fundamentals of Soil Behavior*, John Wiley & Sons, Inc., New York (1976).

Morse, P.M. and Feshbach, H., *Methods of Theoretical Physics*, McGraw-Hill, New York (1953).

Muskat, M., *The Flow of Homogeneous Fluids through Porous Media*, New York (1937).

Nash, J.E., Eagleson, P.S., Philip, J.R., and Van der Molen, W.H., The education of hydrologists, *Hydrol. Sci. Journ.*, Vol. 35, No. 6, pp. 597-607 (1990).

Nawalany, M., Numerical model for the transport velocity representation of groundwater flow, VIth Int. Conf. on Finite Elements in Water Resources, Lisbon (1986a).

Nawalany, M., Environmental applications of the transport velocity representation for groundwater flow, Envirosoft '86, Los Angeles (1986b).

Nawalany, M., FLOSA-3D, Introduction to system analysis of three-dimensional groundwater flow, Report TNO Institute of Applied Geoscience OS 86-07, Delft (1986c).

Nawalany, M., Applications of FLOSA-3FE simulator in solute transport models, Report TNO Institute of Applied Geoscience OS 88-42, Delft (1988).

Nawalany, M., FLOSA-FE, User's manual, TNO Institute of Applied Geoscience OS 89-12, Delft (1989a).

Nawalany, M., FLOSA-FE finite element model of the three-dimensional groundwater velocity field, Report TNO Institute of Applied Geoscience OS 89-47, Delft (1989b).

Nawalany, M., Regional vs. local computations of groundwater flow, VIIIth Int. Conf. on Computational Methods in Water Resources, Venice,June (1990).

Nawalany, M. Decoupling the velocity oriented groundwater flow equations for the third type boundary conditions, IXth Int. Conf. on Computational Methods in Water Resources, Denver, June 9-12 (1992).

Nawalany, M., Loch, J., and Sinicyn, G., Active isolation of waste disposal sites by hydraulic means, Part II – Models, Report TNO Institute of Applied Geoscience OS 91-42c, Delft (1991).

Nayfeh, A., *Perturbation Methods*, John Wiley & Sons, Inc., New York (1973).

Nieuwenhuizen, R., Flow pattern analysis for a well defined by point sinks, MSc. Thesis, TNO Institute of Applied Geoscience, Report No. OS-92-52A (1992).

Pagels, H., *The Dream of Reason: The Computer and the Rise of Sciences of Complexity*, Simon and Schuster, New York (1988).

Peters, A., Romunde, B., and Sartoretto, F., Vectorized implementation of some MCG codes for FE solution of large groundwater flow problems, Int. Conf. on Computational Methods in Flow Analysis, Okayama, September 5-8 (1988).

Pfeffer, R., Heat and mass transport in multiparticle systems, *I&EC Fundamentals*, Vol. 3, No. 4, pp. 380-383 (1964).

Popper, K.R., *Quantum Theory and the Schism in Physics; Vol. 3 of the postscript to The Logic of Scientific Discovery*, Hutchinson & Co. Ltd., London (1982).

Prigogine, I. and Stengers, I., *La nouvelle alliance. Métamorphose de la science*, Éditions Gallimard, Paris (1979).

Quintard, M. and Whitaker, S., Écoulement monophasique en milieu poreux: effet des hétérogénéïtés locales, *Journal de mécanique théorique et appliquée*, Vol. 6, No. 5, pp. 691-726 (1987).

Quintard, M. and Whitaker, S., Two-phase flow in heterogeneous porous media, the method of large-scale averaging, *Transp. Porous Media*, Vol. 3, pp. 357-413 (1988).

Quintard, M. and Whitaker, S., Two-phase flow in heterogeneous porous media I: The influence of large spatial and temporal gradients, *Transp. Porous Media*, Vol. 5, pp. 341-379 (1990a).

Quintard, M. and Whitaker, S., Two-phase flow in heterogeneous porous media II: Numerical experiments for flow perpendicular to a stratified system, *Transp. Porous Media*, Vol. 5, pp. 429-472 (1990b).

Reddy, R., Foundations and grand challenges of artificial intelligence, *AI Magazine*, pp. 9-19 (1988).

Rikitake, T., Sato, R., and Hagiwara, Y, *Applied Mathematics for Earth*

Scientists, Terra/Reidel, Dordrecht (1987).

Rodríquez-Iturbe, I., Febres de Power, B., Sharifi, M.B., and Georgakakos, K.P., Chaos in rainfall, *Water Resour. Res.*, Vol. 25, No. 7, pp. 1667-1676 (1989).

Shanks, D., Nonlinear transformations of divergent and slowly convergent sequences, *J. Math. Phys.*, Vol. 34, pp. 1-42 (1955).

Shephard, M.S. and Law, K.H., The modified quad-tree mesh generator and efficient adaptive analysis, Proc. Int. Conf. on Accuracy Estimates and Adaptive Refinements in Finite Element Computations (ARFEC), Lisbon, June 19–22 (1984).

Silberberg, I.H. and McKetta, J.J., Learning how to use dimensional analysis, *Petroleum Refiner*, Part I, Vol. 32, No. 4, pp. 179-183; Part II, Vol. 32, No. 5, pp. 147-150; Part III, Vol. 32, No. 6, pp. 101-103; Part IV, Vol. 32, No. 7, pp. 130-134 (1953).

Stam, J.M.T. and Zijl, W., Modeling permeability in imperfectly layered porous media II: A two-dimensional application of block-scale permeability, *Math. Geology*, Vol. 24, No. 8, pp. 884-904 (1992).

Stam, J.M.T., Zijl, W., and Turner, A.K., Determination of hydraulic parameters from the reconstruction of alluvial stratigraphy, 4th Int. Conf. on Computational Methods and Experimental Measurements, Capri, May 23-26 (1989).

Stewart, G.W., *Introduction to Matrix Computations*, Academic Press, New York (1973).

Strack, O., *Groundwater Mechanics*, Prentice Hall, Inc., Englewood Cliffs, N.J. (1989).

Strang, G., Wavelets and dilation equations: a brief introduction, *SIAM Review*, Vol. 31, No. 4, pp. 614-627 (1989).

Stumm, W. and Morgan, J.M., *Aquatic Chemistry; An Introduction Emphasizing Chemical Equilibria in Natural Waters*, John Wiley & Sons, New York (1981).

Tetzlaff, D.M. and Harbaugh, J.W., *Simulating Clastic Sedimentation*, Van Nostrand Reinhold Company (1989).

Tikhonov, A. and Arensin, V., *Solutions of Ill-Posed Problems*, John Wiley & Sons, New York (1977).

Todd, D.K., *Groundwater Hydrology*, John Wiley & Sons, New York (1980).

Tóth, J., A theoretical analysis of groundwater flow in small drainage basins, *J. Geophys. Res.*, Vol. 68, No. 16, pp. 4795-4812 (1963).

Tritton, D.J., *Physical Fluid Dynamics*, Van Nostrand Reinhold Company,

New York (1977).

Vafai, K. and Sozen, M., Analysis of energy and momentum transport for fluid flow through a porous bed, *J. Heat Transf.*, Vol. 112, pp. 690-699 (1990).

Van Dyke, M., *Perturbation Methods in Fluid Mechanics*, The Parabolic Press, Stanford (1975).

Van Duijn, C.J. and Knabner, P., Solute transport in porous media with equilibrium and non-equilibrium multiple-site adsorption: travelling waves, *J. Reine Angew. Math.*, Vol. 415, pp. 1-49 (1991).

Van Duijn, C.J. and Van der Zee, S.E.A.T.M., Solute transport parallel to an interface separating two different porous materials, *Water Resour. Res.*, Vol. 22, No. 13, pp. 1779-1789 (1986).

Van Geer, F.C., Te Stroet, C.B.M., and Zhou, Y., Using Kalman filtering to improve and quantify the uncertainty of numerical groundwater simulations I: The role of system noise and its calibration, *Water Resour. Res.*, Vol. 27, No. 8, pp. 1987-1994 (1991).

Van Veldhuizen, M., Bervoets, A.F., and Zijl, W., Dichotomy of a special recurrence relation from the earth sciences, *J. Comput. Appl. Math.* (accepted) (1992a).

Van Veldhuizen, M., Nieuwenhuizen, R., and Zijl, W., A note on log scale Hankel Transforms, *J. Comput. Phys.* (accepted) (1992b).

Van der Vorst, H.A., Preconditioning by incomplete decompositions, Ph.D. Thesis, Rijksuniversiteit Utrecht, Utrecht (1982).

Watson, G.N., *A Treatise on the Theory of Bessel Functions*, Cambridge University Press, Cambridge (1966).

Weast, R.C., *Handbook of Chemistry and Physics, A Ready-Reference Book of Chemical and Physical Data*, CRC Press, Cleveland (1975).

Weber, K.J. and Van Geuns, L.C., Framework for constructing clastic reservoir simulation models, *J. Petrol. Techn.*, Vol. 34, pp. 439-453 (1990).

Whitaker, S., Flow in porous media I: A theoretical derivation of Darcy's law, *Transp. Porous Media*, Vol. 1, No. 1, pp. 3-25 (1986).

Whitehead, A.N., *Science and the Modern World*, The Free Press, New York (1967) (first edition in 1925).

Whitehead, A.N., *An Introduction to Mathematics*, Oxford University Press, London (1982) (first edition in 1911).

Wong, H.Y., *Handbook of Essential Formulae and Data on Heat Transfer for Engineers*, Longman Group Ltd., London (1977).

Yerry, M.A. and Shephard, M.S., A modified quad-tree approach to finite element mesh generation, *IEEE Comp. Graph. and Appl.*, Vol. 3, No. 1 (1983).

Zhou, Y., Te Stroet, C.B.M., and Van Geer, F.C., Using Kalman filtering to improve and quantify the uncertainty of numerical groundwater simulations II: Application to monitoring network design, *Water Resour. Res.*, Vol. 27, No. 8, pp. 1995-2006 (1991).

Zijl, W., Finite-element methods based on a transport velocity representation for groundwater motion, *Water Resour. Res.*, Vol. 20, No. 1, pp. 137-145 (1984).

Zijl, W., Three-dimensional flow systems analysis: Basic equations and numerical modelling, In: *Topics in Groundwater Modelling*, Course Notes Delft University of Technology, Fac. of Civil Engineering (1988).

Zijl, W. and Nawalany, M., Vector field approach for finite differences and first-order finite elements, Expert Meeting on New Developments in Groundwater Modelling, Delft, Sept. 14-16 (1988).

Zijl, W. and Nawalany, M., Robustness and accuracy of groundwater flux computations in large-scale shallow sedimentary basins, Proc. 2nd Int. Conf. on Reliability and Robustness of Engineering Software, Milan, April 22-24 (1991).

Zijl, W., Nawalany, M., and Pasveer, F., Numerical simulation of fluid flow in porous media, Colloquium on Numerical Aspects of Vector and Parallel Processors, 28 Febr. (1986).

Zijl, W. and Stam, J.M.T., Modeling permeability in imperfectly layered porous media I: Derivation of block-scale permeability tensor for thin grid-blocks, *Math. Geology*, Vol. 24, No. 8, pp. 865-883 (1992).

Appendix A

From the Navier-Stokes equations to Darcy's law

A.1. Introduction

The theory of flow in natural porous media is devoted to the description of flow and transport through soils consisting of sand, clay, shale, etc. Typical applications are in the fields of petroleum reservoir engineering and groundwater hydrology.

A porous medium is defined as a solid phase in which a connected pore space exists through which a fluid can flow. Fluid motion in the pore space is governed by the fundamental continuum equations of fluid dynamics expressing conservation of mass and momentum in general, and the generalized Navier-Stokes equations in particular. These equations must be supplemented by appropriate boundary conditions at the fluid-solid interfaces. However, from a practical standpoint, it is hopeless to try to apply these basic laws directly even to small-scale flow problems in porous media. As a first step to small-scale applications, the concept of a 'mixed fluid-solid continuum' is employed, *where the boundary conditions on the internal fluid-solid interfaces are embedded in continuously distributed control volume parameters*. In this continuum approach, the approximate velocity $\bar{\boldsymbol{v}}$ should resemble the exact pore space velocity $\boldsymbol{v}$ as closely as possible; i.e., a suitably chosen norm $\| \bar{\boldsymbol{v}} - \boldsymbol{v} \|$ should be equal to zero or as small as possible. The small-scale equations obtained in this way may be applied as a basis for the derivation of large-scale equations; see Appendix D.

Much effort has been devoted in recent years to developing rigorous local averaging techniques to find the basic fluid-solid continuum equations governing flow in porous media. The various aspects of this macroscopization procedure by spatial averaging techniques have been discussed extensively in the literature to which the reader may be referred here; see the references at the end of this Appendix. In these theories, volume-averaged quantities (e.g., the intrinsic phase-averaged fluid velocity $<\boldsymbol{v}>^f$) play an important role. However, the approximate quantities (e.g., $\bar{\boldsymbol{v}}$) introduced in this Appendix are not necessarily equal to these volume-averaged quantities.

To give the reader a feeling for the most important assumptions underlying the basic fluid-solid continuum equations, without going into the abundance of mathematical details and subtleties, an approach based on arguments derived from generally known fluid dynamic theory will be presented. In this way, the equations governing flow in porous media, viz., the continuity equation and Darcy's law, are obtained for small scales (say for the scale of centimeters).

A.2. Basic equations

The basic equations to be solved for the fluid in the pore space are the partial differential equations expressing conservation of mass and linear momentum:

$$\partial\rho/\partial t + \mathrm{div}(\rho\boldsymbol{v}) = 0 \qquad \text{(mass conservation)}, \tag{A.1}$$

$$\rho\left(\partial\boldsymbol{v}/\partial t + \boldsymbol{v}\boldsymbol{\cdot}\underline{\mathbf{grad}\boldsymbol{v}}\right) = -\,\mathbf{grad}\,p + \rho\boldsymbol{g} + \mathbf{div}\underline{\boldsymbol{S}} \qquad \text{(momentum conservation)}. \tag{A.2a}$$

For a Newtonian fluid the tensor $\underline{\boldsymbol{S}}$ is given by:

$$\underline{\boldsymbol{S}} = \mu\left[\underline{\mathbf{grad}\boldsymbol{v}} + (\underline{\mathbf{grad}\boldsymbol{v}})^T\right] + (\xi - 2/3\,\mu)(\mathrm{div}\,\boldsymbol{v})\,\underline{\boldsymbol{I}}, \tag{A.2b}$$

where the superscript T means the transpose of the tensor; ρ $[kg{\cdot}m^{-3}]$ is the fluid density, $\boldsymbol{v}$ $[m{\cdot}s^{-1}]$ is the fluid velocity in the pore space, $\boldsymbol{g}$ $[m{\cdot}s^{-2}]$ is the gravitational acceleration, p $[Pa]$ is the fluid pressure in the pore space, $\underline{\boldsymbol{S}}$ $[Pa]$ is the viscous stress tensor, μ $[Pa{\cdot}s]$ is the fluid dynamic viscosity, and ξ $[Pa{\cdot}s]$ is the fluid bulk viscosity (see Chorin and Marsden, 1979, pp. 2-4, 43-46). It is remarked here that the generalized Navier-Stokes equations (A.2a), (A.2b), do not hold only for pure Newtonian fluids like water, air, oil and natural gas, but also for Newtonian mixtures, i.e., for fluids in which a solute is dissolved (e.g., salt dissolved in water). Equation (A.1) is often referred to as the continuity equation.

The restriction to Newtonian fluids needs some comment. It is commonly believed that water, water-salt mixtures, oil, etc. are Newtonian fluids, except near walls where fluid is attracted to form an adsorption layer with thickness of several molecule dimensions. This attracted fluid does not satisfy expressions (A.2), which means that fluid in the adsorption layer surrounding the solid constituents of the porous medium is excluded from the description. When considering saturated flow this is a justifiable restriction, since only a minor amount of the water is attracted to the solids. In the unsaturated zone, however, this might be a severe restriction, especially in the description of water flow in almost dry soils. There, most of the water is attracted to the adsorption layer and a description based on the equations (A.2) will break down. For further discussion on this topic, see Gray and Hassanizadeh (1991a, 1991b).

In addition to the four equations (A.1, A.2) for the seven unknowns $\boldsymbol{v}, p, \rho, \mu$ and ξ, the equation of state and expressions for the viscosities must also be given:

$$\rho = F(\rho, \omega_i), \quad i = 0, 1, \ldots, \quad \text{(equation of state)} \tag{A.3a}$$

$$\mu = G(p, \omega_i), \quad i = 0, 1, \ldots, \tag{A.3b}$$

$$\xi = H(p, \omega_i), \quad i = 0, 1, \ldots, \tag{A.3c}$$

where ω_0 $[K]$ is the temperature, ω_1 $[kg \cdot kg^{-1}]$ is the mass fraction of miscible component 1 dissolved in the fluid, ω_2 is the mass fraction of component 2, etc.

A.3. Porous media continuity equation

It will be clear from the way in which the fundamental equations (A.1, A.2) of fluid dynamics are derived (see Chorin and Marsden, 1979, pp. 1-62; Meyer, 1971), that the mixed fluid-solid continuum equations must show some similarity to the fundamental equations, since both sets of equations stem from the basic physical principles of conservation of mass and momentum in continuous media. More specifically, in a fluid-solid continuum approximation, conservation of mass should be expressed by an equation similar to equation (A.1):

$$\partial/\partial t\ (\bar{\rho}\theta) + \text{div}\ (\bar{\rho}\theta\bar{\boldsymbol{v}}) = 0 \quad \text{(continuity equation)}. \tag{A.4}$$

In equation (A.4), $\bar{\boldsymbol{v}}$ is the approximation to the exact fluid velocity in the pore space $\boldsymbol{v}$, $\bar{\rho}$ is the approximation to the exact density in the pore space ρ, and θ is the effective porosity, i.e., the volume fraction of the 'fluid-solid mixture' available for fluid flow. In this way, the volume fraction occupied by dead-end pores and by the stagnant fluid in the adsorption layer is excluded.

The effective volume porosity is defined as the fraction of a control volume occupied by the flowing fluid. The effective surface porosity is the fraction of a control surface covered by the flowing fluid. If we consider the averaged value of the effective surface porosity over the linear dimensions of the control volume, it is simple to prove that this latter value is equal to the previously defined effective volume porosity.

The appropriate size of the control volume, or representative elementary volume (REV), is discussed in many publications; e.g., see Bachmat and Bear (1986).

The presence of the effective porosity θ in the continuity equation (A.4) replaces the need to satisfy the internal fluid-solid boundary condition for the *normal* velocity component. This normal velocity boundary condition expresses the fact that the solid constituents are impervious. This means that at places outside the pore space, i.e., at places where the solid constituents are present (or where the fluid is stagnant), there is no normal component of fluid flow (i.e., $\boldsymbol{n} \cdot \boldsymbol{v} = 0$ at places where solid constituents or stagnant fluid is present).

However, the approximate velocity $\bar{\boldsymbol{v}}$ is nonzero everywhere, which represents a very poor approximation at places where the solid constituents are present. We will, however, be satisfied if $\bar{\boldsymbol{v}}$ is a good approximation to $\boldsymbol{v}$ in the pore space, regardless of the goodness of approximation outside the pore space. In order to take into account the absence of flow outside the pore space, the mass flow rate per unit surface area is taken equal to $\bar{\rho}\theta\bar{\boldsymbol{v}}$ (remember that θ is that part of the surface area through which fluid is flowing). In the same way, the fluid mass contents per unit volume is taken equal to $\bar{\rho}\theta$ (remember that θ is that part of the volume in which flowing fluid is present).

A.4. Porous media constitutive equations

Of course, the equation of state (A.3a) and the equations for the viscosities (A.3b, A.3c) should also be satisfied by the approximations, i.e.:

$$\bar{\rho} = F(\bar{p}, \bar{\omega}_i), \quad i = 0, 1, \ldots, \tag{A.5a}$$

$$\bar{\mu} = G(\bar{p}, \bar{\omega}_i), \quad i = 0, 1, \ldots, \tag{A.5b}$$

$$\bar{\xi} = H(\bar{p}, \bar{\omega}_i), \quad i = 0, 1, \ldots, \tag{A.5c}$$

where $\bar{p}$, $\bar{\mu}$, $\bar{\xi}$ and $\bar{\omega}_i$ are approximations to p, μ, ξ and ω_i, respectively.

A.5. Porous media momentum equations

Similar to conservation of mass, conservation of momentum results in:

$$\partial/\partial t\ (\bar{\rho}\theta\bar{\boldsymbol{v}}) + \mathrm{div}(\bar{\rho}\theta\bar{\boldsymbol{v}}\bar{\boldsymbol{v}}) = \boldsymbol{\psi} \tag{A.6a}$$

where $\boldsymbol{\psi}$ $[N{\cdot}m^{-3}]$ is the force distribution per unit volume acting on the fluid with macroscopic momentum flow rate per unit surface area $\bar{\rho}\bar{\boldsymbol{v}}\theta\bar{\boldsymbol{v}}$ and macroscopic momentum per unit volume $\bar{\rho}\theta\bar{\boldsymbol{v}}$. This force distribution must be chosen in such a way that the approximations $\bar{\boldsymbol{v}}$, $\bar{p}$, etc. resemble as much as possible the exact values $\boldsymbol{v}$, p, etc. Similar to the definition of porosity, the macroscopic force per unit volume is also defined in a control volume.

The role of the force distribution is to replace the internal *tangential* boundary conditions $\boldsymbol{n} \times \boldsymbol{v} = \boldsymbol{0}$ at the fluid-solid interfaces by a more tractable description without internal boundaries.

For reasons which will become clear in the subsequent discussion, another force distribution per unit volume, $\boldsymbol{F}$ $[N{\cdot}m^{-3}]$, is defined by $\boldsymbol{F} = -(\boldsymbol{\psi}/\theta) - \mathbf{grad}\,\bar{p} + \bar{\rho}\boldsymbol{g}$, where $\bar{p}$ is the approximate fluid pressure in the void space.

After some rearrangements, and making use of continuity equation (A.4), momentum balance (A.6a) can be written as:

$$\bar{\rho}\left(\partial\bar{\boldsymbol{v}}/\partial t + \bar{\boldsymbol{v}}{\cdot}\underline{\mathbf{grad}\bar{\boldsymbol{v}}}\right) = -\,\mathbf{grad}\,\bar{p} + \bar{\rho}\boldsymbol{g} - \boldsymbol{F}. \tag{A.6b}$$

The force distribution $\boldsymbol{F}$ must be related to other approximate variables. From the exact equations (A.1, A.2) it will be seen that the exact solutions for $\boldsymbol{v}$ and p depend on the density ρ and on the viscosities μ and ξ. *Only by means of ρ, μ and ξ the influences of temperature ω_0 and mass fractions of dissolved matter ω_i, $i = 1, 2, \ldots$, enter the equations.* Of course, this should also be reflected in the approximate equations. This means that $\boldsymbol{F}$ may directy depend on the scalar fields $\bar{\rho}$, $\bar{\mu}$ and $\bar{\xi}$, but not directly on the fields $\bar{\omega}_i$, $i = 0, 1, \ldots$.
Consequently, in general:

$$\boldsymbol{F} = \left(L_v(\bar{\boldsymbol{v}}),\ L_p(\bar{p}),\ L_\rho(\bar{\rho}),\ L_\mu(\bar{\mu}),\ L_\xi(\bar{\xi})\right), \tag{A.7}$$

where L may represent a function or a differential or integral operator.

Under the condition that there is no flow, i.e., $\boldsymbol{v} = \boldsymbol{0}$, the generalized Navier-Stokes equations (A.2) simplify to the well-known equations for the hydrostatic pressure:

$$\mathbf{grad}\, p = \rho \boldsymbol{g}. \tag{A.8a}$$

From equation (A.6b) it follows that, under no-flow conditions, the following equation holds:

$$\mathbf{grad}\, \bar{p} = \bar{\rho} \boldsymbol{g} - \boldsymbol{F}. \tag{A.8b}$$

In the no-flow situation we require the approximations to be equal to the exact values. That is, if there is no flow, then $\boldsymbol{v} = \bar{\boldsymbol{v}} = \boldsymbol{0}$, $\bar{p} = p$ and $\bar{\rho} = \rho$. From equations (A.8a) and (A.8b) it will be seen that this requirement is satisfied if $\boldsymbol{F} = \boldsymbol{0}$ when $\bar{\boldsymbol{v}} = \boldsymbol{0}$, or:

$$\boldsymbol{F}\left(L_v(\boldsymbol{0}),\ L_p(\bar{p}),\ L_\rho(\bar{\rho}),\ L_\mu(\bar{\mu}),\ L_\xi(\bar{\xi})\right) = \boldsymbol{0}. \tag{A.9}$$

A.6. Steady Navier-Stokes flow in the control volume

The problem of finding an expression for $\boldsymbol{F}$ will be more tractable if we restrict the discussion to quasi-steady flow, i.e., to flow for which the term $\partial \boldsymbol{v}/\partial t$ in momentum equation (A.2a) is negligible. This means that no propagation of acoustical waves is considered (for a discussion of wave propagation phenomena, see e.g., Van der Grinten, 1987). Since $\bar{\boldsymbol{v}}$ is an approximation to $\boldsymbol{v}$, the term $\partial\bar{\boldsymbol{v}}/\partial t$ in the approximate momentum equation (A.6b) is also negligible in this case. Furthermore, only small fluid velocities will be considered, or, more specifically, velocities for which holds: $Ma^2 << minimum\ (Re, 1)$, where Ma = Mach number = v/c, with v the magnitude of the fluid velocity and c the fluid velocity of sound, Re = Reynolds number = $\rho v d/\mu$, with d the dimension of the pore space. For a more detailed discussion involving dimensionless numbers, see Van der Weiden (1988, pp. 27-32).

Under these conditions the continuity equation (A.1) and the generalized Navier-Stokes equations (A.2) applied on the scale of the earlier defined control volume may be simplified to the requirement that the velocity field is solenoidal (divergence-free), and to the steady Boussinesq equations (see Tritton, 1977, pp. 56-59):

$$\operatorname{div} \boldsymbol{v} = 0 \qquad \text{(solenoidal velocity field)}, \tag{A.10}$$

$$\boldsymbol{v}\cdot\underline{\mathbf{grad}\boldsymbol{v}} = -\,\mathbf{grad}\,\pi + (\rho'/\rho_0)\,\boldsymbol{g} + \nu_0\,\boldsymbol{\nabla}^2\boldsymbol{v} \qquad \text{(steady Boussinesq equations)}, \tag{A.11}$$

where $\rho = \rho_0 + \rho'$, ρ_0 is constant, $\rho' = \rho'(\omega_i)$ is independent of p, $\nu_0 = \mu_0/\rho_0$, μ_0 is constant, and $\pi = p/\rho_0 - gz_v$, where g is assumed to be constant; z_v is the vertical coordinate in the direction of the gravitational acceleration $\boldsymbol{g}$, and g is the magnitude of $\boldsymbol{g}$. This means that the dependence of $\boldsymbol{F}$ on $\bar{\mu}$ and $\bar{\xi}$ in equation (A.7) disappears.

If we also assume that variations in density on the scale of the control volume are negligible, the steady Boussinesq equations (A.11) simplify to the steady Navier-Stokes equations:

$$\boldsymbol{v}\cdot\underline{\mathbf{grad}\boldsymbol{v}} = -\,\mathbf{grad}\,\pi + \nu_0\,\boldsymbol{\nabla}^2\boldsymbol{v} \qquad \text{(steady Navier-Stokes equations)}. \tag{A.12}$$

From fluid dynamic theory we know that tangential boundary conditions are sources of vorticity production (vorticity is fluid rotation = $\mathbf{curl}\,\boldsymbol{v}$). Furthermore, flow governed by the steady Navier-Stokes equations (A.12) and by continuity equation (A.10) ($\operatorname{div} \boldsymbol{v} = 0$) is irrotational ($\mathbf{curl}\,\boldsymbol{v} = \mathbf{0}$) and without circulation if, and only if, there are no obstacles bounding the flow field (see Meyer, 1971). For irrotational and solenoidal flow, the terms $\nu_0\ \nabla^2 \boldsymbol{v} \equiv -\nu_0\ \mathbf{curl}(\mathbf{curl}\,\boldsymbol{v}) + \nu_0\ \mathbf{grad}(\operatorname{div}\boldsymbol{v})$ disappear in equations (A.12), and this means that there is no viscous dissipation of mechanical energy (see Batchelor, 1974). For steady flow this results in the fact that there is no drag force, a fact which is known in fluid dynamic theory under the name d'Alembert's paradox (Shinbrod, 1973; Batchelor, 1974). If, in addition, circulation is absent, there is also no lift force.

In conclusion: flow governed by equations (A.10), (A.12) will experience no resistance force (= lift + drag) if, and only if, there are no obstacles bounding the flow field.

If there are no solid obstacles, there are no internal boundary conditions to be satisfied and the approximations $\bar{\boldsymbol{v}}$, $\bar{p}$ and $\bar{\rho}$ are required to be equal to the exact values, $\boldsymbol{v}$, p and ρ. It follows then from equations (A.12) and (A.6b) that $\boldsymbol{F} = \mathbf{0}$ in the absence of a solid phase. On the other hand, $\boldsymbol{F}$ will be nonzero if there is interaction of the fluid with a solid phase.

In conclusion: flow governed by equations (A.10), (A.12) will experience a macroscopic force distribution $\boldsymbol{F} \neq \mathbf{0}$ if, and only if, there are solid constituents in the flow field.

Furthermore, $\boldsymbol{F}$ should be given in such a way that it replaces the tangential boundary conditions $\boldsymbol{n} \times \boldsymbol{v} = \mathbf{0}$ at the internal fluid-solid interfaces. Of course, both the internal tangential boundary conditions at the solid constituents and the tangential boundary conditions at the boundary of the porous medium domain act as sources of vorticity. For naturally occurring porous media, which are small-porosity media, there is no physical reason to take into account the vorticity production caused by the tangential conditions at the domain boundary when, at the same time, vorticity production caused by the tangential conditions at the solid-fluid interfaces of the solid constituents is neglected. This means that we have to avoid a formulation in which tangential boundary conditions are needed. For that reason, $\boldsymbol{F}$ should not contain a term like $\mu_0\ \nabla^2 \bar{\boldsymbol{v}}$, since such a second-order term requires interfacial tangential boundary conditions (see Meyer, 1971; Shinbrod, 1973; Batchelor, 1974; and Chorin and Marsden, 1979). In addition, it can be proved that in most types of naturally occurring porous media the intrinsic permeability $\mu_0\,\underline{\boldsymbol{k}}$ is sufficiently small to neglect the term $\mu_0\ \nabla^2 \bar{\boldsymbol{v}}$ without loss of accuracy (see Nield, 1991).

Therefore, it is appropriate to identify $\boldsymbol{F}$ with the resistance force of the control volume under consideration; this force accounts for the vorticity-producing interaction of the fluid with the solid, and it solely depends on

$\bar{\boldsymbol{v}}$ and not on derivates of $\bar{\boldsymbol{v}}$. In this way, $\boldsymbol{F}$ replaces the internal tangential boundary conditions.

If spatial variations in density on the scale of the control volume are not negligible, it also follows from equation (A.11) that $\mathbf{grad}\,\bar{\rho}$ enters into the expression for $\boldsymbol{F}$. This has been worked out by Hassanizadeh (1986, 1988).

A.7. Steady Stokes flow in the control volume

For low Reynolds number flow ($Re << 1$), the advective acceleration $\underline{\bar{\boldsymbol{v}}\cdot\mathbf{grad}\bar{\boldsymbol{v}}}$ in momentum equation (A.6b) may be neglected. In combination with equation (A.9), the resulting momentum equations for this type of flow become:

$$\boldsymbol{F} = -\,\mathbf{grad}\,\bar{p} + \bar{\rho}\boldsymbol{g}, \quad \bar{\boldsymbol{v}} = \mathbf{0} \Leftrightarrow \boldsymbol{F} = \mathbf{0}. \tag{A.13}$$

For low Reynolds number flow, the steady Navier-Stokes equations (A.12) are often replaced by the steady Stokes equations:

$$\boldsymbol{\nabla}^2\boldsymbol{v} = \mathbf{grad}(\pi/\nu_0) \quad \text{(steady Stokes equations).} \tag{A.14}$$

For an analysis whether the steady Stokes equations (A.14) really govern low Reynolds number flow in porous media, an important question is: what would be the limiting character of the fluid motion obtained from the Navier-Stokes equations (A.12) for $Re \to 0$? That it should be identical with the result obtained from the Stokes equations (A.14), i.e., obtained by assuming $Re = 0$ *ab initio*, is not so obvious, as we can learn from examples in the literature on fluid dynamics where solutions to the steady Navier-Stokes equations (A.12) are considered (Olmstead, 1968; Shinbrod, 1973). In this context it is also worth mentioning that even if $\partial\bar{\boldsymbol{v}}/\partial t$ is negligible in equation (A.6b), the unsteady inertia term $\partial\boldsymbol{v}/\partial t$ is not always negligible in equation (A.2a); see Shinbrod (1973); Dybbs and Edwards (1984). However, in the 'geoscience community' it is generally accepted that, under the conditions described above, the effects of nonlinear and unsteady inertia will be negligibly small for low Reynolds number flow. Perhaps the explanation of linearity and steadiness is that the local effects of nonlinearity and unsteadiness are distributed randomly in such a way that they cancel in the approximate values which are, in fact, average values at a larger scale.

A consequence of linearity and steadiness is that a linear relationship $F_i = \beta_{ij}\,\bar{v}_j$ holds. Since we accept that the steady Stokes equations (A.14) may be considered as a good approximation to the fundamental fluid dynamic equations, Lorentz's reciprocity theorem holds (see Olmstead, 1968). From this latter theorem it follows that, in the limit of a vanishingly small representative elementary volume, $\beta_{ij} = \beta_{ji}$, i.e., the tensor $\underline{\boldsymbol{\beta}}$ is symmetric. This means that there are at least three mutually orthogonal axes (the principal axes) along which the resistance force $\boldsymbol{F}$ and the velocity $\bar{\boldsymbol{v}}$ have the same direction. However, on larger scales the symmetry property is lost (see Appendix D). It can also be proved that the tensor $\underline{\boldsymbol{\beta}}$ is positive definite, i.e., $\bar{v}_i\,\beta_{ij}\,\bar{v}_j > 0$ if $\bar{v}_i \neq 0$ and $\bar{v}_j \neq 0$.

A.8. Darcy's law and 'goodness of approximation'

On the basis of the foregoing discussion it follows that equation (A.13) yields:

$$\bar{\boldsymbol{v}} = -(1/\theta\bar{\mu})\;\underline{\boldsymbol{\kappa}}\boldsymbol{\cdot}(\mathbf{grad}\,\bar{p} - \bar{\rho}\boldsymbol{g}), \tag{A.15}$$

where $\underline{\boldsymbol{\kappa}} = \bar{\mu}\,\theta\,\underline{\boldsymbol{\beta}}^{-1}$.

Equation (A.15) represents Darcy's law on the 'point scale,' in which $\underline{\boldsymbol{\kappa}}$ is the intrinsic permeability tensor, a symmetric positive definite tensor which is independent of the dynamic viscosity $\bar{\mu}$ (see Lamb, 1974, pp. 608-616).

After this introduction to the basic principles underlying the equations of continuity and motion (A.4), (A.15) for the description of flow in porous media, we return to the question where most papers dealing with the derivation of the equations governing flow in porous media start. It is the question of how good the approximations $\bar{\boldsymbol{v}}$, $\bar{p}$ and $\bar{\rho}$ are.

A measure of 'goodness of approximation' is needed. Let $\|\cdot\|$ be a measure of the 'size' of a function, so that the smallness of $\|\bar{\boldsymbol{v}} - \boldsymbol{v}\|$ can serve as a measure of how well $\bar{\boldsymbol{v}}$ approximates $\boldsymbol{v}$ (see Esch, 1974). For example, one may take the norm $\|\cdot\|_{psa} = \left|\left\{\int\int\int(\cdot)d^3\tau\right\}/V\right|$, where the volume integral is defined over the pore space in the control volume, or representative elementary volume V (the indices *psa* denote 'pore space averaged'). The above-defined norm $\|\cdot\|_{vsa}$ is generally accepted in the literature (see Whitaker, 1966; Slattery, 1967; Gray, 1975; Sposito, 1978; Hassanizadeh, 1979, 1986, 1988; Lehner, 1979; Hassanizadeh and Gray,

1979a, 1979b, 1980; Cushman, 1983; Baveye and Sposito, 1984; Bachmat and Bear, 1986; Whitaker, 1986a, 1986b, 1986c) and the abundance of complicated mathematics in these papers is intended to prove that, or at least to list the necessary conditions under which, $\| \bar{\boldsymbol{v}} - \boldsymbol{v} \|_{vsa}$ is vanishingly small.

If single-phase steady Stokes flow is assumed, $\| \bar{\boldsymbol{v}} - \boldsymbol{v} \|_{vsa} = \mathbf{0}$ can be proved under some physically plausible sufficient conditions, which means that the approximate velocity $\bar{\boldsymbol{v}}$ is equal to the intrinsic phase-averaged value of the exact velocity $\boldsymbol{v}$, i.e., $\bar{\boldsymbol{v}} = \langle \boldsymbol{v} \rangle^f$ (e.g., see Lehner, 1979). A similar result holds for the approximation $\bar{p}$ to p. However, it turns out to be almost impossible to list the necessary conditions, and in more general cases it turns out to be impossible even to derive physically plausible sufficient conditions for volume-averaged quantities (e.g., see Hassanizadeh and Gray, 1987, 1988). Perhaps norms different from the norm $\| \cdot \|_{vsa}$ to measure the 'goodness of approximation' will be more appropriate.

A.9. Bibliography

Bachmat, Y. and Bear, J., Macroscopic modeling of transport phenomena in porous media. I: The continuum approach, *Transp. Porous Media*, Vol. 1, pp. 213-240 (1986).

Batchelor, G.K., *An Introduction to Fluid Dynamics*, Cambridge University Press, Cambridge (1974).

Baveye, P. and Sposito, G., The operational significance of the continuum hypothesis in the theory of water movement through soils and aquifers, *Water Resour. Res.*, Vol. 20, No. 5, pp. 521-530 (1984).

Chorin, A.J. and Marsden, J.E., *A Mathematical Introduction to Fluid Mechanics*, Springer-Verlag, New York (1979).

Cushman, J.H., Multiphase transport equations, *Transp. Theory and Statist. Phys.*, Vol. 12, No. 1, pp. 35-71 (1983).

Dybbs, A. and Edwards, R.V., A new look at porous media fluid mechanics – Darcy to turbulent, In: *Fundamentals of Transport Phenomena in Porous Media*, J. Bear and M. Yavus Corapcioglu (eds.), Martinus Nijhoff Publ. (1984).

Esch, R., Functional approximation, In: *Handbook of Applied Mathematics*, C.E. Pearson (ed.), Van Nostrand Reinhold Company, New York (1974).

Gray, W.G., A derivation of the equations for multi-phase transport, *Chem. Eng. Sci.*, Vol. 30, pp. 229-233 (1975).

Gray, W.G. and Hassanizadeh, S.M., Paradoxes and realities in unsaturated flow theory, *Water Resour. Res.*, Vol. 27, No. 8, pp. 1847-1854 (1991a).

Gray, W.G. and Hassanizadeh, S.M., Unsaturated flow theory including interfacial phenomena, *Water Resour. Res.*, Vol. 27, No. 8, pp. 1855-1863 (1991b).

Van der Grinten, J.G.M., An Experimental Study of Shock-Induced Wave Propagation in Dry, Water-Saturated, and Partially Saturated Porous Media, Ph.D. Thesis, Eindhoven University of Technology, Eindhoven (1987).

Hassanizadeh, S.M., Macroscopic Description of Multi-Phase systems – A Thermodynamic Theory of Flow in Porous Media, Ph.D. Thesis, Princeton University, pp. VI.1-VI.5 (Appendix A) (1979).

Hassanizadeh, S.M., Derivation of basic equations of mass transport in porous media, Part 2. Generalized Darcy's and Fick's laws, *Adv. in Water Resour.*, Vol. 9, pp. 207-222 (1986).

Hassanizadeh, S.M., Modeling species transport by concentrated brine in aggregated porous media, *Transp. Porous Media*, Vol. 3, pp. 299-318 (1988).

Hassanizadeh, S.M. and Gray, W.G., General conservation equations for multi-phase systems: I. Averaging procedure, *Adv. Water Resour.*, Vol. 2, pp. 131-144 (1979a).

Hassanizadeh, S.M. and Gray, W.G., General conservation equations for multi-phase systems: II. Mass, momentum, energy and entropy equations, *Adv. Water Resour.*, Vol. 2, pp. 191-203 (1979b).

Hassanizadeh, S.M. and Gray, W.G., General conservation equations for multi-phase systems: III. Constitutive theory for porous media flow, *Adv. Water Resour.*, Vol. 3, pp. 25-40 (1980).

Hassanizadeh, S.M. and Gray, W.G., High velocity flow in porous media, *Transp. Porous Media*, Vol. 2, pp. 521-531 (1987).

Hassanizadeh, S.M. and Gray, W.G., Reply to comments by Barak on 'High velocity flow in porous media' by Hassanizadeh and Gray, *Transp. Porous Media*, Vol. 3, pp. 319-321 (1988).

Lamb, H., *Hydrodynamics*, Cambridge University Press, London (1974).

Lehner, K., A derivation of the field equations for slow viscous flow through a porous medium, *Ind. Eng. Chem. Fundam.*, Vol. 18, No. 1, pp. 41-45 (1979).

Meyer, R.E., *Introduction to Mathematical Fluid Dynamics*, Wiley-Interscience, New York (1971).

Mitchell, J.K., *Fundamentals of Soil Behavior*, John Wiley & Sons, Inc., New York (1976).

Morse, P.M. and Feshbach, H., *Methods of Theoretical Physics*, McGraw-Hill, New York (1953).

Nield, D.A., The limitations of the Brinkman-Forchheimer equation in modeling flow in a saturated porous medium and at an interface, *Int. J. Heat and Fluid Flow*, Vol. 12, No. 3, pp. 269-272 (1991).

Olmstead, W.E., Force relationships and integral representations for the viscous hydrodynamical equations, *Archive for Rational Mech. and Anal.*, Vol. 31, pp. 380-390 (1968).

Shinbrod, M., *Lectures on Fluid Mechanics*, Gordon & Breach, New York (1973).

Slattery, J.C., Flow of viscoelastic fluids through porous media, *AICHE J.*, Vol. 13, No. 6, pp. 1066-1077 (1967).

Sposito, G., The statistical mechanical theory of water transport through unsaturated soil: I. The conservation laws, II. Derivations of the Buckingham-Darcy flux law, *Water Resour. Res.*, Vol. 14, No. 3, pp. 474-484 (1978).

Tritton, D.J., *Physical Fluid Dynamics*, Van Nostrand Reinhold Company, New York (1977).

Van der Weiden, R.M., Boundary integral equations for the computational modeling of three-dimensional steady groundwater flow problems, Report Et/EM 1988-14, Delft University of Technology, Faculty of Electrical Engineering, Laboratory of Electromagnetic Research, and Faculty of Civil Engineering, Water Management Group (1988) (appeared also as a Ph.D. Thesis).

Whitaker, S., The equations of motion in porous media, *Chem. Eng. Sci.* Vol. 21, pp. 291-300 (1966).

Whitaker, S., Flow in porous media I: A theoretical derivation of Darcy's law, *Transp. Porous Media*, Vol. 1, No. 1, pp. 3-25 (1986a).

Whitaker, S., Flow in porous media II: The governing equations for immiscible, two-phase flow, *Transp. Porous Media*, Vol. 1, No. 2, pp. 105-125 (1986b).

Whitaker, S., Flow in porous media III: Deformable media, *Transp. Porous Media*, Vol. 1, No. 2, pp. 127-154 (1986c).

Appendix B

Derivation of the vector field equations

B.1. Three coupled Laplace-type equations

In Cartesian coordinates x, y, z, which do not necessarily coincide with the horizontal axes x_h, y_h and the vertical axis z_v, the condition that the volumetric flow rate vector $\boldsymbol{q} = (q_x, q_y, q_z)$ is solenoidal ($\operatorname{div} \boldsymbol{q} = 0$) yields:

$$\partial q_x/\partial x + \partial q_y/\partial y + \partial q_z/\partial z = 0 \tag{B.1}$$

and the condition that the driving force density vector $\boldsymbol{e} = (e_x, e_y, e_z)$ is irrotational ($\mathbf{curl}\, \boldsymbol{e} = \mathbf{0}$) yields:

$$\partial e_z/\partial y - \partial e_y/\partial z = 0 \tag{B.2a}$$

$$\partial e_x/\partial z - \partial e_z/\partial x = 0 \tag{B.2b}$$

$$\partial e_y/\partial x - \partial e_x/\partial y = 0 \tag{B.2c}$$

If the local-scale conductivity tensor $\underline{\boldsymbol{k}}$, which is symmetric, has its principal axes along the x-, y- and z-axes, the relationship between $\boldsymbol{q}$ and $\boldsymbol{e}$, i.e., Darcy's law, is given by:

$$q_x = k_x e_x \tag{B.3a}$$

$$q_y = k_y e_y \tag{B.3b}$$

$$q_z = k_z e_z. \tag{B.3c}$$

The equations (B.1, B.2, B.3) are not always well-suited for direct solution of $\boldsymbol{q}$ and $\boldsymbol{e}$. Therefore, the equations will be transformed to equivalent equations that are sometimes more convenient.

With the aid of equation (B.3a), equation (B.1) is transformed to:

$$k_x \, \partial e_x/\partial x + q_x \, k_x^{-1} \, \partial k_x/\partial x + \partial q_y/\partial y + \partial q_z/\partial z = 0 \tag{B.4}$$

Differentiation of equation (B.4) with respect to x yields:

$$\begin{aligned} &\partial/\partial x \, (k_x \, \partial e_x/\partial x) + \partial/\partial y \, (\partial q_y/\partial x) + \partial/\partial z \, (\partial q_z/\partial x) + \\ &\quad + \partial q_x/\partial x \left(k_x^{-1} \, \partial k_x/\partial x\right) + q_x \, \partial/\partial x \left(k_x^{-1} \, \partial k_x/\partial x\right) = 0. \end{aligned} \tag{B.5}$$

With the aid of equation (B.3b), equation (B.2c) is transformed to:

$$k_y^{-1} \, \partial q_y/\partial x - \partial e_x/\partial y + q_y \, \partial k_y^{-1}/\partial x = 0. \tag{B.6a}$$

Similarly, with the aid of equation (B.3c), equation (B.2b) is transformed to:

$$k_z^{-1} \, \partial q_z/\partial x - \partial e_x/\partial z + q_z \, \partial k_z^{-1}/\partial x = 0. \tag{B.6b}$$

Substitution of expressions (B.6a) and (B.6b) for $\partial q_y/\partial x$ and $\partial q_z/\partial x$ in (B.5) yields:

$$\begin{aligned} &\partial/\partial x \, (k_x \, \partial e_x/\partial x) + \partial/\partial y \, (k_y \partial e_x/\partial y) + \partial/\partial z \, (k_z \, \partial e_x/\partial z) + \\ &\quad + \left(k_x^{-1} \, \partial k_x/\partial x\right) \, \partial q_x/\partial x + \left(k_y^{-1} \, \partial k_y/\partial x\right) \, \partial q_y/\partial y + \\ &\quad + \left(k_z^{-1} \, \partial k_z/\partial x\right) \, \partial q_z/\partial z + q_x \, \partial/\partial x \left(k_x^{-1} \, \partial k_x/\partial x\right) + \\ &\quad + q_y \, \partial/\partial y \left(k_y^{-1} \, \partial k_y/\partial x\right) + q_z \, \partial/\partial z \left(k_z^{-1} \, \partial k_z/\partial x\right) = 0. \end{aligned} \tag{B.7a}$$

In a similar way, equations for e_y and e_z can be derived.

In index notation ($x = x_1$, $y = x_2$, $z = x_3$, etc.) equation (B.7a) and the equations for e_y and e_z can be written as:

$$\sum_{j=1}^{3} \partial/\partial x_j \left(k_j \, \partial e_k/\partial x_j\right) + (\partial\lambda_j/\partial x_k)\, \partial q_j/\partial x_j + \\ + q_j \, \partial^2\lambda_j/\partial x_j \partial x_k = 0, \quad k = 1,2,3 \tag{B.7b}$$

where $\lambda_j = \ln(k_j/k_G)$; k_G is a constant reference conductivity.

By substitution of $q_k = k_k e_k$ into equation (B.7b), equations for q_k may be derived yielding:

$$\sum_{j=1}^{3} \partial/\partial x_j \left(k_j \, k_k^{-2} \, \partial q_k/\partial x_j\right) + k_k^{-1}(\partial\lambda_j/\partial x_k)\, \partial q_j/\partial x_j + \\ - e_k \, \partial/\partial x_j \left(k_j \, k_k^{-1} \, \partial\lambda_k/\partial x_j\right) + \\ + q_j \, k_k^{-1} \, \partial^2\lambda_j/\partial x_j \partial x_k = 0, \quad k = 1,2,3. \tag{B.8}$$

B.2. Three uncoupled Laplace-type equations

An interesting simplification arises if k_x, k_y and k_z vary in such a way that:

$$k_x = f_1(z)\exp(c_1 x)\exp(c_2 y) \tag{B.9a}$$

$$k_y = f_2(z)\exp(c_1 x)\exp(c_2 y) \tag{B.9b}$$

$$k_z = f_3(z)\exp(c_1 x)\exp(c_2 y) \tag{B.9c}$$

where c_1 and c_2 are constants, and f_1, f_2 and f_3 are arbitrary functions of z. In that case equations (B.7b) for the x- and y-directions ($k = 1,2$) simplify to:

$$\sum_{j=1}^{3} \partial/\partial x_j \left(k_j \, \partial e_k/\partial x_j\right) = 0, \quad k = 1,2. \tag{B.10a}$$

For the z-direction ($k = 3$) let us consider equation (B.8) which simplifies to:

$$\sum_{j=1}^{3} \partial/\partial x_j \left(k_j\, k_3^{-2}\, \partial q_3/\partial x_j \right) + \\ + k_3^{-1} \left(\partial \lambda_j/\partial x_3 \right) \partial q_j/\partial x_j = 0. \tag{B.10b}$$

This latter equation can further be simplified by considering the situation $k_1 = k$, $k_2 = \beta k$, where β is a constant (quasi-isotropy in the x-y plane).

Making use of the continuity equation (B.1), the latter equation then simplifies to:

$$\partial/\partial x_1 \left(k_3^{-1}\, \partial q_3/\partial x_1 \right) + \partial/\partial x_2 \left(\beta\, k_3^{-1}\, \partial q_3/\partial x_2 \right) + \\ + \partial/\partial x_3 \left(k^{-1}\, \partial q_3/\partial x_3 \right) = 0. \tag{B.10c}$$

The set of equations (B.10a)-(B.10c) represents three uncoupled Laplace-type equations for e_x, e_y and q_z, respectively. If $k = k_h$ and $k_3 = k_z$ are discontinuous functions of z, it follows from the weak form of $\operatorname{div} \boldsymbol{q} = 0$ that q_z is continuous over the discontinuities in k_h and k_z, and it follows from the weak form of $\mathbf{curl}\, \boldsymbol{e} = \mathbf{0}$ that e_x and e_y are continuous over the discontinuities in k_h and k_z.

Consequently, the set of Laplace-type equations (B.10a)-(B.10c) describes continuously varying variables e_x, e_y and q_z, even if k_h and k_z are discontinuous functions of z.

If $c_1 = 0$, $c_2 = 0$, the porous medium is said to be perfectly layered in the x-y planes.

B.3. Physical and auxiliary boundary conditions

Now the question arises whether the solutions of equations (B.7b) and (B.8) really are solutions of the continuity equation $\operatorname{div} \boldsymbol{q} = 0$ and Darcy's law $\mathbf{curl}\, \boldsymbol{e} = \mathbf{0}$. To insure that this is really the case we have to specify the boundary conditions for equations (B.7b), (B.8) in the right way.

Let us consider a three-dimensional rectangular region, or block, where $0 \leq x \leq a$, $0 \leq y \leq b$, $0 \leq z \leq c$. As an example, let us choose the boundary conditions $e_x(x, y, 0) = -f_x(x, y)$ and $e_y(x, y, 0) = -f_y(x, y)$ on the boundary $z = 0$, $0 \leq x \leq a$, $0 \leq y \leq b$ where f_x and f_y are derived from a potential $\phi(x, y, 0)$ specified on the boundary; i.e., $f_x(x, y) = \partial\phi(x, y, 0)/\partial x$, $f_y = \partial\phi(x, y, 0)/\partial y$. These boundary conditions may be considered as physical boundary conditions for equations (B.7b) for e_x and e_y.

To find the auxiliary boundary condition for equation (B.8) for q_z we consider $\operatorname{div} \boldsymbol{q} = 0$ on the boundary:

$$\begin{aligned} \partial q_z/\partial n = -\partial q_z/\partial z = -\partial/\partial x\,[k_x(x, y, 0)\ f_x(x, y)] \\ -\,\partial/\partial y\,[k_y(x, y, 0)\ f_y(x, y)] \\ \text{on } z = 0,\ 0 \le x \le a,\ 0 \le y \le b. \end{aligned} \tag{B.11a}$$

As another example, let us choose $q_z(x, y, c) = B(x, y)$ on $z = c$, $0 \le x \le a$, $0 \le y \le b$ (q_z is a specified normal flux $= -k_z\ \partial\phi/\partial z$). This boundary condition may be considered as a physical boundary condition for equation (B.8) for q_z.

To find the auxiliary boundary conditions for equation (B.7b) we consider $(\mathbf{curl}\, \boldsymbol{e})_y = 0$ and $(\mathbf{curl}\, \boldsymbol{e})_x = 0$ on the boundary:

$$\begin{aligned} \partial e_x/\partial n = \partial e_x/\partial z = \partial/\partial x\,[B(x, y)/k_z(x, y, c)] \\ \text{on } z = c,\ 0 \le x \le a,\ 0 \le y \le b \end{aligned} \tag{B.11b}$$

and

$$\begin{aligned} \partial e_y/\partial n = \partial e_y/\partial z = \partial/\partial y\,[B(x, y)/k_z(x, y, c)] \\ \text{on } z = c,\ 0 \le x \le a,\ 0 \le y \le b \end{aligned} \tag{B.11c}$$

Also for the other four boundaries, we can apply either $\operatorname{div} \boldsymbol{q} = 0$ or $\boldsymbol{n} \times \mathbf{curl}\, \boldsymbol{e} = \mathbf{0}$ to find the additional auxiliary boundary condition(s). The above procedure does not lead to a well-posed problem for the Laplace-type equations (B.7b), (B.8) if e_x, e_y and q_z are not specified on at least one point of the closed boundary.

For instance, if only $\partial q_z/\partial n$ is specified on the closed boundary, the solution q_z is unique up to the addition of a constant. If we specify the potential ϕ on the two boundaries $z = 0$ and $z = c$, and we specify $\partial\phi/\partial n$ (flow rates) on the other four boundaries, the boundary conditions for e_x and e_y on $z = 0$ and $z = c$ must be derived from $e_x = -\partial\phi/\partial x$, $e_y = -\partial\phi/\partial y$. If we add a constant C_1 to the value of ϕ on $z = 0$, and a constant $C_2 \neq C_1$ to the value of ϕ on $z = c$, we find the same boundary conditions for e_x and e_y as before. This means that information is lost. This loss of information is expressed by the fact that the solution q_z is not unique. If we connect the two boundaries where ϕ is specified with another boundary, say $x = a$, where ϕ is also specified, the uniqueness will be restored, since in this latter case q_z is specified on $x = a$. From the above discussion we conclude that, if mixed boundary conditions are specified, one cannot specify ϕ on two or more disjunct parts of the boundary.

Summarizing the results discussed above yields:

1. If the normal flux component $\boldsymbol{n} \cdot \boldsymbol{q}$ is specified, then $\boldsymbol{n} \times \mathbf{curl}\, \boldsymbol{e} = \mathbf{0}$ must also be given. (B.12a)

2. If the tangential components $\boldsymbol{n} \times \boldsymbol{e}$ are specified, then $\mathrm{div}\, \boldsymbol{q} = 0$ must also be given. (B.12b)

3. If mixed boundary conditions are given, i.e., if (B.12a) is specified on parts of the boundary and (B.12b) is specified on the remaining parts, then the problem is properly posed only if boundary conditions (B.12b) are specified on a connected region of the boundary. (B.12c)

Now the question whether $\boldsymbol{q}$ and $\boldsymbol{e}$ satisfy the continuity equation and Darcy's law can be answered.

The vector functions $\boldsymbol{q}'$ and $\boldsymbol{e}'$ satisfying $\mathrm{div}\, \boldsymbol{q}' = 0$, $\mathbf{curl}\, \boldsymbol{e}' = \mathbf{0}$ and $\boldsymbol{q}' = \underline{\boldsymbol{k}} \cdot \boldsymbol{e}'$ form a subset of all possible solutions $\boldsymbol{q}$ and $\boldsymbol{e}$ of equations (B.7b) and (B.8). However, the boundary conditions (B.12) for equations (B.7b) and (B.8) are chosen in such a way that nowhere on the boundary $\boldsymbol{q} \neq \boldsymbol{q}'$, $\boldsymbol{e} \neq \boldsymbol{e}'$, $\mathrm{div}\, \boldsymbol{q} \neq \mathrm{div}\, \boldsymbol{q}'$ and $\boldsymbol{n} \times \mathbf{curl}\, \boldsymbol{e} \neq \boldsymbol{n} \times \mathbf{curl}\, \boldsymbol{e}'$ are specified.

Since the solutions of the Laplace-type equations (B.7b), (B.8) with the above-presented boundary conditions (B.12) exist and are unique, it follows that $\boldsymbol{q} = \boldsymbol{q}'$ and $\boldsymbol{e} = \boldsymbol{e}'$, in other words, $\mathrm{div}\, \boldsymbol{q} = 0$, $\mathbf{curl}\, \boldsymbol{e} = \mathbf{0}$, $\boldsymbol{q} = \underline{\boldsymbol{k}} \cdot \boldsymbol{e}$. Consequently, the choice of boundary conditions (B.12) forces $\boldsymbol{q}$ and $\boldsymbol{e}$ to satisfy the continuity equation and Darcy's law.

Appendix C

Flow caused by horizontal gradients in fluid density

C.1. Introduction

Flow systems analysis was for the first time applied by Tóth (1963). Originally, flow systems analysis was devoted to the study of groundwater flow driven only by spatial variations in the topography of the water table. Tóth's theoretical analysis shows that the flow field can be subdivided in flow systems resulting in a hierarchy of nested flow paths. This subdivision in flow systems plays an important role in the conceptual model underlying practical applications of flow systems analysis.

Applications deal with migration of hydrocarbons in sedimentary basins (Tóth, 1980), genesis of strata-bound ore deposits (Garven and Freeze, 1984), and environmental protection (Engelen, 1984; Stuurman *et al.*, 1989).

A potential application of flow systems analysis is related to the storage of highly toxic and/or radioactive waste in deep subsurface formations. The concept of flow systems analysis looks promising to assess the environmental impact of a hypothetical release of this waste out of its containment into the groundwater. However, when considering deep on-shore storage in flat coastal areas or deep off-shore storage, spatial variations in the water table are not likely to be the only driving mechanism for groundwater flow. Therefore, the concept of flow systems analysis will be generalized also to account for groundwater flow driven by spatial differences in the water-salt mixture density. This means that the combination of the following two mechanisms driving groundwater flow will be considered:

(i) Spatial variations in the topography of the water table and well-abstractions causing forced convection.
(ii) Horizontal components in the gradient of the water-salt mixture density causing free convection.

The combination of forced convection with free convection is generally called mixed convection.

In the following discussion gravity will be the only external force field under consideration; in other words, only buoyancy-driven flow is considered. This means that forced convection caused by groundwater well discharge will not be discussed.

Groundwater flow is governed by linear field equations, but the coupling with the dynamic and kinematic boundary conditions at the water table makes the flow problem nonlinear. However, in many practical cases this nonlinearity may be neglected and then the solutions describing groundwater flow to wells may be superimposed on the solutions developed in this Appendix.

To describe the combined effect of the two above-mentioned driving mechanisms, dimensionless quantities and dimensionless numbers will be introduced. The introduction of dimensionless numbers has been inspired by the theory of dimensional analysis (see Silberberg and McKetta, 1953), and they turn out to be useful for further mathematical analysis by means of the perturbation method (see Van Dyke, 1975).

Dimensional analysis is based upon the notion that the physical dimensions of the quantities relevant to the problem under consideration can be arranged in such a way that dimensionless numbers are formed. The introduction of dimensionless quantities strongly depends on the existence of characteristic quantities which are representative for the flow systems under consideration. For instance, when a groundwater basin has a depth d (with dimension 'length' expressed in the unit 'meter' $[m]$), it seems natural to scale all vertical coordinates z (dimension 'length' $[m]$) with the aid of this characteristic vertical length d resulting in a dimensionless vertical coordinate $Z = z/d$.

A similar procedure holds for the horizontal coordinates x, y $[m]$, which can be scaled with the aid of a characteristic horizontal length l_C $[m]$ to obtain the dimensionless horizontal coordinates $X = x/l_C$, $Y = y/l_C$. However, a proper choice for the horizontal characteristic length is more difficult than choosing the vertical characteristic length. For a flow systems analysis, it is proposed that the spatial variations of the water table be decomposed into Fourier modes. Then, for each Fourier mode, the characteristic horizontal length will be chosen (approximately) equal to the wavelength divided by 2π of the Fourier mode under consideration.

Proper scaling of subsurface phenomena is even more difficult because of the heterogeneity of the subsurface. For instance, it is much more difficult to define characteristic values, representative for the whole groundwater basin under consideration, of the horizontal and vertical conductivities than it is to define representative characteristic lengths. The same difficulty holds for the heterogeneous effective porosity and specific storage-coefficient. However, the problem can be simplified when considering an approximately layered basin with sufficiently often differentiable horizontal heterogeneities.

For this model, application of the perturbation method will lead to expressions for the characteristic conductivities. With the aid of the latter characteristic horizontal and vertical conductivities, k_{hc} and k_{zc}, respectively (with dimension 'length^2•pressure^{-1}•time^{-1}' expressed in the SI unit 'meter^2•Pascal^{-1}•second^{-1}' $[m^2 \cdot Pa^{-1} \cdot s^{-1}]$), the conductivities k_h and k_z $[m^2 \cdot Pa^{-1} \cdot s^{-1}]$ can be scaled to obtain the dimensionless conductivities $K_h = k_h/k_{hc}$ and $K_z = k_z/k_{zc}$. In this way, the equations governing groundwater flow can be written in dimensionless quantities, resulting in equations in which dimensionless numbers, which may have any order of magnitude, occur as parameters. For the equations governing groundwater flow in situations where only the two above-mentioned driving mechanisms are accounted for, only one such dimensionless number can be found.

The perturbation method is based upon the notion that the solution of the dimensionless equations depends on the parameters, i.e., on the dimensionless numbers occurring in the equations. Therefore, one seeks to find the solution as a power series in one of the dimensionless numbers, say ϵ, in such a way that the coefficients in the series can readily be obtained by analytical mathematics.

The zeroth-order approximation to the solution is obtained in the limit $\epsilon \rightarrow 0$. In the groundwater flow problem under consideration the zeroth-order solution turns out to be equal to the well-known Dupuit approximation. The terms in ϵ, ϵ^2, etc. are corrections, or perturbations, to this zeroth-order solution.

Characteristic conductivities are important not only in perturbation methods, but also in numerical modeling studies, where characteristic conductivities have to be assigned to the grid blocks. Each grid block is considered as an 'equivalent homogeneous porous medium.' Furthermore, dimensionless numbers can be applied in order of magnitude estimations, for instance to estimate under what circumstances free convective flow is negligible with respect to forced convective flow.

Conclusions derived in this Appendix are:

(i) In perfectly layered groundwater basins with exactly horizontal layers, vertical flow depends only on the horizontal second derivatives of the water table and water-salt mixture density. In the sharp fresh-saline interface approximation, vertical flow depends only on the horizontal second derivatives of the interface. This dependence on second derivatives, and not on first derivatives, explains why vertical transport is much more difficult to determine than horizontal transport. Yet, to investigate whether waste dissolved in the groundwater travels either to the biosphere or to deeper geologic layers, knowledge of the vertical flow component is essential.

(ii) As far as forced convection is considered, an 'equivalent homogeneous porous medium' exists. However, for free convection, an 'equivalent homogeneous porous medium' does not generally exist.

(iii) The long response times of the water-salt mixture density distribution to palaeo-geological or palaeo-climatological disturbances make it evident that both forced convection and free convection caused by a nonstagnant density distribution are important transport mechanisms when considering the transport of highly toxic and/or radioactive matter to the biosphere. The characters of forced convective flow and stable free convective flow are similar, both showing nested flow systems.

C.2. Governing equations and boundary conditions

The starting point for a flow systems analysis is the set of basic equations governing flow in porous media: the continuity equation and Darcy's law. In the context of free convection the approximate form of the continuity equation, the Boussinesq approximation (see Turner, 1973), is given by:

$$\operatorname{div} \boldsymbol{q} = \partial q_x/\partial x + \partial q_y/\partial y + \partial q_z/\partial z = 0. \tag{C.1}$$

In its most general form Darcy's law is given by:

$$\boldsymbol{q} = -\underline{\boldsymbol{k}} \cdot (\mathbf{grad}\, p - \rho \boldsymbol{g}). \tag{C.2}$$

In the following presentation Darcy's law will be simplified by only considering perfectly layered basins, i.e.:

$$\underline{\boldsymbol{k}} = \begin{bmatrix} k_{xx} & k_{xy} & k_{xz} \\ k_{yx} & k_{yy} & k_{yz} \\ k_{zx} & k_{zy} & k_{zz} \end{bmatrix} = \begin{bmatrix} k_h(z,t) & 0 & 0 \\ 0 & k_h(z,t) & 0 \\ 0 & 0 & k_z(z,t) \end{bmatrix} \tag{C.3}$$

where k_h and k_z may be piecewise continuous functions of z. Furthermore, x and y are exactly horizontal coordinates, and z is the exactly vertical coordinate.

In expression (C.3) it is tacitly assumed that the viscosity of the water-salt mixture is independent of the mass fraction of dissolved salt. The time-dependence of the hydraulic conductivities plays only a role on very long geological time scales, which are important when studying the transport of released highly toxic and/or radioactive waste to the biosphere, but can be neglected in most engineering applications.

On the upper plane of the flow domain, $z = 0$, the pressure $p(x, y, 0, t)$ [Pa; $dbar$] of the water-salt mixture is specified and equal to:

$$p(x, y, 0, t) = p_{atm}(t) - g \int_0^{-h_f(x,y,t)} \rho(z)\, \mathrm{d}z, \tag{C.4}$$

where $-h_f(x, y, t)$ [m] represents the topography of the phreatic groundwater level with respect to the fixed level $z = 0$ ($g = |\boldsymbol{g}|$). In the dynamic boundary condition (C.4) it is assumed that in the interval $z = 0$ to $z = -h_f(x, y, t)$ the water-salt mixture density ρ has no horizontal gradient components ($\partial\rho/\partial x = \partial\rho/\partial y = 0$) and that the vertical pressure gradient is hydrostatic ($\partial p/\partial z = \rho g$).

The lower boundary of the basin, $z = d$, is chosen as an impervious base, i.e., as a plane where the vertical component of the volumetric flow rate, $q_z(x, y, d, t)$ is equal to zero.

The above-discussed equations and boundary conditions lead to the following three uncoupled sets of field equations and boundary conditions for the three quantities $e_x(x, y, z, t) = q_x(x, y, z, t)/k_h(z, t)$, $e_y(x, y, z, t) =$

$= q_y(x,y,z,t)/k_h(z,t)$, and $q_z(x,y,z,t)$:

$$\left.\begin{aligned}
&\partial/\partial x(k_h\ \partial e_x/\partial x) + \partial/\partial y(k_h\ \partial e_x/\partial y) + \\
&\quad + \partial/\partial z\left[k_z(\partial e_x/\partial z + g\ \partial\rho/\partial x)\right] = 0 && , 0 < z < d \\
&e_x = -\rho g\ \partial h_f/\partial x && , z = 0 \\
&\partial e_x/\partial z = -g\ \partial\rho/\partial x && , z = d
\end{aligned}\right\} \quad (C.5a)$$

$$\left.\begin{aligned}
&\partial/\partial x(k_h\ \partial e_y/\partial x) + \partial/\partial y(k_h\ \partial e_y/\partial y) + \\
&\quad + \partial/\partial z\left[k_z(\partial e_y/\partial z + g\ \partial\rho/\partial y)\right] = 0 && , 0 < z < d \\
&e_y = -\rho g\ \partial h_f/\partial y && , z = 0 \\
&\partial e_y/\partial z = -g\ \partial\rho/\partial y && , z = d
\end{aligned}\right\} \quad (C.5b)$$

$$\left.\begin{aligned}
&\partial/\partial x\left(k_z^{-1}\ \partial q_z/\partial x - g\ \partial\rho/\partial x\right) + \\
&\quad + \partial/\partial y\left(k_z^{-1}\ \partial q_z/\partial y - g\ \partial\rho/\partial y\right) + \\
&\quad + \partial/\partial z\left(k_h^{-1}\ \partial q_z/\partial z\right) = 0 && , 0 < z < d \\
&\partial q_z/\partial z = k_h \rho g\ (\partial^2/\partial x^2 + \partial^2/\partial y^2)\ h_f && , z = 0 \\
&q_z = 0 && , z = d
\end{aligned}\right\} \quad (C.5c)$$

The proof of the above equations is similar to that given in Appendix C.

In the above three uncoupled sets of Poisson-type equations and boundary conditions (C.5a), (C.5b) and (C.5c), two buoyancy-driven mechanisms for flow are included.

Firstly, the combination of gravity and spatial variations in the topography of the water table are a driving force per unit volume for flow: $\rho g\ \partial h_f(x,y,t)/\partial x$ [$Pa{\cdot}m^{-1}$; $dbar{\cdot}m^{-1}$] occurs in the top boundary condition for $e_x(x,y,z,t)$, $\rho g\ \partial h_f(x,y,t)/\partial y$ [$Pa{\cdot}m^{-1}$; $dbar{\cdot}m^{-1}$] occurs in the top boundary condition for $e_y(x,y,z,t)$, and $\rho g \left\{\partial^2 h_f(x,y,t)/\partial x^2 + \right.$ $\left. + \partial^2 h_f(x,y,t)/\partial y^2\right\}$ [$Pa{\cdot}m^{-1}$; $dbar{\cdot}m^{-1}$] occurs in the top boundary condition for $q_z(x,y,z,t)$. These top boundary conditions cause the topography-driven flow, or forced convection.

Secondly, the combination of gravity and horizontal components of the water-salt mixture density gradient, $g\ \partial\rho(x,y,z,t)/\partial x$ [$Pa{\cdot}m^{-2}$; $dbar{\cdot}$ ${\cdot}m^{-2}$] and $g\ \partial\rho(x,y,z,t)/\partial y$ [$Pa{\cdot}m^{-2}$; $dbar{\cdot}m^{-2}$], occur both in the bottom boundary conditions and in the field equations. These latter components cause the density-driven flow, or free convection. The equations (C.5) hold for a perfectly layered subsurface, but an extension to approximately layered subsurfaces is straightforward. (An approximately layered porous medium has lateral heterogeneities that are 'smooth,' i.e., that are sufficiently often differentiable.)

In this Appendix, further analytical mathematics will be presented, but the equations (C.5) are also promising for further numerical analysis. The more common approach is to base numerical analysis on the equation:

$$\operatorname{div}(\underline{\boldsymbol{k}} \cdot \mathbf{grad}\, p) - g\ \partial(k_z \rho)/\partial z = 0 \tag{C.6}$$

to calculate p, and to evaluate $\boldsymbol{q}$ afterwards from Darcy's law (C.2). However, when the conforming finite element method with piecewise linear interpolations of p and ρ is applied to solve equation (C.6), there will be an inconsistency in the approximation of $\boldsymbol{q}$ which gives rise to artificial vertical velocity components. This inconsistency is due to the numerical differentiation of p resulting in piecewise constant values of $\mathbf{grad}\, p$. The combination in Darcy's law of these piecewise constant values of $\mathbf{grad}\, p$ with the piecewise linear values of ρ gives rise to artificial vertical velocities within one element (for more details, see Voss and Souza, 1987).

The same conforming finite element method, with piecewise linear interpolations of e_x, e_y, q_z and ρ, applied to solve equations (C.5) does not give rise to the above-discussed inconsistency and resulting artificial vertical velocity components.

The transport of salt by convection and dispersion causes the density field $\rho(x, y, z, t)$ to vary with time. Both convection and dispersion strongly depend on the transport velocity field $\boldsymbol{v} = (v_x, v_y, v_z) = (k_h e_x/\theta,\ k_h e_y/\theta,\ q_z/\theta)$. As can be observed from equations (C.5), the velocity field depends on the density field ρ. For a given velocity field the convection-dispersion equation is linear, and for a given density field the flow equations (C.5) are linear. Yet, the combined problem is nonlinear because of the coupling between the two equations. Also in the sharp fresh-saline interface approximation, the coupling of the interfacial kinematic condition, which replaces the convection-dispersion equation, with the flow equations makes the total flow problem nonlinear.

Consider the motion of an element of fluid displaced a small distance $z^*(t)$ vertically from its stagnant position in an environment with vertical components of the density gradient $\partial\rho/\partial z = c$. The vertical component of Darcy's law (C.2), neglecting small pressure fluctuations, together with the dispersion-free approximation $\partial\rho/\partial t + \boldsymbol{v}\boldsymbol{\cdot}\, \mathbf{grad}\, \rho = 0$ yields $\partial z^*/\partial t = -(k_z gc/\theta)\, z^*$ with the solution $z^* = z_0^* \exp[-(k_z gc/\theta)\, t]$. This solution expresses that, in an environment where the density increases with increasing depth ($c > 0$), the fluid element is driven back to the stagnant position. See also Horton and Rogers (1945) who show that diffusion and dispersion also have a stabilizing effect. However, the characteristic time needed to restore the disturbance may be relatively long. For instance, for $k_z = 0.0864\ m^2{\cdot}dbar^{-1}{\cdot}d^{-1} = 10^{10}\ m^2{\cdot}Pa^{-1}{\cdot}s^{-1}$, $c = 25\ kg{\cdot}m^{-4}$, $\theta = 0.25$ and $g = 10\ m{\cdot}s^{-2}$ the characteristic time $(k_z gc/\theta)^{-1} = 10^7\ s \approx 120\ d$. Much longer response times to palaeogeological or palaeo-climatological disturbances causing flow in deeper layers, where k_z is orders of magnitude smaller, make it evident that free convection caused by a nonstagnant density distribution is an important transport mechanism when considering the travel of dangerous matter to the biosphere.

The occurrence of the nonlinear coupling makes fresh-saline problems very difficult to solve, especially in unstable environments (c sufficiently negative), where fresh water is below saline water. However, in this Appendix unstable transport of salt is not considered and only the linear flow problem given by equations (C.5) will be discussed further.

C.3. Dimensionless vector field equations

A useful procedure in a flow systems analysis is to decompose the spatial variations of the water table topography in a sum or an integral of Fourier modes. A Fourier mode of the water table is given by $h_f(x,y,t) = \left[h_f^*(w_x, w_y, t)/4\pi^2\right] \exp\left[i(w_x x + w_y y)\right]\, dw_x dw_y$, where $h_f^*(w_x, w_y, t)$ $[m^3]$ is the Fourier Transform of the water table, and w_x and w_y are the wave numbers $[m^{-1}]$ of the particular Fourier mode under consideration. The characteristic horizontal length scale l_c $[m]$ belonging to this Fourier mode is $l_c = 1/\sqrt{(w_x^2 + w_y^2)}$.

Now define the dimensionless horizontal coordinates $X = x/l_c$ and $Y = y/l_c$, and the dimensionless vertical coordinate $Z = z/d$. Also, define the dimensionless water table height $F(X,Y,t) = h_f(x,y,t)/d$, and the dimensionless water-salt mixture density $P(X,Y,Z,t) = \rho(x,y,z,t)/$

$/\rho_0$, where ρ_0 is the density on $z = 0$. Furthermore, we also need dimensionless conductivities $K_h(z) = k_h(z)/k_{hc}$ and $K_z(z) = k_z(z)/k_{zc}$, where k_{hc} and k_{zc} are, respectively, the horizontal and vertical characteristic conductivities. The question how to choose the characteristic conductivities will be considered in Section C.6. Finally, define the dimensionless quantities $E_x(X,Y,Z,t) = e_x(x,y,z,t)\ l_c/(\rho_0 g\ d)$, $E_y(X,Y,Z,t) = e_y(x,y,z,t)\ l_c/(\rho_0 g\ d)$, and $Q_z(X,Y,Z,t) = q_z(x,y,z,t)\ l_c^2 / \left(\rho_0 g\ \mathrm{d}^2 k_{hc}\right)$. The three decoupled sets of field equations and boundary conditions (C.5a), (C.5b) and (C.5c) written in dimensionless variables are now given as:

$$\left.\begin{array}{ll} \epsilon\ \partial/\partial X\ (K_h \partial E_x/\partial X) + & \\ \quad + \epsilon\ \partial/\partial Y\ (K_h \partial E_x/\partial Y) + & \\ \quad + \partial/\partial Z\ [K_z\ (\partial E_x/\partial Z + \partial P/\partial X)] = 0 & ,\ 0 < Z < 1 \\ E_x = -\partial F/\partial X & ,\ Z = 0 \\ \partial E_x/\partial Z = -\partial P/\partial X & ,\ Z = 1 \end{array}\right\} \qquad (C.7a)$$

$$\left.\begin{array}{ll} \epsilon\ \partial/\partial X\ (K_h\ \partial E_y/\partial X) + & \\ \quad + \epsilon\ \partial/\partial Y\ (K_h\ \partial E_y/\partial Y) + & \\ \quad + \partial/\partial Z\ \left[K_z\ (\partial E_y/\partial Z + \partial P/\partial Y)\right] = 0 & ,\ 0 < Z < 1 \\ E_y = -\partial F/\partial Y & ,\ Z = 0 \\ \partial E_y/\partial Z = -\partial P/\partial Y & ,\ Z = 1 \end{array}\right\} \qquad (C.7b)$$

$$\left.\begin{array}{ll} \partial/\partial X\ (\epsilon K_z^{-1}\ \partial Q_z/\partial X - \partial P/\partial X) + & \\ \quad + \partial/\partial Y\ (\epsilon K_z^{-1}\ \partial Q_z/\partial Y - \partial P/\partial Y) + & \\ \quad + \partial/\partial Z\ (K_h^{-1}\ \partial Q_z/\partial Z) = 0 & ,\ 0 < Z < 1 \\ \partial Q_z/\partial Z = K_h\ (\partial^2/\partial X^2 + \partial^2/\partial Y^2)\ F & ,\ Z = 0 \\ Q_z = 0 & ,\ Z = 1 \end{array}\right\} \qquad (C.7c)$$

Since E_x, E_y and Q_z are dimensionless quantities they may be considered as the three Cartesian components of a dimensionless vector. The three decoupled sets of field equations (C.7a), (C.7b) and (C.7c) are then called the vector field equations. In the vector field equations (C.7) the following dimensionless number occurs:

$$\epsilon = (d/l_c)^2\ (k_{hc}/k_{zc}), \qquad (C.8)$$

Since in the above-presented sets of equations the time t is only a parameter, its explicit notation will be omitted.

C.4. Perturbation series in the ϵ-number

The three vector field components E_x, E_y and Q_z in the three sets of vector field equations (C.7a), (C.7b) and (C.7c) depend on the parameter ϵ. Therefore, the three vector field components are expanded in an infinite power series in the ϵ-number:

$$E_x = E_{x0} + \epsilon E_{x1} + \epsilon^2 E_{x2} + \ldots \epsilon^n E_{nx} + \ldots, \tag{C.9a}$$

$$E_y = E_{y0} + \epsilon E_{y1} + \epsilon^2 E_{y2} + \ldots \epsilon^n E_{ny} + \ldots, \tag{C.9b}$$

$$Q_z = Q_{z0} + \epsilon Q_{z1} + \epsilon^2 Q_{z2} + \ldots \epsilon^n Q_{nz} + \ldots. \tag{C.9c}$$

Now the three series (C.9a), (C.9b) and (C.9c) are substituted into the three sets of vector field equations (C.7a), (C.7b) and (C.7c) yielding three sets of equations in which powers of ϵ occur. Since these equations must hold for all values of ϵ, each coefficient of ϵ must vanish independently because powers of ϵ are linearly independent. Equating the coefficients to zero results in a hierarchy of ordinary differential equations which can easily be solved successively. The resulting hierarchy is given by:

Zeroth-order equations.

$$\left.\begin{array}{ll} \partial/\partial Z\ [K_z\ (\partial E_{x0}/\partial Z + \partial P/\partial X)] = 0 & ,\ 0 < Z < 1 \\ E_{x0} = -\partial F/\partial X & ,\ Z = 0 \\ \partial E_{x0}/\partial Z = -\partial P/\partial X & ,\ Z = 1 \end{array}\right\} \tag{C.10a}$$

$$\left.\begin{array}{ll} \partial/\partial Z\ [K_z\ (\partial E_{y0}/\partial Z + \partial P/\partial Y)] = 0 & ,\ 0 < Z < 1 \\ E_{y0} = -\partial F/\partial Y & ,\ Z = 0 \\ \partial E_{z0}/\partial Z = -\partial P/\partial Y & ,\ Z = 1 \end{array}\right\} \tag{C.10b}$$

$$\left.\begin{array}{ll}\partial/\partial Z\ (K_h^{-1}\ \partial Q_{z0}/\partial Z)+ & \\ \quad -(\partial^2/\partial X^2+\partial^2/\partial Y^2)\ P=0 & ,\ 0<Z<1 \\ \partial Q_{z0}/\partial Z = K_h\ (\partial^2/\partial X^2+\partial^2/\partial Y^2)\ F & ,\ Z=0 \\ Q_{z0}=0 & ,\ Z=1\end{array}\right\} \quad \text{(C.10c)}$$

Higher-order equations ($n>0$).

$$\left.\begin{array}{ll}\partial/\partial X\ (K_h\ \partial E_{xn-1}/\partial X)+ & \\ \quad +\partial/\partial Y\ (K_h\ \partial E_{xn-1}/\partial Y)+ & \\ \quad +\partial/\partial Z\ (K_z\ \partial E_{xn}/\partial Z)=0 & ,\ 0<Z<1 \\ E_{xn}=0 & ,\ Z=0 \\ \partial E_{xn}/\partial Z=0 & ,\ Z=1\end{array}\right\} \quad \text{(C.11a)}$$

$$\left.\begin{array}{ll}\partial/\partial X\ (K_h\ \partial E_{yn-1}/\partial X)+ & \\ \quad +\partial/\partial Y\ (K_h\ \partial E_{yn-1}/\partial Y)+ & \\ \quad +\partial/\partial Z\ (K_z\ \partial E_{yn}/\partial Z)=0 & ,\ 0<Z<1 \\ E_{yn}=0 & ,\ Z=0 \\ \partial E_y/\partial Z=0 & ,\ Z=1\end{array}\right\} \quad \text{(C.11b)}$$

$$\left.\begin{array}{ll}\partial/\partial X\ (K_z^{-1}\ \partial Q_{zn-1}/\partial X)+ & \\ \quad +\partial/\partial Y\ (K_z^{-1}\ \partial Q_{zn-1}/\partial Y)+ & \\ \quad +\partial/\partial Z\ (K_h^{-1}\ \partial Q_{zn}/\partial Z)=0 & ,\ 0<Z<1 \\ \partial Q_{zn}/\partial Z=0 & ,\ Z=0 \\ Q_{zn}=0 & ,\ Z=1\end{array}\right\} \quad \text{(C.11c)}$$

Since the above hierarchy consists of ordinary differential equations, the solutions can easily be found by integration.

C.5. Solutions

The solutions of the three sets of zeroth-order equations (C.10a), (C.10b) and (C.10c) are given by:

Zeroth-order solutions.

$$E_{x0}(X,Y,Z) = -\partial\Phi(X,Y,Z)/\partial X, \quad \text{(C.12a)}$$

$$E_{y0}(X,Y,Z) = -\partial\Phi(X,Y,Z)/\partial Y, \tag{C.12b}$$

$$Q_{z0}(X,Y,Z) = -T_0\left[\partial^2\Phi(X,Y,Z)/\partial X^2 + + \partial^2\Phi(X,Y,Z)/\partial Y^2;\ Z\right], \tag{C.12c}$$

where:

$$\Phi(X,Y,Z) = F(X,Y) + \int_0^Z P(X,Y,Z')\,\mathrm{d}Z', \tag{C.13}$$

and where $T_0[\psi(Z);\ Z]$ is an integral operator operating on the function $\psi(Z)$ with integration from Z to 1:

$$T_0[\psi(Z);\ Z] = \int_Z^1 \psi(Z')\,K_h(Z')\,\mathrm{d}Z'. \tag{C.14}$$

The above-presented zeroth-order solutions are the well-known Dupuit approximations, which are sufficiently accurate for shallow groundwater basins, i.e., for basins in which $\epsilon << 1$. Equation (C.12c) is presented in such a way that it yields the correct solution also in the sharp fresh-saline interface approximation, in which $\partial\Phi/\partial X$ and $\partial\Phi/\partial Y$ are discontinuous at the interface; see Section C.7. Only in situations where $\partial\Phi/\partial X$ and $\partial\Phi/\partial Y$ are continuously differentiable functions of X and Y, the order of differentiation ($\partial/\partial X$, $\partial/\partial Y$) and integration ($T_0[\psi(Z);\ Z]$) may be interchanged.

Especially in combination with the sharp fresh-saline interface approximation (see Section C.7), the Dupuit approximations (C.12a), (C.12b) and (C.12c) form the basis of many geohydrological model codes (see Verruijt, 1987). However, the shallow groundwater basin condition $\epsilon << 1$ is not always satisfied and therefore the higher-order corrections to the Dupuit approximations will be considered.

Higher-order corrections ($n > 0$).

$$E_{xn}(X,Y,Z) = -(\partial^2/\partial X^2 + \partial^2/\partial Y^2)^n \cdot \cdot R_n[\partial\Phi(X,Y,Z)/\partial X;\ Z], \tag{C.15a}$$

$$E_{yn}(X,Y,Z) = -(\partial^2/\partial X^2 + \partial^2/\partial Y^2)^n \cdot \\ \cdot R_n[\partial\Phi(X,Y,Z)/\partial Y;\ Z], \tag{C.15b}$$

$$Q_{zn}(X,Y,Z) = -(\partial^2/\partial X^2 + \partial^2/\partial Y^2)^{(n+1)} \cdot \\ \cdot T_n[\Phi(X,Y,Z);\ Z], \tag{C.15c}$$

where the integral operators $R_n[\psi(Z);\ Z]$ and $T_n[\psi(Z);\ Z]$ are given by:

$$R_n[\psi(Z);\ Z] = \int_0^Z T_{n-1}[\psi(Z');\ Z']/K_z(Z')\ \mathrm{d}Z', \tag{C.16a}$$

$$T_n[\psi(Z);\ Z] = \int_Z^1 R_n[\psi(Z');\ Z']\ K_h(Z')\ \mathrm{d}Z'. \tag{C.16b}$$

Contrary to the Dupuit approximations (C.12a), (C.12b) and (C.12c), the higher-order corrections (C.15a), (C.15b) and (C.15c) do not hold in the sharp fresh-saline interface approximation. The basic reason is that the equations (C.5) do not hold on the fresh-saline interface. This is so because equations (C.5) are derived under the assumption that $\mathbf{curl}\,\mathbf{grad}\,p = \mathbf{0}$ (**curl** and **grad** in the meaning of differential operators; see Section 2.5). However, $\mathbf{grad}\,p$ is not continuously differentiable on the interface, and therefore $\mathbf{curl}\,\mathbf{grad}\,p \neq 0$. Instead, the weak-form meanings of $\mathrm{div}\,\boldsymbol{q} = 0$ and $\mathbf{curl}\,\mathbf{grad}\,p = \mathbf{0}$, namely, the interfacial conditions continuous normal component of $\boldsymbol{q}$ and continuous tangential components of $\mathbf{grad}\,p$ must be applied (see Section 2.5).

The total driving mechanism for flow depends on the lateral derivatives of $\Phi(X,Y,Z)$. According to expression (C.13), $\Phi(X,Y,Z)$ is composed of the sum of two driving mechanisms. The first term in the right-hand side of (C.13) represents the driving mechanism for forced convection, and the second term in the right-hand side of (C.13) represents the driving mechanism for free convection.

The situation in which $Q_z(X,Y,Z) = 0$, but in which $E_x(X,Y,Z)$ and $E_y(X,Y,Z)$ have finite values, is possible. From the above-presented solutions it is observed that this is the case if and only if:

$$(\partial^2/\partial X^2 + \partial^2/\partial Y^2)\ \Phi(X,Y,Z) = 0. \tag{C.17a}$$

Since it follows from equation (C.13) that $P(X,Y,Z) = \partial\Phi(X,Y,Z)/\partial Z$, a necessary condition for only horizontal flow is:

$$(\partial^2/\partial X^2 + \partial^2/\partial Y^2)\ P(X,Y,Z) = 0. \tag{C.17b}$$

Condition (C.17b), a condition for the water-salt mixture density field, is very restrictive. However, in the sharp fresh-saline interface approximation, where a sharp interface $Z = H(X,Y)$ separates two domains with constant densities in each domain, expression (C.17b) is satisfied in each domain. When, furthermore, the interface satisfies $(\partial^2/\partial X^2 + \partial^2/\partial Y^2)\ H(X,Y) = 0$, condition (C.17a) is also satisfied everywhere (see Section C.7). For example, in two-dimensional flow situations, where $H = H(X)$, it follows that condition (C.17a) is satisfied if $H(X)$ is linear in X.

The conclusion of the above-presented discussion is that, in horizontally and perfectly layered groundwater basins, vertical flow depends only on the horizontal second derivatives of the water table and water-salt mixture density. In the fresh-saline sharp interface approximation, vertical flow depends only on the horizontal second derivatives of the interface. The dependence on second derivatives, and independence of first derivatives, explains why vertical transport is much more difficult to determine than horizontal transport. Yet, for an investigation whether released waste travels either to the biosphere or to deeper geologic layers, knowledge of the vertical flow component is essential.

C.6. Characteristic conductivities and equivalent homogeneous porous medium

Let us consider the zeroth-order solution and first-order correction for the vertical component of the volumetric flow rate $Q_z(X,Y,0)$ on the upper plane $Z = 0$:

$$\begin{aligned}-Q_z(X,Y,0) &= T_0[1;0]\ (\partial^2/\partial X^2 + \partial^2/\partial Y^2)\ F(X,Y) + \\ &+ \epsilon T_1[1;0]\ (\partial^2/\partial X^2 + \partial^2/\partial Y^2)^2\ F(X,Y) + \\ &+ (\partial^2/\partial X^2 + \partial^2/\partial Y^2)\ T_0 \left[\int_0^Z P(X,Y,Z')\, dZ'; 0 \right] +\end{aligned}$$

$$+\epsilon\,(\partial^2/\partial X^2 + \partial^2/\partial Y^2)^2\; T_1 \left[\int_0^Z P(X,Y,Z')\,\mathrm{d}Z';0\right] + \qquad \text{(C.18)}$$
$$+\,O(\epsilon^2).$$

For a homogeneous porous medium with horizontal conductivity k_h^o and vertical conductivity k_z^o, the natural choice for the characteristic conductivities is, of course, $k_{hc} = k_h^o$ and $k_{zc} = k_z^o$, which means that $K_h^o = K_z^o = 1$. From (C.14) and (C.16b) it follows then that $T_0[1;0] = 1$ and $T_1[1;0] = 1/3$.

For a perfectly layered porous medium with $T_0[1;0] = 1$ and $T_1[1;0] = 1/3$, the first two terms on the right-hand side of expression (C.18), i.e., the terms describing forced convection, give the same contribution to the dimensionless $Q_z(X,Y,0)$ as a homogeneous porous medium. When choosing $T_0[1;0] = 1$ and $T_1[1;0] = 1/3$, the following expressions are found for the dimensionful characteristic conductivities:

$$k_{hc} = d^{-1}\int_0^d k_h(z')\,\mathrm{d}z', \qquad \text{(C.19a)}$$

$$k_{zc} = d^3\, k_{hc}^2 / \left[3t_1(0)\right] \qquad \text{(C.19b)}$$

where:

$$t_0(z) = \int_z^d k_h(z')\,\mathrm{d}z', \qquad \text{(C.19c)}$$

$$r_1(z) = \int_0^z t_0(z')/k_z(z')\,\mathrm{d}z', \qquad \text{(C.19d)}$$

$$t_1(z) = \int_z^d r_1(z')\,k_h(z')\,\mathrm{d}z'. \qquad \text{(C.19e)}$$

Since, according to expression (C.19a), $k_{hc} = k_h^o$ for a homogeneous porous medium with horizontal conductivity k_h^o, and since by definition $Q_z(X,Y,Z) = q_z(x,y,z)\, l_c^2/(\rho_0 g\, d^2 k_{hc})$, the zeroth- and first-order terms describing forced convection for a perfectly layered porous medium give the same contribution to $q_z(x,y,0)$ as a homogeneous porous medium. By choosing the characteristic conductivities according to the above expressions (C.19), the dimensionless terms in the series expansions for the forced convection have an order of magnitude of one, except for the dimensionless number ϵ, which may have any order of magnitude.

In many practical studies one wants to replace the perfectly layered porous medium by an 'equivalent homogeneous porous medium.' Here an equivalent homogeneous porous medium is defined as a homogeneous porous medium in which the recharge/discharge pattern $q_z(x,y,0)$ is similar to that of the perfectly layered porous medium under consideration. From the preceding discussion it can be seen that such an equivalent homogeneous porous medium exists for the forced convective part of the flow, provided that terms with orders of magnitude ϵ^2, ϵ^3, etc. are negligibly small. The values of the conductivities of the equivalent homogeneous porous medium are then equal to k_{hc} and k_{zc} as defined in expressions (C.19).

An important practical question is now: does an equivalent homogeneous porous medium also exist for the free convective part of the solution? Let us first assume that $K_h(Z)$ and $K_z(Z)$ are nonconstant functions of Z, i.e., $K_h(Z) \neq 1$ and $K_z(Z) \neq 1$. From the third and fourth terms in the right-hand side of equation (C.18) we observe that characteristic conductivities can be defined if and only if:

$$(\partial^2/\partial X^2 + \partial^2/\partial Y^2) \int_0^Z P(X,Y,Z')\, \mathrm{d}Z' = \alpha(X,Y), \tag{C.20}$$

where α is any function independent of Z. Condition (C.20) yields condition (C.17b) as a necessary condition for the water-salt mixture density field $P(X,Y,Z)$. As has already been discussed in Section C.5, this is a very restrictive condition which is hardly ever met in real situations. When a sharp interface $Z = H(X,Y)$ separates two domains with a constant density in each domain, expression (C.17b) is satisfied in each domain separately, and also condition (C.20) is satisfied in each domain. However, the characteristic conductivities thus derived are dependent on the position of the fresh-saline interface. Since, in general, this position depends on the horizontal coordinates, these characteristic conductivities cannot be used to define an equivalent homogeneous porous medium. In many practical cases, however, it is possible to project the spatial variations in the interface $H(X,Y)$ on a horizontal plane $Z = H_0$; this results in an 'equivalent homogeneous porous medium' (see Section C.7).

The conclusion from the above-presented discussion is that, as far as free convection is considered, the notion of an 'equivalent homogeneous porous medium' does not generally exist, but it exists when a sharp fresh-saline interface with relatively small spatial variations around a mean depth is a reasonable approximation. The situation on the scale of grid blocks, which is a smaller scale than the basin scale discussed before, is somewhat more favorable for the existence of an 'equivalent homogeneous porous block.' On the scale of a grid block we may assume that in each grid block condition (C.17b) is satisfied (i.e., $P(X,Y,Z)$ is linear in X and Y within a grid block); the deviations from (C.17b) are then supposed to occur on the inter-block boundaries.

C.7. Order of magnitude estimations

Let us consider a sharp fresh-saline interface $Z = H(X,Y)$ and a density distribution given by:

$$P(X,Y,Z) = \begin{cases} 1 \quad , & \text{for } 0 \le Z < H(X,Y) \\ P_1 \quad , & \text{for } H(X,Y) < Z \le 1 \end{cases} \tag{C.21a}$$

where it is assumed that $P_1 > 1$.

From equation (C.13) it follows then that:

$$\Phi(X,Y,Z) = F(X,Y) + \\ + \begin{cases} Z, & \text{for } 0 \le Z < H(X,Y) \\ -(P_1 - 1)\ H(X,Y) + P_1 Z, & \text{for } H(X,Y) < Z \le 1. \end{cases} \tag{C.21b}$$

The zeroth-order solutions (C.12a), (C.12b) and (C.12c) are:

$$E_{x0}(X,Y) = -\partial F(X,Y)/\partial X + \\ + \begin{cases} 0, & \text{for } 0 \le Z < H(X,Y) \\ (P_1 - 1)\ \partial H(X,Y)/\partial X, & \text{for } H(X,Y) < Z \le 1. \end{cases} \tag{C.22a}$$

$$E_{y0}(X,Y) = -\partial F(X,Y)/\partial Y + \\ + \begin{cases} 0, & \text{for } 0 \le Z < H(X,Y) \\ (P_1 - 1)\ \partial H(X,Y)/\partial Y, & \text{for } H(X,Y) < Z \le 1. \end{cases} \tag{C.22b}$$

$$Q_{z0}(X,Y,Z) = -T_0[1;Z]\left[(\partial^2/\partial X^2 + \partial^2/\partial Y^2)\,F(X,Y) + \right.$$
$$\left. -(P_1-1)\,(\partial^2/\partial X^2 + \partial^2/\partial Y^2)\,H(X,Y)\right] \qquad \text{(C.22c)}$$
$$\text{for } H(X,Y) \le Z \le 1.$$

$$Q_{z0}(X,Y,Z) = Q_{z0}(X,Y,H(X,Y)) +$$
$$-(T[1;Z] - T[1;H(X,Y)])\,\cdot$$
$$\cdot\,(\partial^2/\partial X^2 + \partial^2/\partial Y^2)\,F(X,Y), \qquad \text{(C.22d)}$$
$$\text{for } 0 \le Z < H(X,Y)$$

The above-presented zeroth-order solutions (C.22) are the well-known Dupuit approximations, which are sufficiently accurate for shallow groundwater basins in which $\epsilon << 1$. The Dupuit approximations form the basis of many geohydrological model codes (see, for instance, Verruijt, 1987). For $F(X,Y) = (P_1-1)\,H(X,Y)$ ($h_f(x,y) = [(\rho_1-\rho_0)/\rho_0]\;h(x,y)$ in dimensional form) the Dupuit approximations simplify to the well-known Badon-Ghyben-Herzberg approximations in which the flow in the saline domain $H(X,Y) < Z \le 1$ is stagnant.

Expressions (C.22) are also useful to estimate under what conditions free convection is negligible with respect to forced convection. In dimensional form the criteria are:

$$[(\rho_1-\rho_0)/\rho_0]\;|\partial h(x,y)/\partial x| << |\partial h_f(x,y)/\partial x|, \qquad \text{(C.23a)}$$

$$[(\rho_1-\rho_0)/\rho_0]\;|\partial h(x,y)/\partial y| << |\partial h_f(x,y)/\partial y|, \qquad \text{(C.23b)}$$

$$[(\rho_1-\rho_0)/\rho_0]\;|\partial^2/\partial x^2 + \partial^2/\partial y^2)\,h(x,y)| <<$$
$$<< |(\partial^2/\partial x^2 + \partial^2/\partial y^2)\,h_f(x,y)|. \qquad \text{(C.23c)}$$

In situations where the shallow groundwater basin condition $\epsilon << 1$ is not met, more or even all terms of the series expansions must be taken into account. The general solution for *forced convection* is given by:

$$E_x(X,Y,Z) = -\partial F_w(X,Y)/\partial X \sum_{n=0}^{\infty} (-\epsilon W^2)^n\;R_n[1;Z], \qquad \text{(C.24a)}$$

$$E_y(X,Y,Z) = -\partial F_w(X,Y)/\partial Y \sum_{n=0}^{\infty}(-\epsilon W^2)^n \; R_n[1;Z], \tag{C.24b}$$

$$\begin{aligned} Q_z(X,Y,Z) = -(\partial^2/\partial X^2 + \partial^2/\partial Y^2)\; F_w(X,Y)\cdot \\ \cdot \sum_{n=0}^{\infty}(-\epsilon W^2)^n \; T_n[1;Z], \end{aligned} \tag{C.24c}$$

where:

$$F_w(X,Y) = A(W_x, W_y)\; \exp\left[i(W_x X + W_y Y)\right] \tag{C.24d}$$

is a Fourier component of the dimensionless water table and

$$W = \sqrt{\left(W_x^2 + W_y^2\right)}.$$

For a *homogeneous* basin the resulting series expansions (C.24) turn out to be equivalent to the series expansions of well-known hyperbolic functions:

$$\begin{aligned} &\sum_{n=0}^{\infty}(-\epsilon W^2)^n \; R_n[1;Z] = \\ &\quad = \cosh(ZW\sqrt{\epsilon}) - \tanh(W\sqrt{\epsilon})\; \sinh(ZW\sqrt{\epsilon}) \\ &\qquad (= \exp(-ZW\sqrt{\epsilon}), \text{ in the limit } W\sqrt{\epsilon} >> 1), \end{aligned} \tag{C.25a}$$

$$\begin{aligned} &\sum_{n=0}^{\infty}(-\epsilon W^2)^n \; T_n[1;Z] = \\ &\quad = -\; \left[\sinh(ZW\sqrt{\epsilon}) - \tanh(W\sqrt{\epsilon})\; \cosh(ZW\sqrt{\epsilon})\right]/(W\sqrt{\epsilon}) \\ &\qquad (= \exp(-ZW\sqrt{\epsilon})/(W\sqrt{\epsilon}), \text{ in the limit } W\sqrt{\epsilon} >> 1). \end{aligned} \tag{C.25b}$$

The character of the solutions (C.24), (C.25), which show nested flow systems when the solutions for many Fourier modes are superimposed, has had a great impact on the conceptual model underlying applications of flow systems analysis. The limit $W\sqrt{\epsilon} >> 1$ is important when deep groundwater basins are considered.

In a similar way, solutions can be given for *free convective* flow in a homogeneous groundwater basin in which the sharp interface approximation holds. For this solution also the assumption is made that the mass of water above a horizontal plane $Z = H_0$ has a hydrostatic pressure distribution. This is a reasonable approximation for a fresh-saline interface $Z = H(X,Y)$ with spatial variations around the mean depth $Z = H_0$, where $|H(X,Y) - H_0|$ is sufficiently small with respect to the thickness of the salt water layer $1 - H_0$. Of course, such a situation can only exist in stable flow situations, i.e., for saline water below fresh water ($P_1 > 1$). Under the above conditions the solutions for $H_0 < Z \leq 1$ are:

$$E_x(X,Y,Z) = (P_1 - 1)\, \partial H_w(X,Y)/\partial X \cdot \\ \cdot \left[\cosh\{(Z - H_0)\, W\sqrt{\epsilon}\} - \tanh\{(1 - H_0)\, W\sqrt{\epsilon}\} \cdot \\ \cdot \sinh\{(Z - H_0)\, W\sqrt{\epsilon}\}\right], \tag{C.26a}$$

$$E_y(X,Y,Z) = (P_1 - 1)\, \partial H_w(X,Y)/\partial Y \cdot \\ \cdot \left[\cosh\{(Z - H_0)\, W\sqrt{\epsilon}\} - \tanh\{(1 - H_0)\, W\sqrt{\epsilon}\} \cdot \\ \cdot \sinh\{(Z - H_0)\, W\sqrt{\epsilon}\}\right], \tag{C.26b}$$

$$Q_z(X,Y,Z) = -(P_1 - 1)\, [1/(W\sqrt{\epsilon})] \cdot \\ \cdot (\partial^2/\partial X^2 + \partial^2/\partial Y^2)\, H_w(X,Y) \cdot \\ \cdot \left[\sinh\{(Z - H_0)\, W\sqrt{\epsilon}\} - \tanh\{(1 - H_0)\, W\sqrt{\epsilon}\} \cdot \\ \cdot \cosh\{(Z - H_0)\, W\sqrt{\epsilon}\}\right], \tag{C.26c}$$

where:

$$H_w(X,Y) = B(W_x, W_y)\, \exp\left[i(W_x X + W_y Y)\right] \tag{C.26d}$$

is a Fourier component of the dimensionless fresh-saline interface level and $W = \sqrt{(W_x^2 + W_y^2)}$.

The driving force was derived by assuming that the fluid mass in $0 \leq Z < H_0$ has a hydrostatic pressure distribution, which is in agreement with the Dupuit approximations (C.22). However, when the flow pattern for $H_0 < Z \leq 1$ given by expressions (C.26) are considered as good approximations, it is also possible to give the solutions for $0 \leq Z < H_0$ by requiring continuity of the vertical flux component $Q_z(X,Y,H_0)$ on the 'projected interface' $Z = H_0$, yielding:

$$\begin{aligned} E_x(X,Y,Z) = &(P_1 - 1)\ \tanh\{(1 - H_0)\ W\sqrt{\epsilon}\} \cdot \\ &\cdot \partial H_w(X,Y)/\partial X \cdot \\ &\cdot\ [\sinh\{(H_0 - Z)\ W\sqrt{\epsilon}\} - \tanh(H_0 W\sqrt{\epsilon}) \cdot \\ &\cdot\ \cosh\{(H_0 - Z)\ W\sqrt{\epsilon}\}], \end{aligned} \tag{C.27a}$$

$$\begin{aligned} E_y(X,Y,Z) = &(P_1 - 1)\ \tanh\{(1 - H_0)\ W\sqrt{\epsilon}\} \cdot \\ &\cdot \partial H_w(X,Y)/\partial Y \cdot \\ &\cdot\ [\sinh\{(H_0 - Z)\ W\sqrt{\epsilon}\} - \tanh(H_0 W\sqrt{\epsilon}) \cdot \\ &\cdot\ \cosh\{(H_0 - Z)\ W\sqrt{\epsilon}\}], \end{aligned} \tag{C.27b}$$

$$\begin{aligned} Q_z(X,Y,Z) = &(P_1 - 1)\ [\tanh\{(1 - H_0)\ W\sqrt{\epsilon}\}/(W\sqrt{\epsilon})] \cdot \\ &\cdot (\partial^2/\partial X^2 + \partial^2/\partial Y^2)\ H_w(X,Y) \cdot \\ &\cdot\ [\cosh\{(H_0 - Z)\ W\sqrt{\epsilon}\} - \tanh(H_0 W\sqrt{\epsilon}) \cdot \\ &\cdot\ \sinh\{(H_0 - Z)\ W\sqrt{\epsilon}\}]. \end{aligned} \tag{C.27c}$$

Solutions (C.27) give $E_x(X,Y,0) = E_y(X,Y,0) = 0$, i.e., the solutions describe only free convective flow.

From the general solutions discussed in Section C.5 it may be concluded that the total solutions may be composed of a linear superposition of the solutions for forced convection (C.24), (C.25) and for free convection (C.26), (C.27). Comparison of the magnitudes of solutions (C.24), (C.25) and (C.26), (C.27) will show at what depth forced convection is dominating free convection.

In stable equilibrium vertical free convection and vertical forced convection balance each other on the fresh-saline interface, i.e., $Q_z(X,Y,H_0) = 0$. Both for $\epsilon \to 0$ and for $H_0 W\sqrt{\epsilon} >> 1$ it follows then from (C.24), (C.25) and (C.26), (C.27) that the following *generalized* Badon-Ghyben-Herzberg relation exists: $b(w_x, w_y) = a(w_x, w_y)\ \exp\left[-h_0 w\sqrt{(k_{hc}/k_{zc})}\right]\ \rho_0/ /(\rho_1 - \rho_0)$. (Note that the conventional Badon-Ghyben-Herzberg relation is obtained for $h_0 w\sqrt{(k_{hc}/k_{zc})} \to 0$, i.e., in the limit of shallow basins.) Under that condition there is no flow at all in the saline water. For example, if $\rho_0 = 1000\ kg{\cdot}m^{-3}$, $\rho_1 = 1025\ kg{\cdot}m^{-3}$, $\rho_0/(\rho_1 - \rho_0) = 40$, $h_0 = 300\ m$ and $\sqrt{(k_{hc}/k_{zc})} = 30$, it follows that, for $l_c = w^{-1} >> 10{,}000\ m$, $b \approx 40\ a$, whereas for $l_c = w^{-1} << 10{,}000\ m$, $b << 40\ a$. This means that relatively local variations in the water table are not even felt at the fresh-saline interface.

Of course, storage of highly toxic and radioactive waste in stagnant groundwater is a safe option. However, due to the very long response times of the water-salt mixture density distribution to palaeo-geological or palaeo-climatological disturbances (see Section C.2), there will hardly ever be equilibrium. This means that both forced and free convection are important transport mechanisms when considering the transport of dangerous matter to the biosphere.

Since the characters of solutions (C.24), (C.25) and (C.26), (C.27) are similar, both showing nested flow systems, the basic thoughts underlying applications of flow systems analysis may also guide us at greater depths.

C.8. Bibliography

Engelen, G.B., Hydrological systems analysis, a regional case study, TNO Institute of Applied Geoscience, Delft, Report OS 94-20 (1984).

Garven, C. and Freeze, N.A., Theoretical analysis of the role of groundwater flow in the genesis of strata-bound ore deposits, *Am. J. Sci.*, Vol. 284, pp. 1085-1125, pp. 1129-1174 (1984).

Horton, C.W. and Rogers, F.T., Jr., Convection currents in a porous medium, *J. Appl. Phys.*, Vol. 16, pp. 367-370 (1945).

Silberberg, I.H. and McKetta, J.J., Learning how to use dimensional analysis, *Petroleum Refiner*, Vol. 42, No. 4, pp. 179-183; No. 5, pp. 147-150; No. 6, pp. 101-103; No. 7, pp. 129-133 (1953).

Stuurman, R.J., Biesheuvel, A., and Van der Meij, J.L., The application of regional hydrological systems analysis in water management, In: *Regional Characterization of Water Quality*, S. Ragone (ed.), Wallingford, *IAHS Publication*, No. 183, pp. 45-57 (1989).

Tóth, J., A theoretical analysis of groundwater flow in small drainage basins, *J. Geophys. Res.*, Vol. 68, pp. 4795-4812 (1963).

Tóth, J., Cross-formational gravity-flow of groundwater: a mechanism of the transport and accumulation of petroleum, In: *Problems of Petroleum Migration*, M.H. Roberts and R.J. Cordel (eds.), *AAPG-Studies in Geology*, Vol. 10, pp. 121-167 (1980).

Turner, J.S., "*Buoyancy Effects in Fluids*," Cambridge University Press, London (1973).

Van Dyke, M., *Perturbation methods in fluid mechanics*, The Parabolic Press, Stanford (1975).

Verruijt, A., A finite element model for interface problems in groundwater flow, In: *Microcomputers in Engineering Applications*, B.A. Schrefler and R.W. Lewis (eds.), John Wiley & Sons, Inc. (1987).

Voss. C.I. and Souza, W.R., Variable density flow and solute transport simulation of regional aquifers containing a narrow freshwater-saltwater transition zone, *Water Resour. Res.*, Vol. 23, pp. 1851-1866 (1987).

Appendix D

From the small-scale to the large-scale Darcy's law

D.1. Introduction

Computer codes to calculate the fluid potential and flux in a saturated porous medium are applied on a routine basis by geohydrologists and petroleum reservoir engineers. For this purpose, model codes based on the finite difference or the finite element methods are frequently used. In these methods the subsurface is divided into grid blocks, or finite elements, for which a homogeneous block-scale conductivity $<\underline{\boldsymbol{k}}>$ must be specified (see Aziz and Settari, 1979). However, in the real subsurface the local-scale conductivity $\underline{\boldsymbol{k}}(\boldsymbol{x})$ is heterogeneous within a block. Therefore, it is necessary to find a block-averaged or equivalent homogeneous conductivity for each particular grid block.

A general derivation of the block-averaged conductivity is possible (see Quintard and Whitaker, 1987), but does not lead to simple expressions. Development of practical expressions for block-averaged conductivity components implies restriction of the type of subsurface to which they can be applied. The type of subsurface would be defined by its geological and stratigraphical characteristics. The depthwise sequence of geological units in a sedimentary basin is a spatial image of the chronological process of the origin and further development of the rocks. Sedimentary basins frequently present a predominantly, but not perfectly, layered structure. This paper is concerned with layered subsurfaces.

It follows from the steady Stokes equations underlying the local-scale Darcy's law that the local-scale conductivity is a symmetric tensor (see Lehner, 1979). Of course, this symmetry does not imply that the block-scale conductivity tensor is symmetric. Due to the symmetry principal axes exist (see Bear and Verruijt, 1987). In a layered subsurface it is assumed that the local-scale conductivity tensor has its principal axes in the direction of sedimentation (the lateral or 'horizontal' directions) and normal to the direction of sedimentation (the 'vertical' direction). Furthermore, in a layered subsurface local-scale conductivity may vary abruptly in the vertical direction, but in the lateral directions variations will be smooth.

Based upon the above assumptions, practical expressions are derived here for the nine components of the block-scale conductivity tensor of a thin grid block. These expressions are derived from the local-scale continuity equation and Darcy's law in a layered subsurface. To show the origin of off-diagonal components, this layering is imperfect. The resulting block-scale Darcy's law is considered as a linear relation between a constant block-scale flux and a constant block-scale potential gradient. To be able to assume a constant block-scale flux, the flow problem is separated in a bottom flux formulation and a top flux formulation; both formulations can be solved in essentially the same way. The bottom flux formulation is then worked out in detail. This formulation is separated in vertical potential difference equations and lateral (or horizontal) potential difference equations. These two types of equations are solved with different approaches specially designed for the types of equations under consideration. In both types of equations a perturbation technique is applied to obtain first-order corrections to the well-known results for a perfectly layered subsurface. Depth-averaged expressions are obtained first. Then the lateral potential difference and vertical potential difference solutions, and the bottom flux and top flux solutions are combined. Finally, block-scale expressions are obtained by making use of the assumption of a constant block-scale flux. It turns out that, in general, the resulting block-scale conductivity tensor is nonsymmetric.

D.2. Basic local-scale equations

The continuity equation and Darcy's law are simplified by neglecting specific storage and density-driven flow; under these assumptions the latter equations are given by:

$$\operatorname{div} \boldsymbol{q}(\boldsymbol{x}) = 0 \tag{D.1}$$

$$\boldsymbol{q}(\boldsymbol{x}) = -\underline{\boldsymbol{k}}(\boldsymbol{x}) \cdot \mathbf{grad}\, \phi(\boldsymbol{x}). \tag{D.2}$$

We now introduce a Cartesian coordinate system with lateral coordinates x and y, and vertical coordinate z.

The top boundary $z = 0$ of the block under consideration is chosen as a plane where the potential is specified:

$$\phi(x, y, 0) = f(x, y). \tag{D.3}$$

The lower boundary $z = d$ of the block with thickness d is chosen as a plane where the vertical flux component is specified:

$$q_z(x, y, d) = w(x, y). \tag{D.4}$$

Since we are considering thin blocks, i.e., blocks in which the length scale of lateral variations in conductivity is large with respect to the thickness of the block, specification of side boundary conditions may be neglected.

By choosing an asymmetrical set of boundary conditions (D.3) and (D.4), with a potential on top and a flux at the bottom, the solution can be found in a relatively simple way. A similar flow formulation with a flux specified on top and a potential specified at the bottom could equally well have been defined. This latter formulation will be called the top flux formulation, whereas the former formulation, as denoted by boundary conditions (D.3) and (D.4), will be called the bottom flux formulation. Combination of the top flux with the bottom flux solution leads to the final solution.

The above-presented equations (D.1-4), together with the local-scale conductivity distribution in a layered subsurface, lead to the following equations for the potential $\phi(x, y, z)$:

$$\left.\begin{array}{ll} \partial/\partial x_i\ (k_h\ \partial\phi/\partial x_i) + \partial/\partial z\ (k_z\ \partial\phi/\partial z) = 0 & ,\ 0 < z < d \\ \phi = f & ,\ z = 0 \\ \partial\phi/\partial z = -k_z^{-1}\ w & ,\ z = d \end{array}\right\} \tag{D.5}$$

The set of equations for the flux $\boldsymbol{q}(x, y, z)$ will be presented in Section D.4. In equation (D.5) x_i, $i = 1, 2$, is short-hand notation for the pair of lateral Cartesian coordinates $x_i = (x_1, x_2) = (x, y)$. Use has been made of the summation convention meaning that, when a subscript occurs twice in the same term, it is to be summed from 1 to 2 (subscripts in the argument of a function, for instance $w(x_i)$, are not subjected to the summation convention).

Due to the linearity of equation (D.5), the potential may be written as the sum of two potentials:

$$\phi(\boldsymbol{x}) = \phi^{(1)}(\boldsymbol{x}) + \phi^{(2)}(\boldsymbol{x}). \tag{D.6}$$

Potential $\phi^{(1)}(\boldsymbol{x})$ is defined as the potential satisfying the equations where flow is driven only by the vertical flux $w(x_i) = w(x_1, x_2)$:

$$\left.\begin{array}{ll} \partial/\partial x_i\ (k_h\ \partial\phi^{(1)}/\partial x_i) + & \\ \quad + \partial/\partial z\ (k_z\ \partial\phi^{(1)}/\partial z) = 0 & ,\ 0 < z < d \\ \phi^{(1)} = 0 & ,\ z = 0 \\ \partial\phi^{(1)}/\partial z = -k_z^{-1}\ w & ,\ z = d \end{array}\right\} \tag{D.7}$$

Potential $\phi^{(2)}(\boldsymbol{x})$ is defined as the potential satisfying the equations where flow is only driven by a lateral gradient in $f(x_j)$:

$$\left.\begin{array}{ll} \partial/\partial x_i\ (k_h\ \partial\phi^{(2)}/\partial x_i) + & \\ \quad + \partial/\partial z\ (k_z\ \partial\phi^{(2)}/\partial z) = 0 & ,\ 0 < z < d \\ \phi^{(2)} = f & ,\ z = 0 \\ \partial\phi^{(2)}/\partial z = 0 & ,\ z = d \end{array}\right\} \tag{D.8}$$

Equations (D.7) describe flow caused by the vertical flux component $w(x_j)$. These equations will be applied to derive the vertical component of the block-scale conductivity $<k_{zz}>$. However, the influence of a vertical potential difference on lateral flow, expressed by $<k_{xz}>$ and $<k_{yz}>$, can also be determined from equations (D.7).

Equations (D.8) describe flow caused by the lateral components of the potential gradient $\partial f(x_j)/\partial x_i$. These equations will be applied to derive the lateral components of the block-scale conductivity $<k_{xx}>$, $<k_{yy}>$ and the components $<k_{xy}>$ and $<k_{yx}>$. However, the influence of a lateral potential difference on vertical flow expressed by $<k_{zx}>$ and $<k_{zy}>$ can also be determined from equations (D.8).

D.3. Solution for vertical potential difference

In this section an approximate solution to equations (D.7) will be presented. To simplify the notation, the superscript $^{(1)}$ is omitted in $\phi^{(1)}(\boldsymbol{x})$.

Now we define the dimensionless lateral coordinates $X_i = x_i/l_L$, where l_L is the characteristic length scale of lateral variations in conductivity, and the dimensionless vertical coordinate $Z = z/d$, where d is the thickness of the block.

Furthermore, we define the dimensionless potential $\Phi(X_i, Z) = \phi(x_i, z)/ /\phi_c$, where ϕ_c is the characteristic potential difference.

We also introduce dimensionless conductivities $K_h(X_i, Z) = k_h(x_i, z)/ /k_{hc}$ and $K_z(X_i, Z) = k_z(x_i, z)/k_{zc}$, where k_{hc} and k_{zc} are characteristic conductivities. Finally, we define the dimensionless flux $W(X_i) = = dw(x_i)/(k_{zc}\,\phi_c)$.

Equations (D.7) written in dimensionless variables are then given by:

$$\left.\begin{array}{ll} \epsilon\, \partial/\partial X_i\ (K_h\ \partial\Phi/\partial X_i) + & \\ \quad + \partial/\partial Z\ (K_z\ \partial\Phi/\partial Z) = 0 & ,\ 0 < Z < 1 \\ \Phi = 0 & ,\ Z = 0 \\ \partial\Phi/\partial Z = -K_z^{-1}\ W & ,\ Z = 1 \end{array}\right\} \tag{D.9}$$

In equation (D.9) all terms have order of magnitude one, except for the dimensionless number ϵ:

$$\epsilon = (d/l_L)^2\ k_{hc}/k_{zc}. \tag{D.10}$$

Since only thin blocks are considered, the ϵ-number is assumed to be much smaller than one. For instance, if $d/l_L = 10^{-1}$ then $(d/l_L)^2 = 10^{-2}$ and, for a basin in which $k_{hc}/k_{zc} = 10$, we find that $\epsilon = 10^{-1}$.

To obtain ordinary differential equations, $\Phi(X_i, Z)$ will be expanded in a power series of the ϵ-number (see Van Dyke, 1975):

$$\Phi(X_i, Z) = \Phi_0(X_i, Z) + \epsilon\ \Phi_1(X_i, Z) + \epsilon^2\ \Phi_2(X_i, Z) + \ldots \tag{D.11}$$

where $\Phi_0(X_i, Z)$, $\Phi_1(X_i, Z)$, $\Phi_2(X_i, Z)$ etc. are the perturbation terms of order 0,1,2, respectively.

Substitution of the series (D.11) into equation (D.9), and equating terms with the same power of ϵ, yields the following infinite hierarchy of equations:

$$\left.\begin{array}{ll} \partial/\partial X_i\,(K_h\,\partial\Phi_{n-1}/\partial X_i)+ & \\ \quad +\,\partial/\partial Z\,(K_z\,\partial\Phi_n/\partial Z)=0 & ,\ 0<Z<1 \\ \Phi_n=0 & ,\ Z=0 \\ \partial\Phi_0/\partial Z=-K_z^{-1}\,W; & \\ \quad \text{for } n>0\ \partial\Phi_n/\partial Z=0 & ,\ Z=1 \end{array}\right\} \tag{D.12}$$

In the above hierarchy n is the order of the perturbation term under consideration; $n = 0, 1, 2, \ldots, \infty$, and $\Phi_{-1} = 0$.

The zeroth-order solution ($n = 0$), rewritten in dimensional form, is given by:

$$\phi_0(x_i, z) = -c_0(x_i, z)\, w(x_i) \tag{D.13a}$$

where $c_0(x_i, z)$ is given by:

$$c_0(x_i, z) = \int_0^z k_z(x_i, z')^{-1}\, dz'. \tag{D.13b}$$

The resistance, well-known in geohydrology, is given by $c_0(x_i, d)$.

Substitution of (the dimensionless form of) solution (D.13a) into the first-order equations of the hierarchy ($n = 1$) yields the first-order correction to the potential. In this first-order correction, terms with $\partial w(x_j)/\partial x_i$ and higher derivatives with respect to x_i occur. However, a block-scale Darcy's law is essentially a linear relationship between a flux and a potential gradient; there is no room for derivatives of the flux and the potential gradient. Therefore, in the vertical potential difference problem under consideration, either the terms with derivatives of $\partial f(x_j)/\partial x_i$, or the terms with derivatives of $w(x_i)$ must be negligibly small. Here the last possibility is chosen, because under that condition a simple procedure for averaging in the lateral directions is obtained, as will be shown in Section D.6.

Under the condition that $\partial w(x_j)/\partial x_i \approx 0$ we find:

$$\phi_1(x_i, z) = -c_1(x_i, z)\, w(x_i) \tag{D.14a}$$

where $c_1(x_i, z)$ is given by:

$$c_1(x_j, z) = \int_0^z k_z(x_j, z')^{-1} \, \partial\alpha_i(x_j, z')/\partial x_i \, dz' \qquad \text{(D.14b)}$$

with:

$$\alpha_i(x_j, z) = \int_z^d k_h(x_j, z') \, \partial c_0(x_j, z')/\partial x_i \, dz'. \qquad \text{(D.14c)}$$

Since it is assumed that the block is thin, i.e., that $\epsilon << 1$, higher-order corrections are negligibly small. In that case we will derive from equations (D.13) and (D.14) approximations to the depth averaged conductivity components $<k_{zz}>(x, y)$, $<k_{xz}>(x, y)$ and $<k_{yz}>(x, y)$. For that purpose the depth-averaged Darcy's law will be considered as a linear relationship between the bottom flux $w(x, y)$ and the potential difference over the thickness of the block. This potential difference is given by $\phi(x, y, d) = = \phi_0(x, y, d) + \phi_1(x, y, d)$. Consequently, it makes sense to define the vertical depth-averaged potential gradient $\Delta<\phi>(x, y)/\Delta z = \phi(x, y, d)/d$, which is equal to the potential difference over the thickness of the block divided by that thickness. The vertical bottom flux component will be denoted by $<q_z>(x, y) = w(x, y)$ and in this way we find from equations (D.13) and (D.14) the depth-averaged Darcy's law:

$$<q_z>(x, y) = -<k_{zz}>(x, y) \; \Delta<\phi>(x, y)/\Delta z \qquad \text{(D.15a)}$$

where the depth-averaged conductivity component $<k_{zz}>(x, y)$ is given by:

$$<k_{zz}>(x, y) = d \, [c_0(x, y, d) + c_1(x, y, d)]^{-1}. \qquad \text{(D.15b)}$$

From equations (D.13) and (D.14) it also follows that:

$$q_i = -k_h \, \partial\phi/\partial x_i = k_h \, (\partial c/\partial x_i \; w + c \; \partial w/\partial x_i) \qquad \text{(D.16)}$$

where $c = c_0 + c_1$. Under the condition that terms with derivatives of $w(x_i)$ are negligible, and with the definition for the depth-averaged lateral flux components:

$$<q_i>(x_j) = d^{-1} \int_0^d q_i(x_j, z)\, \mathrm{d}z \tag{D.17}$$

we find:

$$<q_x>(x, y) = -<k_{xz}>(x, y)\ \Delta<\phi>(x, y)/\Delta z \tag{D.18a}$$

$$<q_y>(x, y) = -<k_{yz}>(x, y)\ \Delta<\phi>(x, y)/\Delta z \tag{D.18b}$$

with:

$$\begin{aligned} <k_{xz}>(x, y) = \ & c(x, y, d)^{-1} \int_0^d k_h(x, y, z) \cdot \\ & \cdot\, \partial c(x, y, z)/\partial x \ \mathrm{d}z \end{aligned} \tag{D.18c}$$

$$\begin{aligned} <k_{yz}>(x, y) = \ & c(x, y, d)^{-1} \int_0^d k_h(x, y, z) \cdot \\ & \cdot\, \partial c(x, y, z)/\partial y \ \mathrm{d}z \end{aligned} \tag{D.18d}$$

Expressions (D.18) describe flow in the two lateral directions caused by a vertical potential difference over the block. The magnitude of this lateral flow is proportional to the magnitudes of the depth-averaged conductivities $<k_{xz}>(x, y)$ and $<k_{yz}>(x, y)$. In the zeroth-order approximation the top flux $q_{z0}(x, y, 0)$ is exactly equal to the bottom flux $w(x, y) =< q_z > (x, y)$, but in the first-order approximation a small difference between $q_z(x, y, 0)$ and $w(x, y)$ may exist. However, if $<k_{xz}>(x, y) = <k_{yz}>(x, y) = 0$ then $<q_x>(x, y) = <q_y>(x, y) = 0$ and, consequently, the top flux $q_z(x, y, 0)$ is equal to the bottom flux $w(x, y)$.

D.4. Solution for lateral potential differences

In this section, first-order correct solutions to equation (D.8) are sought. The superscript $^{(2)}$ is omitted in $\phi^{(2)}(\boldsymbol{x})$ to simplify the notation. Equations (D.2) and (D.8) are equivalent to the following coupled equations:

$$\left.\begin{array}{ll} \partial e_i/\partial z - \partial/\partial x_i\,(k_z^{-1}\,q_z) = 0 & ,\ 0 < z < d \\ e_i = -\partial f/\partial x_i & ,\ z = 0 \end{array}\right\} \tag{D.19a}$$

$$\left.\begin{array}{ll} \partial x_i/\partial\,(k_h\,e_i) + \partial q_z/\partial z = 0 & ,\ 0 < z < d \\ q_z = 0 & ,\ z = d \end{array}\right\} \tag{D.19b}$$

In equation (D.19) e_i, $i = 1, 2$, is short-hand notation for the pair of Cartesian components $(e_1, e_2) = k_h^{-1}\,(q_x, q_y)$.

After scaling, the following set of dimensionless equations is found:

$$\left.\begin{array}{ll} \partial E_i/\partial Z - \epsilon\,\partial/\partial X_i\,(K_z^{-1}\,Q_z) = 0 & ,\ 0 < Z < 1 \\ E_i = -\partial F/\partial X_i & ,\ Z = 0 \end{array}\right\} \tag{D.20a}$$

$$\left.\begin{array}{ll} \partial/\partial X_i\,(K_h\,E_i) + \partial Q_z/\partial Z = 0 & ,\ 0 < Z < 1 \\ Q_z = 0 & ,\ Z = 1 \end{array}\right\} \tag{D.20b}$$

The scaling is similar to that described in Section D.3, except that $Q_z(X_i, Z) = q_z(x_i, z)\,l_L^2/(\phi_c\,d\,k_{hc})$, and $E_i(X_j, Z) = l_L\,e_i(x_j, z)/\phi_c$.

To obtain ordinary differential equations, $E_i(X_j, Z)$ and $Q_z(X_j, Z)$ in equation (D.20) will be expanded in a power series of the ϵ-number:

$$\begin{array}{l} E_i(X_j, Z) = E_{i0}(X_j, Z) + \epsilon\,E_{i1}(X_j, Z) + \\ \quad + \epsilon^2\,E_{i2}(X_j, Z) + \ldots \end{array} \tag{D.21a}$$

$$\begin{array}{l} Q_z(X_j, Z) = Q_{z0}(X_j, Z) + \epsilon\,Q_{z1}(X_j, Z) + \\ \quad + \epsilon^2\,Q_{z2}(X_j, Z) + \ldots \end{array} \tag{D.21b}$$

Substitution of the series (D.21) into equations (D.20), and equating terms with the same power of ϵ, yields the following infinite hierarchy of equations:

$$\left.\begin{array}{ll} \partial E_{in}/\partial Z - \partial/\partial X_i\,(K_z^{-1}\,Q_{zn-1}) = 0 & ,\ 0 < Z < 1 \\ E_{i0} = -\partial F/\partial X_i;\ \text{for } n > 0\ E_{in} = 0 & ,\ Z = 0 \end{array}\right\} \tag{D.22a}$$

$$\left.\begin{array}{ll} \partial/\partial X_i\,(K_h\,E_{in}) + \partial Q_{zn}/\partial Z = 0 & ,\ 0 < Z < 1 \\ Q_{zn} = 0 & ,\ Z = 1 \end{array}\right\} \tag{D.22b}$$

In the above hierarchy n is the order of the perturbation term under consideration; $n = 0, 1, 2, \ldots, \infty$, and $E_{i-1} = Q_{z-1} = 0$.

Rewritten in dimensional form, the zeroth-order solutions to the hierarchy of equation (D.22) are given by:

$$q_{i0}(x_j, z) = -k_h(x_j, z)\ \partial f(x_j)/\partial x_i \tag{D.23a}$$

$$q_{z0}(x_j, z) = -\partial/\partial x_i\ [t_0(x_j, z)\ \partial f(x_j)/\partial x_i] \tag{D.23b}$$

where $t_0(x_i, z)$ is defined as:

$$t_0(x_i, z) = \int_z^d k_h(x_i, z')\ \mathrm{d}z'. \tag{D.23c}$$

Equations (D.23) represent the Dupuit approximation in which $t_0(x_i, 0)$ is the transmissivity. This approximation is very popular in geohydrology.

It will now be clear why, instead of solving the equations for the potential $\phi(\boldsymbol{x})$, the equations for $e_i(\boldsymbol{x})$ and $q_z(\boldsymbol{x})$ were applied. If the equations for $\phi(\boldsymbol{x})$ had been solved, it would have been found that $\phi_0(x_i, z) = f(x_i)$ resulting in $q_{z0}(x_i, z) = -k_z(x_i, z)\ \partial\phi_0(x_i, z)/\partial z = 0$, which is not a useful result.

In equation (D.23b) the terms $(\partial t_0/\partial x)\,(\partial f/\partial x)$, $(\partial t_0/\partial y)\,(\partial f/\partial y)$ and $t_0\ \partial^2 f/\partial x^2$, $t_0\ \partial^2 f/\partial y^2$ occur. However, a block-scale Darcy's law is a relationship between a flux and a potential gradient, with no room for derivatives of the flux and the potential. Therefore, in the lateral potential differences problem under consideration, the terms with second derivatives must be negligibly small. After substitution of the dimensionless equivalents of the zeroth-order solutions (D.23) into equations (D.22), and neglecting terms in which second derivatives in f occur, the following solutions, reformulated in dimensional form, are found:

$$q_{i1}(x_k, z) = -k_h(x_k, z)\ \mu_{ij}(x_k, z)\ \partial f(x_k)/\partial x_j \tag{D.24a}$$

$$q_{z1}(x_k, z) = -\partial/\partial x_i\ [t_{1ij}(x_k, z)]\ \partial f(x_k)/\partial x_j \tag{D.24b}$$

with:

$$t_{1ij}(x_k,z) = \int_z^d k_h(x_k,z')\,\mu_{ij}(x_k,z')\,\mathrm{d}z' \tag{D.24c}$$

$$\mu_{ij}(x_k,z) = \partial a_j(x_k,z)/\partial x_i \tag{D.24d}$$

$$a_j(x_k,z) = \int_0^z k_z(x_k,z')^{-1}\,\partial t_0(x_k,z')/\partial x_j\,\mathrm{d}z'. \tag{D.24e}$$

One might wonder whether solutions (D.24) really satisfy the continuity equation and Darcy's law. This can be checked by noting that equations (D.24) satisfy the continuity equation (D.19b) and $\partial e_2/\partial x_1 - \partial e_1/\partial x_2 = 0$ exactly. Equation (D.19a) is satisfied approximately, with first-order accuracy. This means that $\mathbf{curl}\,\boldsymbol{e}(\boldsymbol{x}) = \mathbf{0}$ also holds with first-order accuracy, where $(e_1,e_2,e_3) = (k_h^{-1}\,q_1,\ k_h^{-1}\,q_2,\ k_z^{-1}\,q_3)$. Since any irrotational vector function can be written as the gradient of a scalar function, it follows that $\boldsymbol{e}(\boldsymbol{x}) = -\,\mathbf{grad}\,\phi(\boldsymbol{x})$ and, consequently, Darcy's law is satisfied with first-order accuracy.

In analogy to the solutions for the vertical potential difference, we derive approximations to the depth-averaged conductivity components $<k_{xx}>\cdot$ $\cdot(x,y)$, $<k_{yy}>(x,y)$, $<k_{xy}>(x,y)$, $<k_{yx}>(x,y)$, $<k_{zx}>(x,y)$ and $<k_{zy}>\cdot$ $\cdot(x,y)$ from equations (D.23) and (D.24). As in the previous section, the depth-averaged vertical potential gradient is defined as the potential difference over the depth of the block, divided by that depth. In the zeroth-order approximation (the Dupuit approximation) this potential difference is equal to zero. In the first-order approximation a small correction term occurs, which is equal to $a_j(x_k,d)\,\partial f(x_k)/\partial x_j$. This means that the potential difference is negligibly small if $\left|a_j(x_k,d)/l_c\right| << 1$ (see also Section D.5 where this condition is relaxed). As a consequence, the depth-averaged vertical potential gradient may be taken equal to zero, and it makes sense to denote the depth-averaged potential as $<\phi>(x,y) = f(x,y)$. Furthermore, the vertical top flux component driven by the lateral components of the potential gradient $(\partial f/\partial x,\ \partial f/\partial y)$ is denoted by $<q_z>(x,y) = q_z(x,y,0)$. The depth-averaged lateral fluxes $<q_x>(x,y)$, $<q_y>(x,y)$ are defined by equation (D.17). In this way we find:

$$<k_{xx}>(x,y) = d^{-1}\,[t_0(x,y,0) + t_{1xx}(x,y,0)] \tag{D.25a}$$

$$<k_{yy}>(x,y) = d^{-1}\ [t_0(x,y,0) + t_{1yy}(x,y,0)] \tag{D.25b}$$

$$<k_{xy}>(x,y) = d^{-1}\ t_{1xy}(x,y,0) \tag{D.25c}$$

$$<k_{yx}>(x,y) = d^{-1}\ t_{1yx}(x,y,0) \tag{D.25d}$$

$$\begin{aligned} <k_{zx}>(x,y) &= \partial t_0(x,y,0)/\partial x + \\ &+ \partial t_{1xx}(x,y,0)/\partial x + \partial t_{1yx}(x,y,0)/\partial y \end{aligned} \tag{D.25e}$$

$$\begin{aligned} <k_{zy}>(x,y) &= \partial t_0(x,y,0)/\partial y + \\ &+ \partial t_{1yy}(x,y,0)/\partial y + \partial t_{1xy}(x,y,0)/\partial x. \end{aligned} \tag{D.25f}$$

With the aid of equations (D.24c), (D.24d) and (D.24e) it can be seen that, in general, $<k_{xy}>(x,y) \neq <k_{yx}>(x,y)$, which means that the depth-averaged conductivity tensor is nonsymmetric. This conclusion is based on a consideration of the lateral potential difference problem only, unbiased by the separation of the total problem in lateral and vertical potential difference equations.

D.5. Combination of bottom flux with top flux solution

Thanks to the asymmetrical boundary conditions (D.3) and (D.4) we were able to separate the flow formulation into 'lateral' (horizontal) and 'vertical' flow equations with fluxes $\boldsymbol{q}^{(1)}(x,y,z)$ and $\boldsymbol{q}^{(2)}(x,y,z)$, respectively. The two types of flow equations could then be solved by two different approaches specially designed for the type of equations under consideration.

The final first-order accurate result is obtained by adding the solutions for $\boldsymbol{q}^{(1)}(x,y,z)$ and $\boldsymbol{q}^{(2)}(x,y,z)$ and their depth-averaged values $<\boldsymbol{q}^{(1)}>\cdot$ $\cdot(x,y)$ and $<\boldsymbol{q}^{(2)}>(x,y)$ for a specified depth-averaged potential 'gradient' $\boldsymbol{gra\Delta}\ <\phi^{(1)}>(x,y) = \boldsymbol{gra\Delta}\ <\phi^{(2)}>(x,y) = \boldsymbol{gra\Delta}\ <\phi>(x,y)$. This superposition results in a depth-averaged Darcy's law given by:

$$<\boldsymbol{q}>(x,y) = -<\underline{\boldsymbol{k}}>(x,y)\cdot\boldsymbol{gra\Delta}\ <\phi>(x,y) \tag{D.26}$$

where the symbol $\boldsymbol{gra\Delta}$ is introduced here to represent a mixed differential-difference operator defined by $\boldsymbol{gra\Delta} = (\partial/\partial x, \partial/\partial y, \Delta/\Delta z)$.

The above solution (D.26) has already been obtained for the bottom flux formulation in which a potential is specified on the top boundary $z = 0$, whereas a vertical flux is specified on the bottom boundary $z = d$. A similar top flux formulation, in which a vertical flux is specified on the top boundary $z = 0$, and a potential is specified on the bottom boundary $z = d$, could also be specified. This latter top flux problem results in exactly the same solution as that of the original problem, however, expressed in the coordinate system (x, y, ζ) where $\zeta = d - z$ and where the flux vector has components (q_x, q_y, q_ζ) with $q_\zeta = -q_z$.

When in the latter top flux solution ζ is replaced by $d - z$ and q_ζ is replaced by $-q_z$, then the same depth-averaged Darcy's law as equation (D.26) is obtained. Also the expressions for $<k_{xx}>(x, y)$, $<k_{yy}>(x, y)$, $<k_{zz}>(x, y)$, $<k_{xy}>(x, y)$ and $<k_{yx}>(x, y)$ are the same, however the integration is over ζ instead of z. Similarly, the expressions for $<k_{xz}>(x, y)$, $<k_{yz}>(x, y)$, $<k_{zx}>(x, y)$ and $<k_{zy}>(x, y)$ are also obtained by integration over ζ instead of z, but they also have a minus sign.

When assuming the same depth-averaged potential 'gradient' $\boldsymbol{gra\Delta} \cdot \cdot <\phi>(x, y)$ for the bottom flux formulation and for the top flux formulation, combination of equation (D.26) for the bottom flux formulation with the equivalent equation for the top flux formulation results in the same depth-averaged Darcy's law as equation (D.26). However, in this combined equation the flux vector $<\boldsymbol{q}>(x, y)$ represents the arithmetic mean value of the flux vector of the bottom flux problem and the top flux problem. Also the depth-averaged conductivity tensor $<\underline{\mathbf{k}}>(x, y)$ represents the arithmetic mean value of the conductivity in the bottom flux problem and the top flux problem, i.e.:

$$2 <\underline{\mathbf{k}}> =$$
$$= \begin{pmatrix} <k_{xx}>_b + <k_{xx}>_t & <k_{xy}>_b + <k_{xy}>_t & <k_{xz}>_b - <k_{xz}>_t \\ <k_{yx}>_b + <k_{yx}>_t & <k_{yy}>_b + <k_{yy}>_t & <k_{yz}>_b - <k_{yz}>_t \\ <k_{zx}>_b - <k_{zx}>_t & <k_{zy}>_b - <k_{zy}>_t & <k_{zz}>_b + <k_{zz}>_t \end{pmatrix} \quad \text{(D.27)}$$

where the subscripts b and t mean that the conductivity component is derived from the bottom flux formulation and the top flux formulation, respectively.

From the above discussion it can be deduced that in equations (D.25e) and (D.25f) the terms $\partial t_0(x_k, 0)/\partial x_i$ cancel in equation (D.27); in other words, in equations (D.25e) and (D.25f) only the first-order terms with

$t_{1ij}(x_k, 0)$ play a role in the final result given by equation (D.27). It also follows that the first-order correction to the vertical potential difference is negligible in the determination of the depth-averaged potential gradient if $\left|\left[a_j(x_k, d)\right]_b - \left[a_j(x_k, d)\right]_t\right| / l_c << 1$ (see Section D.4).

D.6. Averaging in the lateral directions

The last step is to average the final form of equation (D.26), with $<\underline{\boldsymbol{k}}>$ given by equation (D.27), over the lateral dimensions of the block under consideration. For this purpose let us consider the meaning of $<\boldsymbol{q}>(x, y)$ in more detail.

When we deal with purely unidirectional lateral block-scale flow, say in the x-direction, it follows immediately from continuity equation (D.1) that the depth-averaged lateral flux $<q_x>(x) = <q_x>$ is constant, i.e., independent of x. This fact should be reflected in the lateral averaging procedure. Furthermore, a block-scale Darcy's law is a relationship between a constant block-averaged flux and a constant block-averaged potential gradient. The question is what to average laterally, the flux assuming a constant potential gradient, the potential gradient assuming a constant flux, or a combination of both. For a block surrounded by a large number of other blocks, where all blocks together constitute the total flow domain, the potential differences over the block will be negligibly small with respect to the total potential difference over the total flow domain. It follows that the flux through one particular block will be almost completely determined by all the neighboring blocks, independent of the conductivity distribution in that particular block. Therefore, it is reasonable to define the block-scale average with the aid of a constant depth-averaged flux in the block. A constant depth-averaged flux means that not only the lateral component, say $<q_x>$, but also the vertical component $<q_z>$ is constant. For that reason, the assumption that $<q_z>$ is constant has also been applied in the form $\partial w(x, y)/\partial x = \partial w(x, y)/\partial y = 0$ for the derivation of equation (D.14a), and it justifies the approach presented here with a combined bottom flux and top flux solution. The externally-determined depth-averaged flux results in vertical potential differences over the thickness of the block that do not vary linearly in the lateral directions. After averaging this varying potential difference over the lateral dimensions of the block, the result will be a block-averaged potential difference.

Application to the depth-averaged Darcy's law (D.26) of the assumption that $<\boldsymbol{q}>(x,y) = <\boldsymbol{q}>$ is constant in the block, leads to:

$$\begin{aligned} &\int_{x-a}^{x+a}\int_{y-b}^{y+b} \boldsymbol{gra\Delta}\ <\phi>(\lambda,\mu)\ \mathrm{d}\lambda\mathrm{d}\mu = \\ &= -\left[\int_{x-a}^{x+a}\int_{y-b}^{y+b} <\underline{\boldsymbol{k}}>(\lambda,\mu)^{-1}\ \mathrm{d}\lambda\mathrm{d}\mu\right] \bullet <\boldsymbol{q}>. \end{aligned} \tag{D.28}$$

In equation (D.28) λ and μ denote the two lateral coordinates, while x and y denote the coordinates of the center of the block; i.e., $x-a \leq \lambda \leq x+a$ and $y-b \leq \mu \leq y+b$ with $2\,a$ and $2\,b$ the lateral dimensions of the block. Integration of the left-hand side of equation (D.28) leads to:

$$\begin{aligned} &\int_{x-a}^{x+a}\int_{y-b}^{y+b} \boldsymbol{gra\Delta}\ <\phi>(\lambda,\mu)\ \mathrm{d}\lambda\mathrm{d}\mu = \\ &= \left[\int_{y-b}^{y+b} [<\phi>(x+a,\mu) - <\phi>(x-a,\mu)]\ \mathrm{d}\mu\ , \right. \\ &\qquad \int_{x-a}^{x+a} [<\phi>(\lambda,\mu+b) - <\phi>(\lambda,\mu-b)]\ \mathrm{d}\lambda, \\ &\qquad \left. \int_{x-a}^{x+a}\int_{y-b}^{y+b} \Delta<\phi>(\lambda,\mu)/\Delta z\ \mathrm{d}\lambda\mathrm{d}\mu\right]. \end{aligned} \tag{D.29}$$

Defining the block-averaged potential $\boldsymbol{<\phi>}$ as:

$$\boldsymbol{<\phi>} = (4\,a\,b)^{-1}\int_{x-a}^{x+a}\int_{y-b}^{y+b} <\phi>(\lambda,\mu)\ \mathrm{d}\lambda\mathrm{d}\mu \tag{D.30}$$

we find from equation (D.29) that:

$$\begin{aligned} &(4\,a\,b)^{-1}\int_{x-a}^{x+a}\int_{y-b}^{y+b} \boldsymbol{gra\Delta}\ <\phi>(\lambda,\mu)\ \mathrm{d}\lambda\mathrm{d}\mu = \\ &= (\partial\boldsymbol{<\phi>}/\partial x,\ \partial\boldsymbol{<\phi>}/\partial y,\ \Delta\boldsymbol{<\phi>}/\Delta z) = \boldsymbol{gra\Delta<\phi>} \end{aligned} \tag{D.31}$$

where the notation $\Delta\boldsymbol{<\phi>}/\Delta z$ is used to denote the laterally averaged value of the depth-averaged vertical potential gradient $\Delta<\phi>/\Delta z$.

Combination of equation (D.28) and equation (D.31) results in the block-scale Darcy's law:

$$\boldsymbol{<q>} = -\boldsymbol{<\underline{k}>} \cdot \boldsymbol{gra\Delta} \; \boldsymbol{<\phi>} \tag{D.32a}$$

with:

$$\boldsymbol{<\underline{k}>} = 4\,a\,b \left[\int_{x-a}^{x+a} \int_{y-b}^{y+b} \boldsymbol{<\underline{k}>}(\lambda,\mu)^{-1} \, \mathrm{d}\lambda \mathrm{d}\mu \right]^{-1} . \tag{D.32b}$$

Until now it has been assumed that the local-scale permeabilities are differentiable functions, up to the third derivatives, in the two lateral directions. In reality, however, discontinuous steps in the local-scale permeabilities often occur. In that case, the third derivatives in the first-order correction terms lead to second derivatives of Dirac-delta functions, a result which is useless for practical applications. Therefore, three integrations in the lateral directions over the correction terms are needed to obtain discontinuous step functions, which is a useful result. The averaging procedure in equation (D.32b) is one such integration. Consequently, to obtain a sufficiently smooth block-averaged conductivity, the averaging procedure must be repeated twice more, resulting in the discontinuous block-averaged conductivity $\boldsymbol{<<<\underline{k}>>>}$.

Finally, it is useful to remark on inverting the order of depth-averaging and lateral averaging. In the above derivation, depth-averaged values were derived first, while the averages in the lateral directions were determined afterwards. This procedure results in a smoothing of the depth-averaged conductivity distribution over the lateral scale. However, for practical applications of the above-presented theory, it is much simpler to invert the order of averaging. In that case, the local-scale permeabilities are smoothed first over the lateral scale. Then these smoothed values are used in the depth-averaging procedure, in which lateral derivatives of the laterally smoothed permeabilities are used. Such a procedure has been applied by Stam and Zijl (1992). Since the depth-averaged permeabilities are nonlinear in the local-scale permeabilities, inverting the order of averaging will not yield exactly the same results, but both results may reasonably be expected to have the same orders of magnitude.

D.7. Bibliography

Aziz, K. and Settari, A., *Petroleum Reservoir Simulation*, Applied Science Ltd., London (1979).

Lehner, K., A derivation of the field equations for slow viscous flow through a porous medium, *Ind. Eng. Chem. Fundam.*, Vol. 18, No. 1, pp. 41-45 (1979).

Quintard, M. and Whitaker, S., Écoulement monophasique en milieux poreux: effet des hétérogénéïtés locales, *Journal de mécanique théorique et appliquée (Journ. Theor. Appl. Mech.)*, Vol. 6, No. 5, pp. 691-726 (1987).

Stam, J.M.T. and Zijl, W., Modelling conductivity in imperfectly layered porous media II: A two-dimensional application of block-scale permeability, *Math. Geology* (accepted) (1992).

Van Dyke, M., *Perturbation Methods in Fluid Mechanics*, The Parabolic Press, Stanford (1975).

Appendix E

Basic functions of the finite element method

Simple tetrahedrons can be used to fill a three-dimensional flow domain Ω with finite elements. Numeration of nodes in every tetrahedron $T = (n_1, n_2, n_3, n_4)$ – see Figure 39 – assures that

$$V^* = \left[\overrightarrow{n_1 n_2},\ \overrightarrow{n_1 n_3},\ \overrightarrow{n_1 n_4}\right]$$

is positive and so is a volume $V = V^*/6$ of the tetrahedron. Tetrahedrons result from covering a three-dimensional flow domain Ω with pentahedrons (see Section 9.1) followed by breaking each pentahedron into three tetrahedrons as shown in Figure 17b.

The triangular sides S_1, S_2, S_3, S_4 of the tetrahedrons also have their vertices numerated in such a way that the corresponding normal vectors $\boldsymbol{N}_1$, $\boldsymbol{N}_2$, $\boldsymbol{N}_3$, $\boldsymbol{N}_4$ are pointing outwards:

$$\begin{aligned} \boldsymbol{N}_1 &\Leftrightarrow S_1 = \{n_2, n_3, n_4\} \\ \boldsymbol{N}_2 &\Leftrightarrow S_2 = \{n_3, n_1, n_4\} \\ \boldsymbol{N}_3 &\Leftrightarrow S_3 = \{n_4, n_1, n_2\} \\ \boldsymbol{N}_4 &\Leftrightarrow S_4 = \{n_1, n_3, n_2\}\,. \end{aligned} \tag{E.1}$$

Approximation to an arbitrary function f at any point (x, y, z) of the tetrahedron T can be written as the linear combination of some basic functions Φ_i $(i = 1, 2, 3, 4)$ defined over Ω:

$$f(x, y, z) = \sum_{i=1}^{4} f_i \cdot \Phi_i(x, y, z) \tag{E.2}$$

where f_i; $i = 1, 2, 3, 4$ is the value of function f at the i-th vertex of the tetrahedron and Φ_i; $i = 1, 2, 3, 4$ is the i-th *basic function* which satisfies the following orthogonality condition:

$$\Phi_i(x(j), y(j), z(j)) = \begin{cases} 1, & \text{for } j = i, \\ 0, & \text{otherwise} \end{cases} \tag{E.3}$$

changes linearly within the tetrahedron and vanishes elsewhere.

The i-th basic function, when restricted to the tetrahedron, is therefore given by the following formula:

$$\Phi_i(x, y, z) = \alpha_i + \beta_i\ x + \gamma_i\ y + \delta_i\ z \tag{E.4}$$

where:

$$\alpha_i = (-1)^{i+1}/V^* \det \begin{pmatrix} x(p(i+1)) & x(p(i+2)) & x(p(i+3)) \\ y(p(i+1)) & y(p(i+2)) & y(p(i+3)) \\ z(p(i+1)) & z(p(i+2)) & z(p(i+3)) \end{pmatrix} \tag{E.5}$$

$$\beta_i = (-1)^i/V^* \det \begin{pmatrix} 1 & 1 & 1 \\ y(p(i+1)) & y(p(i+2)) & y(p(i+3)) \\ z(p(i+1)) & z(p(i+2)) & z(p(i+3)) \end{pmatrix} \tag{E.6}$$

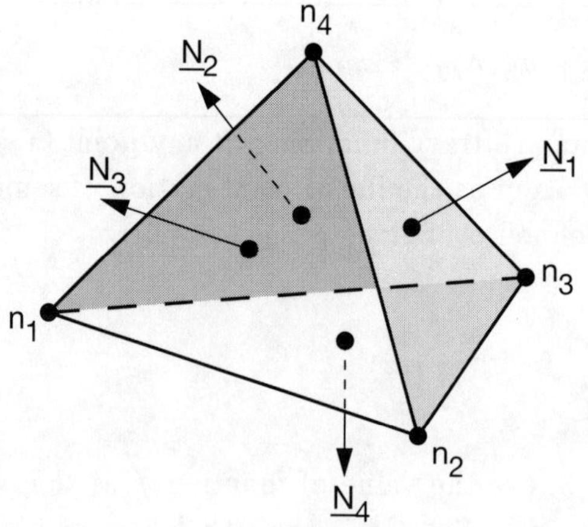

Figure 39. A tetrahedral finite element.

$$\gamma_{i+1} = (-1)^{i+1}/V^* \det \begin{pmatrix} 1 & 1 & 1 \\ x(p(i+1)) & x(p(i+2)) & x(p(i+3)) \\ z(p(i+1)) & z(p(i+2)) & z(p(i+3)) \end{pmatrix} \quad \text{(E.7)}$$

$$\delta_i = (-1)^i/V^* \det \begin{pmatrix} 1 & 1 & 1 \\ x(p(i+1)) & x(p(i+2)) & x(p(i+3)) \\ y(p(i+1)) & y(p(i+2)) & y(p(i+3)) \end{pmatrix} \quad \text{(E.8)}$$

$$V^* = 6V = \alpha_1 + \beta_1 \, x(1) + \gamma_1 \, y(1) + \delta_1 \, z(1) \quad \text{(E.9)}$$

where p is the integer function returning a cyclic permutation of numbers 1, 2, 3 and 4:

$$p = \begin{pmatrix} 1 & 2 & 3 & 4 & 5 & 6 & 7 \\ 1 & 2 & 3 & 4 & 1 & 2 & 3 \end{pmatrix} \quad \text{(E.10)}$$

and $x(j)$, $y(j)$, $z(j)$ are the coordinates of the j-th vertex of the tetrahedron; $j = 1, \ldots, 4$. Part of the space Ω on which a basic function Φ_i does not vanish is called support of Φ_i and is abbreviated as:

$$\Omega_i = \text{supp } \Phi_i. \quad \text{(E.11)}$$

The common part of the supports for two basic functions Φ_i and Φ_j is usually abbreviated as:

$$\Omega_{ij} = \Omega_i \cap \Omega_j. \quad \text{(E.12)}$$

INDEX

Y

Z